Reviews of Environmental Contamination and Toxicology

VOLUME 255

More information about this series at http://www.springer.com/series/398

Reviews of Environmental Contamination and Toxicology Volume 255

Glyphosate

Editor
James B. Knaak

Volume 255

Springer

ISSN 0179-5953 ISSN 2197-6554 (electronic)
Reviews of Environmental Contamination and Toxicology
ISBN 978-3-030-68482-2 ISBN 978-3-030-68483-9 (eBook)
https://doi.org/10.1007/978-3-030-68483-9

This Springer imprint is published by the registered company Springer Nature Switzerland AG
The registered company address is: Gewerbestrasse 11, 6330 Cham, Switzerland

Special Foreword

This Volume of Reviews in Environmental Contamination and Toxicology (RECT) is devoted to an assessment of the chemistry, toxicology and uses of the herbicide glyphosate (N-phosphonomethyl)glycine, CAS # 1071-83-6)). This herbicide is a broad spectrum and highly translocated foliar herbicide. Glyphosate inhibits 5-enolpyruvylshikimate-3-phosphate synthase (EPSPS, 2.5.1.26) in the shikimate biosynthetic pathway that produces the essential amino acids (tryptophan, tyrosine and phenylalanine) and plant phenolics. Glyphosate was originally registered in 1974 as a noncrop herbicide and in 1996 through biotechnology on glyphosate-resistant (GR) soybeans and cotton. This was achieved by inserting CP4 EPSPS gene into the plant genome.

As of today, glyphosate is registered on nine GR crops: soybeans, cotton, corn, Argentine canola, Polish canola, alfalfa, sugar beets, creeping bent grass and wheat. Genetically modified plants are regulated by various US agencies pursuant to the Coordinated Framework for Regulation of Biotechnology., 51 Fed. Reg. 23, 302 (June 26, 1986).

The USEPA indirectly regulates genetically modified plants through FIFRA, 7 U.S.C. 136-136y, which governs the use, sale and labeling of glyphosate.

In January 2020, USEPA under FIFRA/FQPA issued a glyphosate Interim Registration Review Decision (Docket Number EPA-HQ-OPP-2009-0351). The registration review covers glyphosate acid (PC code 417300) and the following salt forms with active pesticide registrations:

1. isopropylamine salt (PC code 10360)
2. ammonium salt (PC code 103604)
3. ethanol amine salt (PC code 1036050
4. diammonium salt (PC code 103607)
5. dimethyl ammonium salt (PC code 103608)
6. potassium salt (PC code 103613)

This document finalizes the agency's draft supporting documents on the following items!

1. Glyphosate Draft Human Health Risk Assessment for Registration Review.

The epidemiological literature was reviewed. The USEPA found there was insufficient evidence to conclude that glyphosate plays a role in any human diseases! Tolerances are established for residues of glyphosate on plant commodities in 40 CFR 180.364. Tolerances range from 0.2 to 400 ppm.

2. Registration-Preliminary Ecological Risk Assessment for Glyphosate and its Salts

The USEPA did not identify potential risks of concern for fish, aquatic invertebrates or aquatic-phase amphibians. The acute adverse effects to adult honey bees were considered low for application rates of up to 5.7 lb a.e./A. The Agency is currently determining whether additional data is needed on honey bees including pollinator studies!

The USFDA regulates genetically modified plants under the Federal Food, Drug, and Cosmetic Act (FFDCA). The FDA's authority is limited to removing adulterated food from the food supply. The FFDCA does not contain any provisions that directly address genetically modified plants.

The Animal and Plant Health Inspection Service (APHIS), US Department of Food and Agriculture, regulates transgenic plants under 7 CFR part 340, "Genetically Engineered Organisms and Products", which was published in1997. APHIS regulates plant pests and noxious weeds under the Plant Protection Act (PPA). The PPA defines a plant pest as any living stage of any of the following:

1. protozoan
2. nonhuman animal
3. parasitic plant
4. bacterium
5. fungus
6. virus or viroid
7. infectious agent or other pathogen

The PPA states the organisms regulated as "plant pests" must be organisms that cause physical harm to plants through injury, damage or disease. Neither the statute nor the regulations indicate that a genetically engineered plant (corn, cotton, etc.) which does not physically damage plants can be considered a plant pest.

Health Canada's Pest Management Regulatory Agency (PMRA) re-evaluated the use of glyphosate (RVD2017-01) in 2017. According to their overall assessment, glyphosate was unlikely a human cancer risk. Dietary, occupational, residential and environmental risks were not of concern provided label instructions were followed!

Glyphosate is currently approved for use in the EU until December 15, 2022 as an active substance in Plant Protection Products (PPPs). The European Commission appointed four Member States (France, Hungary, the Netherlands and Sweden) as the next Assessment Group on Glyphosate (AGG). The Glyphosate Renewal Group in December 2019 applied for renewal of Glyphosate in the EU. The application for renewal past 2022 was sent to the AGG, the other Member States, the European

Food Safety Authority (EFSA) and the European Commission. Glyphosate is not classified by the European Chemicals Agency and the EFSA as a carcinogen!

Glyphosate is regulated and used in most countries of the world except Iceland, Greenland, and a small island in the Pacific. In March 2015, the World Health Organization's (WHO) International Agency for Research on Cancer (IARC) classified glyphosate as a probable carcinogen. The IARC classification is a hazard classification and not a health risk assessment, because the level of human exposure was not taken into account by IARC.

The editor wishes to thank Dr. Stephen Duke for organizing the RECT Scientific Volume 255 on Glyphosate by recruiting Drs. Solomon, Green and Tranel, suggesting the titles of each chapter and writing the first chapter. The editor also wishes to thank the first authors of Chapters Evolution of Glyphosate-Resistant Weeds (p 93) and Ecotoxicology of Glyphosate, Its Formulants, and Environmental Degradation Products (p 129), Drs. Y. Baek and J.L. Rodriguez-Gil for their contribution to those chapters and Mr. Daniel Siehl for the kinetic parameters of EPSPS variants found in GR Weeds and Crops in Chapter History and Outlook for Glyphosate-Resistant Crops (p 67)!

The information provided in these chapters will give regulatory agencies, agriculturists and plant scientists easy access to a concise and informative review of glyphosate!

Fort Myers, FL, USA James B. Knaak

Foreword

International concern in scientific, industrial, and governmental communities over traces of xenobiotics in foods and in both abiotic and biotic environments has justified the present triumvirate of specialized publications in this field: comprehensive reviews, rapidly published research papers and progress reports, and archival documentations These three international publications are integrated and scheduled to provide the coherency essential for nonduplicative and current progress in a field as dynamic and complex as environmental contamination and toxicology. This series is reserved exclusively for the diversified literature on "toxic" chemicals in our food, our feeds, our homes, recreational and working surroundings, our domestic animals, our wildlife, and ourselves. Tremendous efforts worldwide have been mobilized to evaluate the nature, presence, magnitude, fate, and toxicology of the chemicals loosed upon the Earth. Among the sequelae of this broad new emphasis is an undeniable need for an articulated set of authoritative publications, where one can find the latest important world literature produced by these emerging areas of science together with documentation of pertinent ancillary legislation.

Research directors and legislative or administrative advisers do not have the time to scan the escalating number of technical publications that may contain articles important to current responsibility. Rather, these individuals need the background provided by detailed reviews and the assurance that the latest information is made available to them, all with minimal literature searching. Similarly, the scientist assigned or attracted to a new problem is required to glean all literature pertinent to the task, to publish new developments or important new experimental details quickly, to inform others of findings that might alter their own efforts, and eventually to publish all his/her supporting data and conclusions for archival purposes.

In the fields of environmental contamination and toxicology, the sum of these concerns and responsibilities is decisively addressed by the uniform, encompassing, and timely publication format of the Springer triumvirate:

Reviews of Environmental Contamination and Toxicology [Vol. 1 through 97 (1962–1986) as Residue Reviews] for detailed review articles concerned with any aspects of chemical contaminants, including pesticides, in the total environment with toxicological considerations and consequences.

Bulletin of Environmental Contamination and Toxicology (Vol. 1 in 1966) for rapid publication of short reports of significant advances and discoveries in the fields of air, soil, water, and food contamination and pollution as well as methodology and other disciplines concerned with the introduction, presence, and effects of toxicants in the total environment.

Archives of Environmental Contamination and Toxicology (Vol. 1 in 1973) for important complete articles emphasizing and describing original experimental or theoretical research work pertaining to the scientific aspects of chemical contaminants in the environment.

The individual editors of these three publications comprise the joint Coordinating Board of Editors with referral within the board of manuscripts submitted to one publication but deemed by major emphasis or length more suitable for one of the others.

Coordinating Board of Editors

Preface

The role of *Reviews* is to publish detailed scientific review articles on all aspects of environmental contamination and associated (eco)toxicological consequences. Such articles facilitate the often complex task of accessing and interpreting cogent scientific data within the confines of one or more closely related research fields.

In the 50+ years since *Reviews of Environmental Contamination and Toxicology* (formerly *Residue Reviews)* was first published, the number, scope, and complexity of environmental pollution incidents have grown unabated. During this entire period, the emphasis has been on publishing articles that address the presence and toxicity of environmental contaminants. New research is published each year on a myriad of environmental pollution issues facing people worldwide. This fact, and the routine discovery and reporting of emerging contaminants and new environmental contamination cases, creates an increasingly important function for *Reviews*. The staggering volume of scientific literature demands remedy by which data can be synthesized and made available to readers in an abridged form. *Reviews* addresses this need and provides detailed reviews worldwide to key scientists and science or policy administrators, whether employed by government, universities, nongovernmental organizations, or the private sector.

There is a panoply of environmental issues and concerns on which many scientists have focused their research in past years. The scope of this list is quite broad, encompassing environmental events globally that affect marine and terrestrial ecosystems; biotic and abiotic environments; impacts on plants, humans, and wildlife; and pollutants, both chemical and radioactive; as well as the ravages of environmental disease in virtually all environmental media (soil, water, air). New or enhanced safety and environmental concerns have emerged in the last decade to be added to incidents covered by the media, studied by scientists, and addressed by governmental and private institutions. Among these are events so striking that they are creating a paradigm shift. Two in particular are at the center of ever increasing media as well as scientific attention: bioterrorism and global warming. Unfortunately, these very worrisome issues are now superimposed on the already extensive list of ongoing environmental challenges.

The ultimate role of publishing scientific environmental research is to enhance understanding of the environment in ways that allow the public to be better informed or, in other words, to enable the public to have access to sufficient information. Because the public gets most of its information on science and technology from internet, TV news, and reports, the role for scientists as interpreters and brokers of scientific information to the public will grow rather than diminish. Environmentalism is an important global political force, resulting in the emergence of multinational consortia to control pollution and the evolution of the environmental ethic. Will the new politics of the twenty-first century involve a consortium of technologists and environmentalists, or a progressive confrontation? These matters are of genuine concern to governmental agencies and legislative bodies around the world.

For those who make the decisions about how our planet is managed, there is an ongoing need for continual surveillance and intelligent controls to avoid endangering the environment, public health, and wildlife. Ensuring safety-in-use of the many chemicals involved in our highly industrialized culture is a dynamic challenge, because the old, established materials are continually being displaced by newly developed molecules more acceptable to federal and state regulatory agencies, public health officials, and environmentalists. New legislation that will deal in an appropriate manner with this challenge is currently in the making or has been implemented recently, such as the REACH legislation in Europe. These regulations demand scientifically sound and documented dossiers on new chemicals.

Reviews publishes synoptic articles designed to treat the presence, fate, and, if possible, the safety of xenobiotics in any segment of the environment. These reviews can be either general or specific, but properly lie in the domains of analytical chemistry and its methodology, biochemistry, human and animal medicine, legislation, pharmacology, physiology, (eco)toxicology, and regulation. Certain affairs in food technology concerned specifically with pesticide and other food-additive problems may also be appropriate.

Because manuscripts are published in the order in which they are received in final form, it may seem that some important aspects have been neglected at times. However, these apparent omissions are recognized, and pertinent manuscripts are likely in preparation or planned. The field is so very large and the interests in it are so varied that the editor and the editorial board earnestly solicit authors and suggestions of underrepresented topics to make this international book series yet more useful and worthwhile.

Justification for the preparation of any review for this book series is that it deals with some aspect of the many real problems arising from the presence of anthropogenic chemicals in our surroundings. Thus, manuscripts may encompass case studies from any country. Additionally, chemical contamination in any manner of air, water, soil, or plant or animal life is within these objectives and their scope.

Manuscripts are often contributed by invitation. However, nominations for new topics or topics in areas that are rapidly advancing are welcome. Preliminary communication with the Editor-in-Chief is recommended before volunteered review manuscripts are submitted. *Reviews* is registered in WebofScience™.

Inclusion in the Science Citation Index serves to encourage scientists in academia to contribute to the series. The impact factor in recent years has increased from 2.5 in 2009 to 7.0 in 2017. The Editor-in-Chief and the Editorial Board strive for a further increase of the journal impact factor by actively inviting authors to submit manuscripts.

Amsterdam, The Netherlands Pim de Voogt
February 2020

Contents

List of Contributors

Yousoon Baek Department of Crop Sciences, University of Illinois, Urbana, IL, USA

Lucas K. Bobadilla Department of Crop Sciences, University of Illinois, Urbana, IL, USA

Stephen O. Duke National Center for Natural Products Research, School of Pharmacy, University of Mississippi, University, MS, USA

Darci A. Giacomini Department of Crop Sciences, University of Illinois, Urbana, IL, USA

Jerry M. Green Green Ways Consulting LLC, Landenberg, PA, USA

Jacob S. Montgomery Department of Crop Sciences, University of Illinois, Urbana, IL, USA

Brent P. Murphy Department of Crop Sciences, University of Illinois, Urbana, IL, USA

Ryan S. Prosser School of Environmental Sciences, University of Guelph, Guelph, ON, Canada

Jose Luis Rodríguez-Gil IISD – Experimental Lakes Area, Winnipeg, MB, Canada

Department of Environment and Geography, University of Manitoba, Winnipeg, MB, Canada

Daniel L. Siehl Sr. Scientist (ret.), Corteva Agriscience, Wilmington, DE, USA

Keith R. Solomon Centre for Toxicology, School of Environmental Sciences, University of Guelph, Guelph, ON, Canada

Patrick J. Tranel Department of Crop Sciences, University of Illinois, Urbana, IL, USA

Glyphosate: Uses Other Than in Glyphosate-Resistant Crops, Mode of Action, Degradation in Plants, and Effects on Non-target Plants and Agricultural Microbes

Stephen O. Duke

Contents

Abstract Glyphosate is the most used herbicide globally. It is a unique non-selective herbicide with a mode of action that is ideal for vegetation management in both agricultural and non-agricultural settings. Its use was more than doubled by the introduction of transgenic, glyphosate-resistant (GR) crops. All of its phytotoxic effects are the result of inhibition of only 5-enolpyruvylshikimate-3-phosphate synthase (EPSPS), but inhibition of this single enzyme of the shikimate pathway results in multiple phytotoxicity effects, both upstream and downstream from EPSPS, including loss of plant defenses against pathogens. Degradation of

S. O. Duke (✉)
National Center for Natural Products Research, School of Pharmacy, University of Mississippi, University, MS, USA
e-mail: sduke@olemiss.edu

J. B. Knaak (ed.), *Reviews of Environmental Contamination and Toxicology Volume 255*, Reviews of Environmental Contamination and Toxicology 255,
https://doi.org/10.1007/398_2020_53

glyphosate in plants and microbes is predominantly by a glyphosate oxidoreductase to produce aminomethylphosphonic acid and glyoxylate and to a lesser extent by a C-P lyase to produce sarcosine and phosphate. Its effects on non-target plant species are generally less than that of many other herbicides, as it is not volatile and is generally sprayed in larger droplet sizes with a relatively low propensity to drift and is inactivated by tight binding to most soils. Some microbes, including fungal plant pathogens, have glyphosate-sensitive EPSPS. Thus, glyphosate can benefit GR crops by its activity on some plant pathogens. On the other hand, glyphosate can adversely affect some microbes that are beneficial to agriculture, such as *Bradyrhizobium* species, although GR crop yield data indicate that such an effect has been minor. Effects of glyphosate on microbes of agricultural soils are generally minor and transient, with other agricultural practices having much stronger effects.

Keywords EPSPS · Glyphosate · Herbicide · Herbicide degradation · Hormesis · Mode of action · Weed management

Abbreviations

2PG	2-Phosphoglycolate
AKR	Aldo-keto reductase
ALA	Acetolactate synthase
AMF	Arbuscular mycorrhizal fungi
AMPA	Aminomethylphosphonic acid
AOPP	L-α-aminooxy-β-phenylpropionic acid
CFU	Colony-forming unit
DAHPS	3-Deoxy-D-arabinoheptulosonate-7-phosphate synthase
DQS	3-Dehydroquinate synthase
E4P	Erythrose-4-phosphate
EPSPS	5-Enolpyruvylshikimate-3-phosphate synthase
EU	European Union
GAT	Glyphosate acyltransferase
GOX	Glyphosate oxidoreductase
GR	Glyphosate-resistant
IAA	Indole acetic-3-acid
IPA	Isopropylamine
MRL	Minimum residue level
NT	No-tillage
PAL	Phenylalanine ammonia-lyase
PDS	Phytoene desaturase
PEP	Phosphoenolpyruvate
PGA	3-Phosphoglycerate
PPO	Protoporphyrinogen oxygenase
RH	Relative humidity
ROS	Reactive oxygen species
RUBISCO	Ribulose-1,5-bisphosphate carboxylase
S3P	Shikimate-3-phosphate
USA	United States of America

1 Introduction

After commercialization in 1974, glyphosate (*N*-(phosphonomethyl)glycine; CAS # 1071-83-6) became the most used herbicide worldwide. According to SciFinder®, in 2020 there were over 23,000 scientific publications, including patents, on glyphosate since 1972. Numerous general reviews (e.g., Baylis 2000; Dill et al. 2010; Duke 1988, 2018a; Duke et al. 2003a) and two entire books (Grossbard and Atkinson 1985; Franz et al. 1997) on glyphosate are available. There have been two special issues of a journal on use of glyphosate as a herbicide (Pest Management Science, April, 2008 and May, 2018) and a special issue of Critical Reviews of Toxicology (supplemental issue of 2016) on glyphosate's toxicological properties. Additionally, there are numerous reviews on specific aspects of glyphosate, such as its metabolic degradation in plants (e.g., Duke 2011), its degradation by microbes (e.g., Zhan et al. 2018), glyphosate extraction and analysis methods (Koskinen et al. 2016), its behavior in soil (Borggaard and Gimsing 2008), human exposure to glyphosate (Solomon 2020), and its environmental toxicology (Geisy et al. 2000). This review will not deal with formulation ingredients used with glyphosate, as these can vary between different products, and can vary with a particular product name between countries and over time. Unfortunately, many published studies are designed so that the effects of glyphosate cannot be differentiated from those of formulation ingredients. Furthermore, the exact ingredients of commercial glyphosate formulations are sometimes proprietary, making it impossible to evaluate some studies done with these products. The ecotoxicology of glyphosate and its formulants are covered by Rodríguez-Gil et al. (2020) in this volume.

The selection of topics covered by this review could be considered eclectic, but they were determined by what was not covered by the three other reviews on glyphosate of this volume. The review of Green and Siehl (2020) is on glyphosate-resistant (GR) crops, that of Rodríguez-Gil et al. (2020) covers the ecotoxicology of glyphosate, its formulants, and degradation products, and Baek et al. (2020) discuss evolved GR weeds. This review covers uses of glyphosate other than on GR crops, mode of action of glyphosate, metabolic degradation of glyphosate in microbes and plants, non-target vegetation effects and indirect effects of agricultural glyphosate use on non-target organisms, and effects of glyphosate on microbes in agriculture. A significant amount of this review is germane to the environmental toxicology of glyphosate, but I have tried to avoid those aspects covered by Rodríguez-Gil et al. (2020). This review emphasizes the more recent significant literature that has not been previously reviewed and will not discuss the burgeoning literature (often questionable toxicology studies) frequently found in predatory or very low impact journals. See Mesnage and Antoniou (2017) for an analysis of some of this questionable literature and its potentially harmful effects.

Glyphosate was an important herbicide when it was introduced, as there was no previous herbicide available that was effective on all weeds (non-selective) that was also considered to have low toxicity to animals, including humans. The only highly effective, non-selective herbicide alternatives at that time were paraquat

(1,1'dimethyl-4,4''-bipyridinium dichloride; CAS # 75365-73-0) and diquat (1,1-'-ethylene--2,2'-bipyridinium dibromide; CAS # 85-00-7), two pyridinium herbicides, both with high acute toxicity to animals. In the USA, paraquat use is much greater than diquat use (United States Geological Survey 2020). Paraquat is so acutely toxic to humans that it has often been used to commit suicide (Onyon and Volans 1987). Furthermore, paraquat and diquat are perhaps the fastest acting herbicides, so there is insufficient time for them to be translocated from sprayed foliage to protected plant meristems before the tissues to which they are applied are killed. Thus, after treated foliage dies, paraquat-treated plants, especially perennials, often regrow from meristems that do not come in contact with the herbicide. Glyphosate is highly systemic, translocating both acropetally and basipetally to metabolic sinks like meristems from treated parts of the plant. In most weed species, glyphosate is metabolized slowly to non-phytotoxic or very weakly phytotoxic compounds (Duke 2011), giving the herbicide time to reach critical metabolic sinks without being metabolized. It is also one of the slowest acting herbicides on most plant species, giving the plant adequate time to translocate it to meristems before translocation is adversely affected by glyphosate. This combination of attributes made it more effective than other herbicides in killing weeds with the potential to regrow, being effective on many perennial weed species.

Glyphosate was significantly more expensive than paraquat, but more effective and much safer. Even before the introduction of GR crops, its use was considerably higher than that of paraquat in agriculture (Fig. 1). The rapid increase in glyphosate use after the introduction of GR crops in the USA (Fig. 1a) did not affect the patterns of paraquat use in agriculture (Fig. 1b), and the use of paraquat went up in cotton (*Gossypium hirsutum*) and soybean (*Glycine max*) production after evolved GR weeds became a major problem in these crops (Fig. 1b). Thus, before the introduction of GR crops, glyphosate captured a strong market for vegetation management in

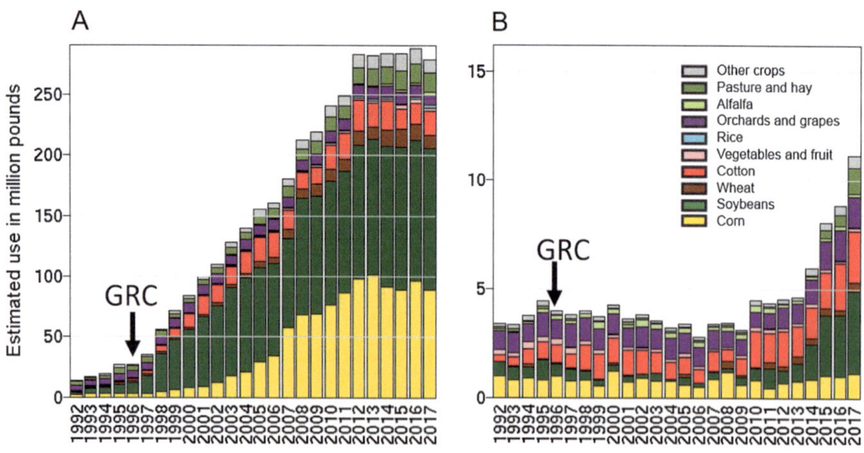

Fig. 1 Glyphosate (**a**) and paraquat (**b**) use in agriculture in the USA. **GRC** designates the introduction of GR crops. From the United States Geological Survey (2020)

situations, other than in most horticultural and agronomic row crops while they are growing, as all crops were susceptible to glyphosate. Before GR crops were available, most herbicide use within growing crops was with highly selective herbicides that do not substantially harm the crop, even when sprayed directly on them; e.g., diclofop-methyl ((RS)-methyl-2-[4-(2,4-dichlorophenoxy)phenoxy]propanoate; CAS # 51338-27-3) on soybean.

Compared to other herbicides introduced since 1974, glyphosate is a high use rate herbicide, requiring 0.5 to 2.0 kg/ha of active ingredient for management of most weeds. Most more recent herbicides, except for bioherbicides, are applied at a few hundred grams or less per hectare. Glyphosate is an anionic compound that is sold as a formulated salt (usually with potassium or isopropylamine (IPA) cations), but the glyphosate anion is the only substantially herbicidal compound in the commercialized formulations. In solution, at physiological pHs, glyphosate exists mostly as a divalent anion (Wauchope 1976). Ions of Ca, Mn, and Zn in tank mixtures of glyphosate can reduce its efficacy (Chahal et al. 2010). A glyphosate product (sometimes called sulfosate) that used a cationic sulfur counterion (trimesium or trimethylsulfonium) was sold at one time, but it was reported to have greater acute human toxicity than a commercial formulation of the IPA salt of glyphosate (Sørensen and Gregersen 1999). The trimesium salt is no longer sold.

Glyphosate's non-selectivity significantly limited its potential market, because it could not be sprayed directly on any growing crop like a selective herbicide. This changed dramatically in the USA with the introduction of transgenic, GR crops in 1996 (Duke 2014) (Fig. 1a). Similar increases in usage occurred in other countries that adopted GR crops, such as Argentina and Brazil. Agricultural use of glyphosate use plateaued in the USA in 2012 (Fig. 1a), probably due to both GR crop market saturation and farmers turning to other herbicides due to the rapid evolution and spread of GR weeds (Heap and Duke 2018). In 2016, about 56% of all glyphosate used globally was estimated to be used on GR crops, and 72% of all glyphosate used globally in its first 40 years of sales was used in the last 10 of those years (Benbrook 2016). The topic of GR crops and glyphosate use in them has been reviewed before (e.g., Duke 2014, 2015) and will be updated in this volume by Green and Siehl (2020). Other uses of glyphosate are briefly reviewed below.

2 Uses of Glyphosate Other Than in GR Crops

Glyphosate was a very successful herbicide for more than 20 years before the introduction of GR crops. Furthermore, it is still extensively used globally for other than weed management in GR crops. Gaines (2018) reviewed the topic of glyphosate use in non-GR crop settings in the USA. Wiese et al. (2018) and Antier et al. (2020a, b) provide good analyses of glyphosate use in Europe, where GR crops are essentially not grown. Even in Europe, glyphosate is the most used herbicide, comprising more about 33% of all herbicide use by volume. Figure 2 provides a

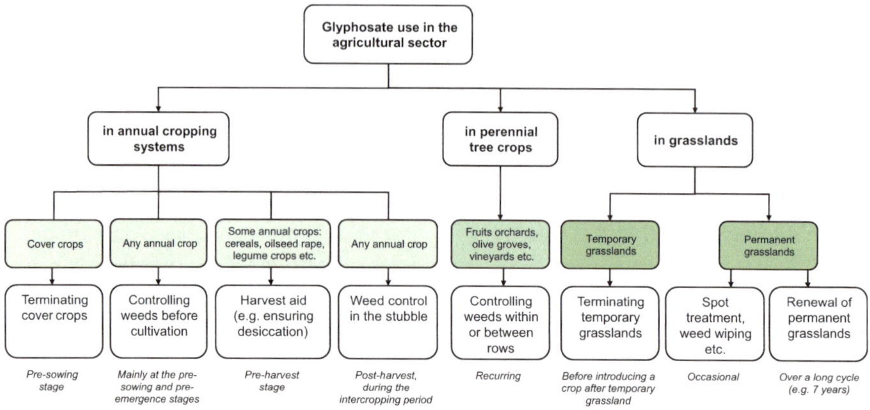

Fig. 2 Glyphosate uses in agriculture in the European Union. With permission from Antier et al. (2020a, b) with slight modification

breakdown of the many uses of glyphosate in agriculture in the European Union (EU). These EU uses are similar to the non-GR crop uses of glyphosate in agriculture throughout the rest of the world.

2.1 Weed Control in Non-Agricultural Situations

In 1995, before GR crops were introduced in the USA in 1996, 31% of the glyphosate used was for non-agricultural uses (Benbrook 2016). This percentage decreased to about 10% after GR crops were introduced in 1996, but the actual amount used for non-agricultural needs had more than doubled by 2014. The main non-selective alternatives for such uses are paraquat, with its toxicity issues discussed above, and glufosinate ((*RS*)-2-amino-4-(hydroxy(methyl)phosphonoyl) butanoic acid; CAS # 51276-47-2), which is less effective, less non-selective, and more expensive than glyphosate in most settings. Glufosinate was first commercialized in 1993, almost 20 years after glyphosate was introduced to the market. It is structurally similar to glyphosate, but has an entirely different molecular target site, glutamine synthetase (EC 6.3.1.2), involved in amino acid metabolism (Takano and Dayan 2020). Glyphosate is an ideal herbicide for total vegetation control in non-crop settings such as roadsides, railroad sidings, and preparation of land for installation of turf. It is used in turf to spot treat weeds (e.g., Burt 1980) or when the desired turf grass is dormant in the winter to kill winter weeds (e.g., Johnson 1976; Binkholder et al. 2011). Glyphosate is virtually inactive in soil and has a relatively short half-life (5.7 to 40.9 days) in moist soil in most climates (Blake and Pallett 2018). Thus, there are no long-lasting effects of these uses, other than indirect effects of killing the unwanted vegetation.

In the USA, glyphosate is used or has been proposed to be used to manage invasive weeds in non-agricultural settings such as *Bromus tectorum* (Sebastian et al. 2017), *Typha* spp. (Linz and Homan 2011), *Oxalis pes-caprae* (Lazzaro et al. 2019), and *Chrysanthemoides monilifera* ssp. *rotundata* (Matarczyk et al. 2002). Glyphosate has been recommended for management of invasive weed species such as *Spartina densiflora* that has become a problem in tidal marshes of southwest Spain (Mateos-Naranjo et al. 2009) and *Bischofia javanica*, an invasive tree species in the Ogasawara Islands (Itou et al. 2015). It is effective in control of invasive Mexican petunia (*Ruellia simplex*) in the state of Florida of the USA (Adams et al. 2014). These are but a few of the uses and proposed uses of glyphosate to manage invasive plant species in non-agricultural ecosystems.

Glyphosate is also used for aquatic weed management (Barrett 1985). There is at least one commercial formulation of glyphosate sold in the USA exclusively for management of aquatic weeds found growing on bodies of water or along shorelines. It is used for macrophyte aquatic weeds with foliage that is not submerged such as water hyacinth (*Eichhonia crassipes*) (e.g., Lopez 1993) and alligator weed (*Alternanthera philoxeroides*) (Bowmer et al. 1993). Many such targeted weed species are invasive, exotic weeds that are harmful to native aquatic vegetation. Glyphosate formulated for aquatic vegetation is sprayed on emergent aquatic vegetation, but it can also be wiped on (e.g., Kay 1995) in order to reduce water contamination. It is not used for submerged macrophytic vegetation such as *Hydrilla verticillata* (Dayan and Netherland 2005) or algae control, as the concentrations required would have to be very high, with potentially harmful environmental effects. On small, floating aquatic plants that have foliage exposed to the atmosphere like duckweed (*Lemna minor*), glyphosate is not effective in the water in which they grow, but it is very effective when sprayed on the foliage (Lockhart et al. 1989).

2.2 Weed Control in Non-GR Crops

Negatively charged glyphosate at soil pH ranges binds soil components (especially the clay fraction and Fe and Al oxides) so tightly (Morillo et al. 2000; Borggaard and Gimsing 2008) that it has no herbicidal effects in most soils. Therefore, it is commonly used in non-GR row crops for weed control before planting. A study meant to simulate effects of the potential accumulation of glyphosate and its main degradation compound, aminomethylphosphonic acid (AMPA; CAS # 1066-51-9) in soil when used at very high rates over multiple years on the growth and development of wheat (*Triticum aestivum*), field peas (*Pisum sativum*), and canola (*Brassica napus*), found no effects at recommended application rates (0.5 to 2.0 kg a.e./ha) (Blackshaw and Harker 2016). They found that application rates of 17.6 to 77 kg a. e./ha would be required to add enough glyphosate to soil to cause any crop injury, depending on the crop and location. The experiment assumed that glyphosate would be retained in the top 2 cm of soil. If glyphosate was distributed throughout a deeper soil profile because of tillage or high rainfall, the application rates required to cause crop injury would be even higher. Another proof of its safety to plants in soil is that dormant turf grasses can be sprayed with glyphosate in the winter to control winter weeds without damage to the dormant grass that regrows in the spring from subterranean meristems.

However, there are a few reports of glyphosate causing crop injury by uptake from sandy soils, especially when phosphate fertilizers are used (e.g., Cornish 1992). Phosphate can displace glyphosate from its soil binding sites in some cases (Gimsing and Borggaard 2001). In sandy loam soil, glyphosate application to weeds, followed by planting of wheat immediately or 1 day after spraying the weeds sometimes reduced wheat growth (Jang et al. 2020). However, in clay loam soil, growth of wheat was sometimes increased by such treatments, perhaps because of glyphosate hormesis (see Sect. 2.5). These effects were influenced by weed densities, target weed species, and soil water conditions. Glyphosate is less commonly used to kill weeds in fields of crops (both GR and non-GR) after harvest. Despite being non-selective, glyphosate is widely used in non-GR crop agriculture, as evidenced by its heavy use in Europe, where GR crops are not grown (e.g., Weise et al. 2018. Antier et al. 2020a, b) and in the USA in non-GR crop settings (Gaines 2018). In the USA in 2014, ca. 12% of the glyphosate use in agriculture was in non-GR crops (Benbrook 2016). The analysis by Gaines (2018) of glyphosate, glufosinate, and paraquat use in various non-GR USA crops showed that glyphosate use predominated, except for peanuts (*Arachis hypogaea*), in which case paraquat use (percent of hectares treated) in 2013 was slightly higher than glyphosate use.

Glyphosate can be safely used in orchards and vineyards to control weeds when crop foliage is high enough to avoid significant spray reaching leaves from directed applications to lower-growing weeds among these crops. The distance between orchard and vineyard crop plants also assists in avoiding contact of the crop foliage with spray. Glyphosate was predicted to end problems with perennial weeds in tree and vine crops soon after it was introduced (Lange et al. 1975). If used properly in vineyards and other perennial, woody crops, there is no crop damage. However, used improperly, drift of glyphosate to foliage can cause crop injury (e.g., Mohseni-Moghadam et al. 2016; Schrübbers et al. 2014). Gaines (2018) reported that in the USA in 2017, glyphosate was used for weed management in 35 to 42% of such crops. Glyphosate has been used so much in some vineyards, that its use has been associated with contamination of nearby surface waters with the herbicide (Daouk et al. 2013). Another evidence of the intensive use of glyphosate in orchard crops is that one of the first cases of evolved resistance of a weed (*Eleusine indica*) to glyphosate was in a fruit orchard in Malaysia (Lee and Ngim 2000). Plants do not evolve resistance to glyphosate easily, as with some herbicides (e.g., the sulfonylureas), as it required more than 20 years for the first report of revolved resistance (Baek et al. 2020), despite its widespread use and resulting strong selection pressure. Glyphosate has been used extensively in conifer silvaculture (Freedman 1990), mostly in the early stages of establishment of the conifer crop. It has also been used to destroy illicit crops, including *Erythroxylum coca* (Solomon et al. 2007; Marshall et al. 2009), marijuana (*Cannabis sativa*) (Lanaro et al. 2015), and opium poppy (*Papaver somniferum*) (Solomon et al. 2007). Glyphosate does not have to kill the *Erythroxylum coca* plant in order to lower the cocaine levels in leaves to uneconomical concentrations (Casale and Lydon 2007).

Another common use of glyphosate is to kill cover crops that are used to prevent soil loss and for suppression of weeds between crops in no-tillage agriculture (e.g., Reddy and Koger 2004; Nascente et al. 2013). The most environmentally damaging weed management option is tillage, as it facilitates erosion of top soil which can take eons to replace. Reduced tillage and plant residue management provide many environmental advantages (Locke and Bryson 1997). Tillage also results in loss of soil moisture (e.g., Blevins et al. 1971). Adoption of GR crops (soybean, maize (*Zea mays*), cotton, canola (*Brassica napus*) and sugar beet (*Beta vulgaris*)) allowed farmers to greatly reduce tillage in these crops (Cerdeira and Duke 2006; Duke and Powles 2009; Givens et al. 2009; Morishita 2018). Use of reduced tillage and cover crops with GR crops can reduce soil erosion, moisture loss, and movement of pre-emergence herbicides from the field (Krutz et al. 2009). Even in non-GR crops, glyphosate use has reduced tillage for weed management both directly (e.g., Melander et al. 2013; Kudsk and Mathiasson 2020) and for facilitation of the use of cover crops that reduce soil erosion (e.g., Weston 1990). Glyphosate is also used extensively in wheat crops before planting and after harvesting to facilitate reduced and no-tillage agriculture (Gaines 2018). Similar practices have been used with glyphosate to facilitate reduced and no-tillage agriculture in Europe, where GR crops are not grown (Wiese et al. 2018; Antier et al. 2020a, b). Furthermore, tillage is a fossil fuel-intensive procedure. Largely due to the reduction of tillage, the use of GR crops in 2016 reduced worldwide fossil fuel use by the equivalent of removing 1.8 million family automobiles from the road for 1 year (Brookes and Barfoot 2018). This figure is only for 1 year and does not include the fuel savings by the reduction of tillage facilitated by glyphosate in non-GR crops.

Some effort has been made to use glyphosate in glyphosate-sensitive row crops by using devices to wipe glyphosate on weeds that are taller than the crop (McWhorter and Derting 1985; Derting 1987; Harrington and Ghanizadeh 2017) and by using shielded or hooded sprayers between rows (e.g., Westerman and Murray 1994). Such methods greatly reduce the amount of herbicide needed per unit area. These approaches have been used with tractor-mounted booms over several crop rows and with hand-held devices for spot treatments. Even with these devices to reduce contact of the crop by glyphosate, crop injury is common. Contact with even one leaf of a plant can cause significant injury or plant death because of glyphosate's ability to translocate well (see Sect. 3.2) once it enters the plant. Although these application technologies were largely developed in the USA, this type of glyphosate application in the USA became rare after the introduction for GR crops. However, methods are being developed to apply herbicides with robotic systems that can differentiate between crops and weeds, applying the herbicide only to the weeds (e.g., Rajaa et al. 2020). Because glyphosate is non-selective, it is ideal for this technology, as the robot would only have to determine if the detected plant is the crop or not. Such technology used with glyphosate would change it from a high use rate herbicide to a very low use rate herbicide.

2.3 Use as a Crop Harvest Aid

After the harvested portion of annual crops are mature, there is an advantage to killing the crop and letting it desiccate so that it can be harvested efficiently with mechanical equipment. Living, green shoots of crops can interfere with harvesting equipment. Also, waiting for the annual crop to die naturally and desiccate so that it can be harvested can delay harvesting until times of the year that are too wet for harvesting (e.g., cotton in the southeast USA). Several herbicides have been used as crop harvest aids to rapidly kill the crop, and glyphosate has become the most commonly used herbicide for this purpose (Griffin et al. 2010). An additional benefit of this practice is that seed-producing weeds that are in the field at the time of application are killed, preventing them from contributing viable seeds to the weed seed bank for future cropping seasons. For example, late season application of glyphosate after seed set of the crop reduced seed production of the weeds *Sesbania herbacea* and *Senna obtusifolia* by 85%, and the *S. herbacea* seeds produced had only 6% viability (Clay and Griffin 2000).

Glyphosate-based herbicides are recommended to be used as a harvest aid at least a week before harvest during the ripe stage of physiological seed maturity. When so used, some shikimic acid ((3R,4S,5R)-(−)-3,4,5-trihydroxy-1-cyclohexenecarboxylic acid; CAS 138-59-0) can accumulate in the grain (see Sect. 3), indicating that some glyphosate translocates to the grain, but no impact on amino acid composition or gluten protein composition is seen, unless glyphosate is applied too early (Malalgoda et al. 2020). Glyphosate applied too early as a harvest aid can result in translocation of enough glyphosate to developing seeds to cause developmental problems. If this occurs, the germination vigor of some or all of these seeds may be compromised (e.g., Jeffery et al. 1981; Whigham and Stoller 1979), and residues of glyphosate and AMPA in the harvested food product will be increased (e.g., Cessna et al. 2002). However, when properly used as a harvest aid in wheat, most of the glyphosate ends up in the straw, with very little in the seed, and relatively little AMPA, the main metabolite of glyphosate, is found (Cessna et al. 1994). Even if there is no translocation, glyphosate residues, but not AMPA, can contaminate harvested food products from use of glyphosate as a harvest aid.

Reports of a few ppm of glyphosate contamination of cereal grain-based foods (e.g., Harris and Gaston 2004) such as beer (e.g., Jansons et al. 2018) and grain-based breakfast foods (e.g., Zoller et al. 2018) are almost certainly due to contamination from use as a harvest aid. How much of the glyphosate is due to translocation to the seed vs contamination from sprayed surfaces is unknown. Residues of glyphosate in these food products are generally below what is permitted by regulatory agencies and are thus not considered to be a health concern by these agencies. In a recent review of the topic, Xu et al. (2019) found that the reported glyphosate levels in grains and other foods were below the residue limits of all regulatory authorities listed in the paper. For example, the maximum residue levels (MRL – called tolerances by the USEPA) for glyphosate in wheat are 30 ppm in the USA and for FAO/WHO, 10 ppm in the EU, and 5 pm in Canada (Xu et al. 2019). The highest level reported by Xu et al. (2019) was 11.1 ppm by Gélinas et al. (2018), but the sample from this study was not from the commercial food supply. This was far higher than most of the other reports that found most wheat-based foods to have glyphosate residues of less than 1 ppm. AMPA was found in some of the samples of the papers reviewed by Xu et al. (2019), indicating that translocated glyphosate was degraded in the grain or at some point in the food supply chain. Similar results were reported by Kolakowski et al. (2020) who found glyphosate residues in a wide range of foods in Canada, but the levels in 99.4% of the almost 8,000 samples tested were lower than Canadian MRLs. No glyphosate was found in dairy and meat samples, and the highest amounts tended to be in grain-derived foods, especially wheat products, likely to be due to glyphosate use as a harvest aid. A recent review by Solomon (2020) of glyphosate levels found in urine of the general public (e.g., in California from 1993–2016 that are assumed to be mostly from dietary exposure (Mills et al. 2017)), concluded that the exposure from this source poses a *de minimis* risk. The results of Mills et al. (2017) indicated an increasing exposure during the time period of the study (1993–2016), a time span when the use of glyphosate in agriculture in the USA grew rapidly until 2012 (Fig. 1a).

2.4 Use as a Sugarcane Ripener

Low application rates (0.16 to 0.47 kg a.i./ha) of glyphosate applied to sugarcane (*Saccharum officinarum*) at 8 weeks before harvest enhances the yield of sucrose (Dalley and Richard 2010; Dusky et al. 1986; Legendre and Finger 1987; Nguyen et al. 2019; Velini et al. 2010). Used in this way, glyphosate is called a ripener. These glyphosate rates are lower than those recommended to kill weeds and are sublethal to sugarcane at the growth stage at which it is treated, yet glyphosate use at these low application rates causes marked increases in shikimic acid (up to 12-fold increases, reaching concentrations of up to 120 ppm) (Carbonari et al. 2014; Viana et al. 2019; Pincelli-Souza et al. 2020), the best biomarker for glyphosate reaching its molecular target site as a herbicide (see Sect. 3). The sucrose yield increase resulting from glyphosate treatment can be more than 10%, depending on the cultivar, weather, treatment timing, application rate of glyphosate, and timing of harvest after treatment (Dalley and Richard 2010). In addition to increasing sucrose yield, low application rates of glyphosate can enhance other growth parameters, such as leaf area and internode numbers (Pincelli-Souza et al. 2020). The low glyphosate application rates used may be sufficient to reduce enough metabolic activity in metabolic sink tissues such as meristems and developing leaves, so that less sucrose is translocated to them. These low application rates, however, do not affect photosynthesis and transport of sucrose from mature leaves to stem internodes. Thus, sucrose accumulates to higher than normal levels in the harvested part of the plant. Some other herbicides with different modes of action (e.g., fluazifop-butyl; butyl-(*R*)-2-(4-{[trifluoromethyl)-2-pyridyl]oxy}phenoxy)propionate; CAS #79241-46-6) cause similar effects, but they are not permitted for this use in the USA. GR sugarcane, as proposed by several groups (e.g., Wang et al. 2017), would render glyphosate ineffective as a sugarcane ripener.

Because glyphosate and sucrose translocate similarly (see Sect. 3.2), glyphosate contamination of sugars from glyphosate-treated sugarcane and GR sugar beet might be expected. However, Barker and Dayan (2019) found that, even with the high application rates of glyphosate for weed control in GR sugar beet (Morishita 2018), processing of the sugar reduced glyphosate levels to below the limit of detection in the refined, crystalline sugar. Similar results should be expected with refined sugarcane sugar, especially since the application rate of glyphosate used as a ripener is much less than that used for weed management in GR sugar beet. A recent study found ca. 1 ppm of glyphosate in a crude extract of juice of sugarcane which had been treated with glyphosate to enhance sugar yields in Vietnam (Nguyen et al. 2019). This level was stated to be below the MRL of 2 ppm allowed by the Vietnamese Ministry of Health.

2.5 Potential Use as a Plant Growth Regulator

Low application rates of glyphosate have been proposed to slow turf growth without unacceptable injury (e.g., Johnson 1990; Fry 1991; Dias et al. 2019). However, glyphosate is not used for this purpose, as the risk of injuring or killing the turf instead of stunting its growth is too great. Transgenic GR turf grasses have been developed (e.g., Blume et al. 2010; Wang and Brummer 2012), and glyphosate-tolerant fescue (*Festuca arundinacea*) has been developed through conventional breeding (Rose-Fricker 2002), although such products have not yet reached the commercial market. Low application rates of glyphosate (up to 0.7 kg/ha) can provide good weed control with some available fineleaf fescue varieties without damage to the turf (Askew et al. 2019). There is concern that glyphosate resistance genes could move from GR or glyphosate-tolerant turf grasses, creating major GR weeds in GR crops (Zapiola and Mallory-Smith 2012). As mentioned earlier, glyphosate can be used in winter to kill weeds without injury to dormant turf grass. Low application rates of glyphosate have been proposed as a plant growth regulator for tomato (*Solanum lycopersicum*) production (Pombo et al. 1985), but this use has not materialized. Later work showed that low application rates of glyphosate can enhance tomato plant photosynthetic rates and growth (Khan et al. 2020).

Hormesis is the stimulatory effect of a subtoxic dose of a toxin (Calabrese et al. 2007). Such an effect is not always beneficial. Very low, subtoxic application rates of herbicides often enhance plant growth (Belz and Duke 2014), but glyphosate is unique, in that its stimulatory effects are the strongest and most consistent among herbicides (Belz and Duke 2017; Brito et al. 2018). Application rates of glyphosate that are effective in stimulation of growth usually range from 1.8 to 32 g/ha (compared to the 500–2,000 g/ha used to kill most weeds) for glyphosate-susceptible plants. Hormetic application rates of glyphosate can increase growth,

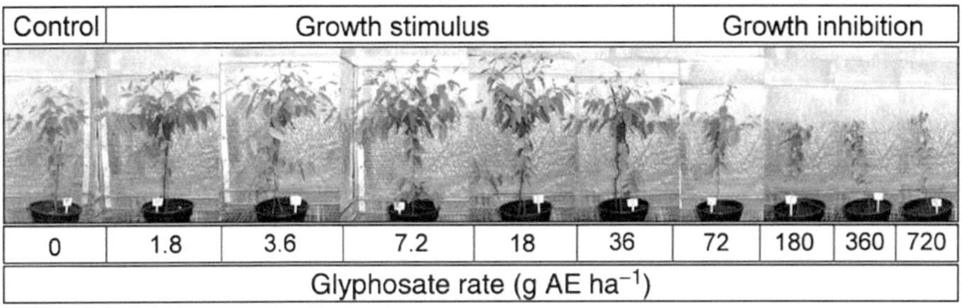

Fig. 3 Effects of different doses of glyphosate on Eucalyptus 60 days after spraying. From Velini et al. (2008) with permission

photosynthesis, seed production, and other developmental parameters. Increases in growth for herbaceous plants are generally 10 to 30% (e.g., Wagner et al. 2003) and sometimes greater (e.g., Sammons et al. 2018), whereas for some woody plants, such as *Eucalyptus* spp., the increase can be 50 to more than 100% increase over untreated plants, depending on the plant part measured (e.g., Velini et al. 2008) (Fig. 3).

The physiological mechanism of glyphosate-caused hormesis is unknown, but the fact that hormesis is not seen in GR crops at glyphosate application rates that cause hormesis in non-GR crops (Velini et al. 2008) indicates that the effect is tied to the herbicidal mode of action of glyphosate. Sammons et al. (2018) found that glyphosate hormesis of GR *Arabidopsis thaliana* lines with one, two, or four copies of a transgene for GR 5-enolpyruvylshikimate-3-phosphate synthase (EPSPS, EC 2.5.1.19), glyphosate's molecular target, occurred at the same and higher glyphosate application rates than hormesis of susceptible plants. Application rates that were very toxic to the wild type were hormetic to the transformants, and the more resistant the transformant, the higher the maximum hormetic application rate. Thus, hormesis might be seen in GR crops at much higher glyphosate application rates than in non-GR crops, because the dose-response curves are shifted to higher application rates by a factor of about fifty (Nandula et al. 2007). Thus, it is possible that the weed-killing rates of glyphosate used on GR crops might sometimes stimulate their growth. I am unaware of any published studies designed to specifically test this hypothesis. However, in a multi-year study with GR maize, ear number, green ear mass, and kernel mass were increased by a recommended glyphosate rate (1.7 kg/ha) for weed management compared to maize kept weed-free without glyphosate use (Williams et al. 2015). Likewise, 1 and 3.33 kg a.e./ha of glyphosate stimulated early growth of GR canola in greenhouse studies in which the plants were not grown full term to harvest (Corrêa et al. 2016).

The hormetic effect of glyphosate has led some to propose that ultralow application rates of glyphosate could be used commercially to increase crop yield (e.g., Abbas et al. 2015, 2016). However, the stimulatory effects on growth are generally transitory and seldom lead to yield increases (Cedergreen 2008; Brito et al. 2018). Nevertheless, the hormetic effects (greater tiller numbers, culm length and dry mass, leaf dry mass, internode numbers, leaf area, and sugar yield) of a low glyphosate dose on sugarcane are sustained until harvest (Pincelli-Souza et al. 2020). For other crops, getting reproducible and predictable results in the field is difficult, as the hormetic dose range is affected by environmental conditions and plant developmental stage, as well as the time between application and harvest. For example, hormetic effects can be reduced by water stress in the weed *Echinochloa colona* (Mollaee et al. 2020). However, in safflower (*Carthamus tinctorius*), a drought-tolerant crop, a glyphosate application rate of 36 g a.e./ha caused hormesis under water stress (dos Santos et al. 2021). The makeup of the microbiome associated with the plant root can have a strong effect on glyphosate hormesis. Ramirez-Villacis et al. (2020) found the presence of a few root growth-inhibiting microbial strains (e.g., *Firmicutes* spp. and *Burkholdia* spp.) could eliminate the hormetic effect of glyphosate on *A. thaliana*. The presence of these soil microbiota could account for the fewer reports of glyphosate hormesis in the field than one would expect. However, the study of Ramirez-Villacis et al. (2020) was not done in soil and, thus, may not extrapolate to the field. Furthermore, in this study, glyphosate was applied to the roots in agar rather than as a foliar application, as it is used as a herbicide in the field.

As far as I know, glyphosate is not being used as a yield enhancer, except for sugarcane. The ripener effect of glyphosate on sugar yield of sugarcane is an atypical example of hormesis. Whether the stimulatory effects of glyphosate on growth of woody plants is a benefit of using the herbicide at the early stage of tree establishment is unclear. For example, the use of glyphosate for weed management in early cultivated *Pinus taeda* forest establishment results in larger tree seedlings (Pehl and Shelnutt 1990), and glyphosate use during the establishment of several tree species resulted in larger trees (Fu et al. 2008). Whether these effects are due to elimination of competition with other, more glyphosate-sensitive vegetation, to hormesis, or to both was not determined in these studies.

Glyphosate-associated hormesis has recently been proposed to facilitate evolution of GR weeds (Belz and Duke 2017; Brito et al. 2018). In the field, drift concentrations of glyphosate can stimulate the growth of glyphosate-susceptible weeds, such as *Urochloa decumbens* (de Moraes et al. 2020). Hormesis can be more pronounced in GR weeds, giving them a growth advantage in a competitive environment (Belz 2014). Furthermore, low application rates of glyphosate can be more advantageous to certain subpopulations of a single plant species than another, altering the makeup of the population (Belz and Sinkkonen 2019) in a way that favors survival of tolerant members of the population.

2.6 Glyphosate Effects on Non-Plant Pests

Phytotoxicity of glyphosate to non-target plant species outside of fields can influence ecosystems, especially if it changes the species composition of an ecosystem. For example, glyphosate could have a harmful effect on an animal species that depends on a plant species that is adversely impacted by glyphosate. This is likely if both species are native to a region in which glyphosate is heavily used. In some cases, glyphosate is used to influence unwanted non-plant species. For example, glyphosate management of invasive cattail (*Typha* spp.) has also had the benefits of reducing the sanctuary of cattail stands for blackbird (*Icteridae*) pests (Linz and Homan 2011). This program reduced blackbird damage to sunflower (*Helianthus annuus*) crops in North and South Dakota of the USA.

Glyphosate elimination of most weeds in agroecosystems should reduce the incidence of pests that use weeds as a food source and/or breeding habitat, but very little has been done to verify this. Elimination of all vegetation, other than the crop, in a GM crop field can also result in disruption of some pest biocontrol technologies, as vegetational diversity is needed for many biological control organisms as a source of habitat and nutritional resources (Lundgren et al. 2009). A few studies have correlated patterns of decline of certain arthropods with glyphosate-killed weeds (e.g., Haughton et al. 2001). There is much more literature on the direct effects of glyphosate (usually as a formulated product) on insects (e.g., Bernal and Dussán 2020) than on the much more severe and long-lasting effects of killing their food sources and habitat.

Desirable insects can be indirectly adversely affected by killing weeds on which they rely on or very near agricultural fields where glyphosate is used. For example, both the monarch butterfly (*Danaus plexippus*) and certain *Asclepias* species upon which this butterfly exclusively depends are found in the parts of North America where glyphosate is heavily used because of GR crop adoption. The decline of this butterfly has been largely attributed to glyphosate use by some (e.g., Pleasants and Oberhauser 2013; Thogmartin et al. 2017). However, an analysis by Boyle et al. (2019) reported that the beginning of the decline of the monarch butterfly predates the adoption of GR crops. Their analysis shows that the decline of both *Asclepias* species and the monarch butterfly in North America began at close to the same time, when there was a widespread shift to synthetic herbicide-based weed management in the middle of the twentieth century. The use of synthetic insecticides also increased dramatically at approximately the same time. With the widespread adoption of GR crops, there was no inflection in the decline plot of either the butterfly nor its host plant (Boyle et al. 2019). Hartzler (2010) found little effect of adoption of GR crops in Iowa (USA) on *Asclepias syriaca*, the main milkweed species host of the monarch butterfly outside of agricultural fields in this area, where insecticides are generally not used. However, in agricultural fields, where insecticides are often sprayed, *A. syriaca* populations were reduced after the introduction of GR crops. *Asclepias* spp. in fields where insecticides are used could be considered an attract and kill situation for the monarch butterfly. Thus, as long as insecticides are sprayed in crops, *Asclepias* spp. growing in such crops could be more of a risk than a benefit to the monarch butterfly. Therefore, glyphosate reducing the milkweed in GR crops, while having almost no effect on this plant species outside of fields where insecticides are not sprayed, might benefit the butterfly. Clearly, more study of the roles of these factors in the decline of the monarch butterfly is warranted. This example illustrates that cause and effect conclusions based on incomplete knowledge of all factors affecting an ecosystem or a species in it can be erroneous.

3 Mode of Action of Glyphosate

3.1 Effect of Glyphosate on 5-Enolpyruvylshikimate-3-Phosphate Synthase

The only molecular target site of glyphosate as a herbicide is EPSPS, an enzyme of the shikimate pathway that produces the three aromatic amino acids (phenylalanine (CAS 63-91-2), tyrosine (CAS 60-18-4), and tryptophan (CAS 73-22-3) required for protein synthesis and for production of compounds required for plant growth and development such as the plant hormone indole acetic-3-acid (IAA, CAS 87-51-4 and plastoquinone (PQ, CAS 4299-57-4) that is essential for photosynthesis and carotenoid synthesis (Fig. 4). Plants, fungi, and bacteria, but not animals, possess EPSPS (Kishore and Shah 1988; Dill et al. 2010). The only exceptions are most of the Apicomplexan parasitic parasites, such as those that cause malaria, which all contain a vestigial plastid, the apicoplast, which is considered to be the result of endosymbiosis of a red alga by a heterotopic, unicellular eukaryote (Arisue and Hashimoto 2015). Even though the apicoplast is not photosynthetic, it contains much of the biosynthetic capability of a plant plastid, including EPSPS that is sensitive to glyphosate (Roberts et al. 1998; McConkey et al. 2004). Glyphosate was once proposed as an antimalarial pharmaceutical with inhibition of EPSPS as its mode of action (Roberts et al. 2002). This has not occurred, but environmentally realistic exposure of mosquito larvae to glyphosate can reduce their infection with *Plasmodium relictum*, a prevalent avian malaria in Europe (Bataillard et al. 2020).

The percent of the carbon in terrestrial plants that passes through the shikimate pathway is estimated to range from 20 to 50% (Tohge et al. 2013), varying largely with the amount of lignin synthesized. Therefore, blocking this pathway has profound effects on plant metabolism. There has been speculation about some of the toxic effects of glyphosate on plants being due to effects unrelated to the shikimate pathway, but the finding that transgenes encoding GR EPSPS render plants approximately 50-fold less sensitive to foliar-applied glyphosate (application rates for 50% growth reduction were 0.47 and 22.8 kg a.e./ha for sensitive and GR soybean, respectively) (Fig. 5) (Nandula et al. 2007) proves EPSPS to be the only herbicide target for glyphosate at the range of recommended application rates used for weed management (0.5–2.0 kg/ha). This supports the view that none of the molecular targets held in common between plants and animals are likely to be affected by the much lower concentrations of glyphosate to which animals are exposed than to which target plants are exposed. For example, some have claimed that because glyphosate can be an in vitro inhibitor of some P450 monooxygenase enzymes (e.g., Xiang et al. 2005), they could cause human toxicity by such a mechanism in gut microbes (e.g., Samsel and Seneff 2013). Because P450 monooxygenases are essential to plants, and GR crops are completely resistant to much higher concentrations (more than 10 kg a.e./ha) of glyphosate than to which they are exposed in the field, such enzymes are highly unlikely to be affected by glyphosate in vivo at recommended application rates for weed management (0.5 to 2.0 kg a.e./ha). Thus,

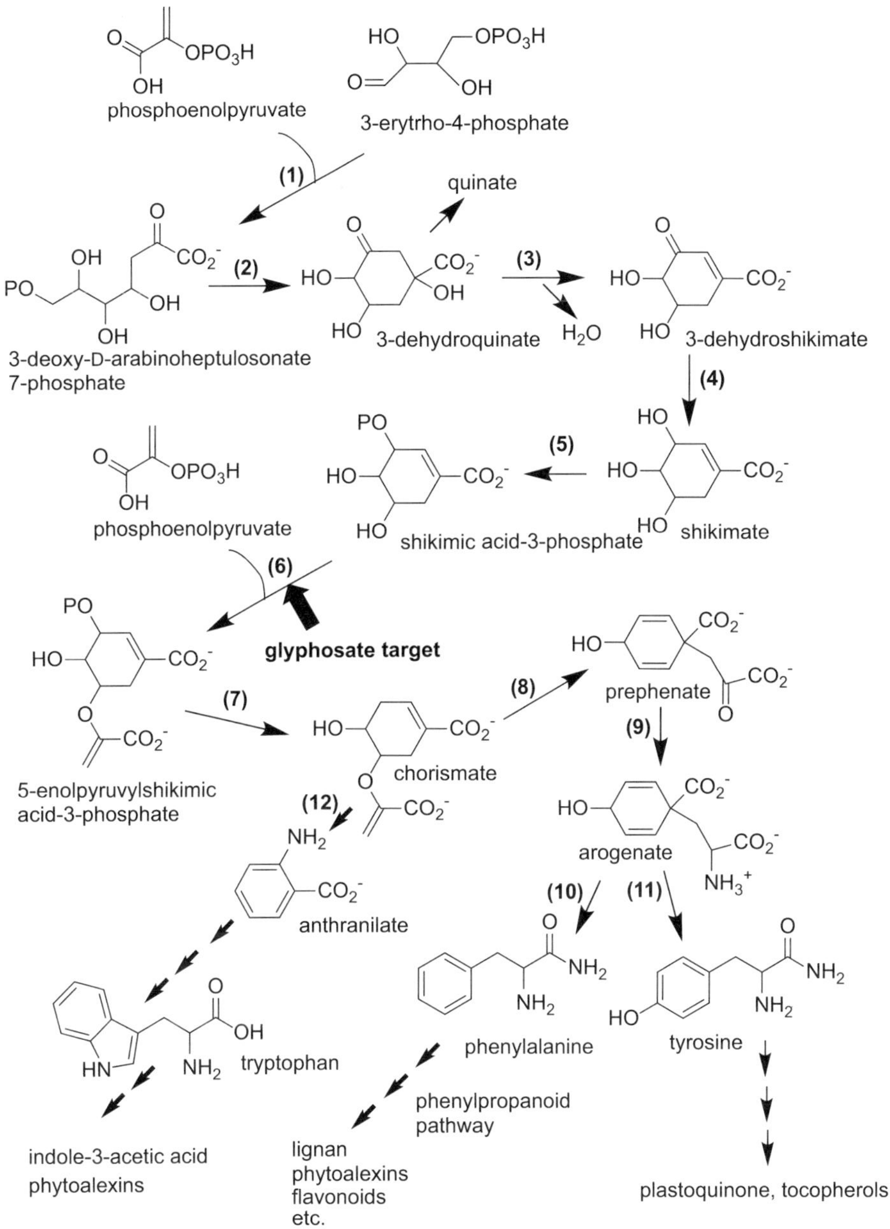

Fig. 4 The shikimate pathway with some of its products. The bold arrow indicates the target of glyphosate (6). Numbered enzymes of the pathway are: (1) 3-deoxy-D-arabinoheptulosonate-7-phosphate synthase (DAHPS); (2) 3-dehydroquinate synthase (DQS); (3) 3-dehydroquinate dehydratase; (4) shikimate dehydrogenase; (5) shikimate kinase; (6) 5-enolpyruvylshikimate-3-phosphate synthase (EPSPS); (7) chorismate synthase; (8) chorismate mutase; (9) prephenate amino transferase; (10) arogenate dehydratase; (11) arogenate dehydrogenase; (12) anthranilate synthase

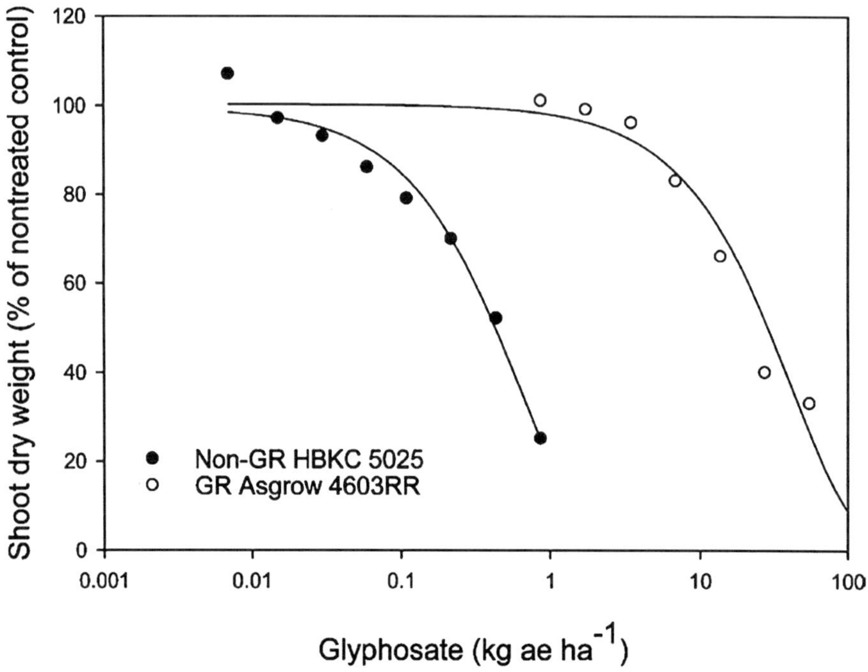

Fig. 5 Growth response of GR soybean (Agrow 460RR) and non-GR soybean (HBKC 5025) 3 weeks after treatment with glyphosate applications to 22-day-old (one- to two-trifoliolate leaf stage) seedlings. The I_{50} values for the GR and non-GR varieties were 22.8 and 0.47 kg a.e./ha, respectively. From Nandula et al. (2007) with permission

the much lower concentrations of glyphosate to which animals are normally exposed (Solomon 2020), compared to those used for weed control, are highly unlikely to affect any P450 monooxygenases of gut microbes of animals, including humans, at concentrations found in the food supply.

Early glyphosate mode of action research findings indicated that it inhibited synthesis of aromatic amino acids (Jaworski 1972). The main clue that led to the discovery of EPSPS as the target of glyphosate (Steinrücken and Amrhein 1980) was the large increase in shikimic acid seen soon after plant exposure to glyphosate (Amrhein et al. 1980). The shikimic acid concentrations in most plant species are exceedingly low and sometimes undetectable. For example, Lydon and Duke (1988) found 0, 0, 0, 5, and 31 nmoles/g dry weight in leaf tissues of *Amaranthus retroflexus*, *Abutilon theophrasti*, soybean, *Lolium perenne*, and *Cyperus esculentus* that had not been treated with glyphosate. Six days after treatment with 10 mM glyphosate, the concentrations were 65, 211, 120, 190, and 135 nmoles/g dry weight, respectively. This rapid, pronounced, and easily measured response to glyphosate is the best biomarker for glyphosate exposure or injury to almost all plants (Harring et al. 1998; Singh and Shaner 1998; Shaner et al. 2005). Even glyphosate exposures which cause no injury or even promote growth (see Sect. 2.5 for discussion of hormesis) can result in shikimate increases (Velini et al. 2008). Hydroxybenzoic acids, such as gallic (CAS 149-91-7), protocatechuic (CAS 99-50-3), and 4-hydroxybenzoic (CAS 99-96-7) acids can also accumulate in glyphosate-treated plants (Lydon and Duke 1988; Becerril et al. 1989) and glyphosate-sensitive microbes (Moorman et al. 1992), apparently derived from shikimate. For example, 6 days of treatment with 10 mM glyphosate caused increases in gallate, protocatechuate, and hydroxybenzoate in soybean leaves from 0.7, 5.9, and 2.3 nmoles/g dry weight to 3.9, 44.6, and 4.8 nmoles/g dry weight, respectively (Lydon and Duke 1988). These biomarkers for glyphosate exposure are not as pronounced as that of shikimate accumulation.

EPSPS is a nuclear-coded enzyme that is located in the plastid. All plant cells contain plastids (green chloroplasts in leaves and other green tissues, chromoplasts (plastids without chlorophyll, but with carotenoids), and leucoplasts (with neither carotenoids nor chlorophyll) such as amyloplasts and etioplasts in roots and other non-green tissues) that are involved in many aspects of plant metabolism other than photosynthesis. Like all other nuclear-coded, plastid enzymes, EPSPS is synthesized in the cytoplasm and enters the plastid by cleavage of a terminal transit peptide in the process of crossing the plastid envelope (della-Cioppa et al. 1986). Unlike other nuclear-coded plastid enzymes, EPSPS is catalytically active in the cytoplasm with its transit peptide (preEPSPS). Furthermore, EPSPS and preEPSPS bind glyphosate with the same affinity. Glyphosate-bound preEPSPS is not processed to EPSPS or taken up by plastids (della-Cioppa and Kishore 1988). Glyphosate has no direct effect on the import of other nuclear-coded enzymes into the plastid.

EPSPS transfers the enolpyruvyl moiety of phosphoenolpyruvate (PEP; CAS 138-08-9) to the 5-hydroxyl of shikimate-3-phosphate (S3P; CAS 63959-45-5) to produce 5-enolpyruvylshikimate-3-phosphate (EPSP; CAS 9068-73-9). The active catalytic site of the enzyme is highly conserved (CaJacob et al. 2003). Glyphosate forms a tight ternary complex with EPSPS and S3P and is competitive with respect to PEP, with a Ki of 1.1 μM, and is an uncompetitive inhibitor with respect to S3P (Boocock and Coggins 1983; Sammons et al. 1995). S3P must bind the enzyme first, followed by either PEP or glyphosate (Anderson et al. 1988; Boocock and Coggins 1983). However, the inhibition is reversible (Boocock and Coggins 1983; Steinrücken and Amrhein 1984). Binding of S3P ligand-free EPSPS causes a large conformational change in the enzyme (Fig. 6a), after which either PEP or glyphosate can bind (Fig. 6b) (Pollegioni et al. 2011). The EPSPS reaction occurs through a tetrahedral intermediate formed between S3P and the carbonation state of PEP, after which inorganic phosphate is released (Anderson and Johnson 1990a, b). The binding interactions of glyphosate and PEP to the same binding site are similar (Eschenburg et al. 2003). The complete enzyme kinetics for each step in the enzymatic production of EPSP from PEP and S3P are discussed in Anderson et al. (1988) and Anderson and Johnson (1990a). The 12 rate constants for EPSPS for the six steps of the EPSPS reaction are provided in Fig. 7. These constants were obtained by analysis of data from a large number of experiments with a computer simulation (modification of KINSIM). The overall equilibrium constant calculated by [EPSP] $[P_i]$/[S3P][PEP] was calculated to be 180 (Anderson and Johnson 1990a).

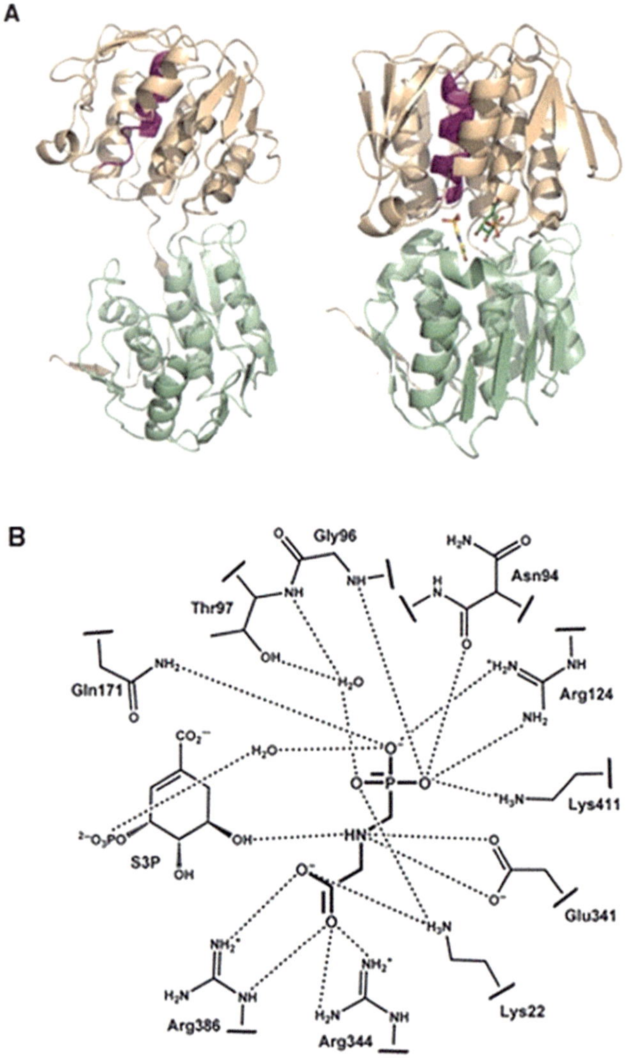

Fig. 6 Molecular binding of glyphosate to EPSPS (**a**) In its ligand-free state, EPSPS exists in the open conformation (left; Protein Data Bank (PDB): 1eps). Binding of S3P induces a large conformational change in the enzyme to the closed state, to which glyphosate or the substrate PEP bind (PDB: 1g6s). The respective crystal structures of the *E. coli* enzyme are shown, with the N-terminal globular domain colored pale green and the C-terminal domain colored brown. The helix containing Pro101 is colored magenta, and the S3P and glyphosate molecules are colored green and yellow, respectively**.** (**b**) Schematic representation of potential hydrogen-bonding and electrostatic interactions between glyphosate and active site residues including bridging water molecules in EPSPS from *E. coli* (PDB: 1g6s). Adapted from Pollegioni et al. (2011)

One of the commercial advantages of glyphosate is that no other inhibitor of EPSPS has been found that is a good herbicide. This is unusual for herbicide target sites, as there are several commercial herbicides targeting most other herbicide

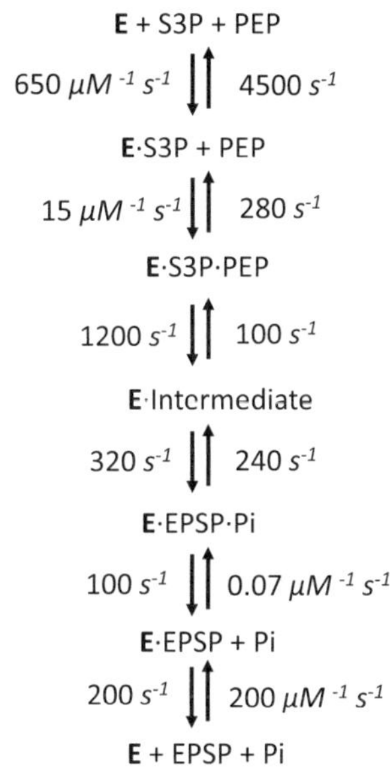

Fig. 7 The six steps of the conversion of S3P and PEP to EPSP and inorganic phosphate by EPSPS (**E**). The tetrahedral intermediate is created by step 3. The equilibrium constants are from Anderson and Johnson (1990a)

targets (Herbicide Resistance Action Committee 2020). Considering the enormous commercial success of glyphosate, it is reasonable to assume that there has been considerable effort to find other herbicides that target EPSPS. Some of these discovery efforts have been published (e.g., Funke et al. 2007; Marzabadi et al. 1996), but none have resulted in a commercial herbicide. Good inhibitors of EPSPS such as *N*-amino-glyphosate (Knowles et al. 1993) have been found, but they have not been commercialized. Good in vitro activity on a molecular target of a herbicide is only one of many characteristics required for commercial viability.

3.2 *Glyphosate Uptake and Translocation*

To have its desired effects, glyphosate must be taken up by the plant and moved to the plastid (in both green and non-green cells), where EPSPS resides in plant cells. Caseley and Coupland (1985) and Duke (1988) reviewed the uptake and translocation of glyphosate decades ago, and little of significance has been added to the literature since then. Of 147 commercial herbicides used in postemergence applications, glyphosate is second to glufosinate as the most hydrophilic (Dayan 2018). Without the help of adjuvants in the solution to be sprayed, glyphosate is poorly taken up by plants compared to the uptake of most other foliar-applied herbicides. A problem with early formulations of glyphosate was that rain within a day or two after application would prevent enough glyphosate from being absorbed by foliage to act effectively as a herbicide. The most efficient formulation (the IPA salt of glyphosate with cationic surfactants, including polyethoxylated tallow amines) studied by Feng et al. (2000) on the common weed *A. theophrasti* resulted in about 15 and 30% of the glyphosate on the leaf being taken up by the plant within 6 and 24 h, respectively. Other less efficient commercial formulations of glyphosate took 24 h for 15% uptake. About 2 and 6% of the glyphosate applied to the foliage had been translocated to the root after 6 and 24 h, respectively, with the most efficient formulation, whereas about 1 and 3.5% was translocated to the root after 6 and 24 h, respectively, with the less efficient formulations. After simulated rainfall at 0.5, 1, and 2 h after application, growth inhibition was doubled by the use of the most efficient formulation over that of the others. The authors concluded that "rainfastness" of the formulation correlates more with the speed (% of glyphosate retained by the leaf after application that is taken up per unit time before a rainfall event) and the quantity of uptake by the foliage than with how much is retained by the leaf surface after a rainfall event. Considerable effort has been exerted in improving the earlier, less rainfast formulations of glyphosate. Unfortunately, some formulation ingredients have proven more toxic than glyphosate (Rodríguez-Gil et al. 2020).

The movement across the cuticle and cell wall is passive, with the rate of diffusion being dependent on many factors such as cuticle composition and thickness, temperature, concentration gradient, and formulation ingredients. With the help of formulation ingredients, sufficient glyphosate for herbicidal effect moves through the leaf cuticle and cell wall to reach the epidermal cell plasma membrane relatively rapidly (e.g., 6 h or less (Feng et al. 2000)). Glyphosate salts (K, Na, NH_4, IPA, and trimethylsulfonium) move across the cuticle better than the free acid of glyphosate, moving in a first order process (Schönherr 2002). For example, at 90% relative humidity (RH), the time for 50% uptake of the free acid was 866 h, whereas that for the IPA salt was ca. 10 h. The time for 50% penetration of the cuticle increased with lower humidity, being ca. 10, 21, and 37 h for the IPA salt at 90, 80, and 70% RH, respectively. The tolerance (not evolved resistance) of some plant species is at least partly due to reduced glyphosate uptake due to low levels of movement from the leaf surface into the plant cells (absorption). For example, Norsworthy et al. (2001) found glyphosate-tolerant *Ipomoea lacunosa* to take up only about 5% of radiolabeled glyphosate in a 0.28 kg a.e./ha glyphosate application 48 h after application. In the same experiment, they found the uptake of glyphosate to be 15 to 62% in three more glyphosate-sensitive species.

With the help of effective formulation ingredients, glyphosate is more readily taken up from sprayed foliage. After traversing the non-living cuticle and cell wall, the herbicide must enter living cells of the leaf and the phloem by crossing the plasma membrane. Early work by Gougler and Geiger (1981) indicated that glyphosate crosses the plasma membrane by passive diffusion, with dependency on glyphosate concentration. After 3 h of exposure, they found a linear relationship between cellular uptake and external glyphosate concentration up to 10 mM with sugar beet leaf discs. However, a later study found uptake through the plasma membrane is first order with respect to extracellular glyphosate concentration, independent of pH and dependent on ATP (Ge et al. 2014). Also, glyphosate does not passively diffuse across semi-permeable membranes such as the plant plasma and vacuolar membranes (Takano et al. 2019). Evidence exists to support the view that phosphate transporters are involved in cellular uptake of glyphosate (Morin et al. 1997; Pereira et al. 2019).

One of the reasons that glyphosate is so effective is that it is a slow-acting herbicide, usually taking several days to kill a plant. It thus has time to be translocated to metabolic sinks such as young, developing leaves and meristems, to which it is translocated in hours (e.g., Gougler and Geiger 1981). Glyphosate moves in both the phloem (symplastic) and xylem (apoplastic) of plants, but its movement in phloem is much greater than xylem movement. Its phloem movement in plants is very much like that of sucrose, with a linear relationship between movement of radiolabeled sucrose and glyphosate from a treated leaf to other parts of a sugar beet plant (Gougler and Geiger 1981; Duke 1988). Gougler and Geiger (1981) found that glyphosate is taken up slowly and released slowly by plant cells, with a plasma membrane permeability of 1.7×10^{-10} m per second, allowing it to accumulate in and be transported by the phloem to plant tissues far from the tissues to which it is applied and taken up before exiting the phloem cell. In a later study from Geiger's lab, glyphosate and CO_2 assimilate accumulated similarly in rhizomes of the perennial weed *Elytrigia repens* (Shieh et al. 1993). McAllister and Haderlie (1985) also found phloem movement of glyphosate and photoassimilate to translocate similarly in *Cirsium arvense*, but they found glyphosate to translocate a little better to roots than photoassimilates. In an analysis of the phloem mobility of all herbicides, based on their pK_a and log K_{ow} values, in Chap. 5 of Devine et al. (1993), glyphosate ranks among the most phloem mobile. Phytotoxic effects on cells that take up glyphosate can limit its movement to phloem cells and translocation in hypersensitive plant species like sugar beet (Geiger and Bestman 1990), but its action in most species is so slow that translocation is initially very good, even at eventually lethal application rates. Some weeds have evolved glyphosate resistance mechanisms based on reduced translocation. This uncommon mechanism of evolved glyphosate resistance is dealt with in the chapter in this volume by Baek et al. (2020).

Vacuolar uptake of glyphosate competes with movement into the phloem and perhaps into the plastid (Ge et al. 2013). In some cases, enhanced vacuolar uptake of glyphosate results in reduced translocation and glyphosate resistance. Grown under similar conditions and treated with the same amount of glyphosate, the fraction of glyphosate that is found in the vacuole varies considerably between species (Ge et al. 2013). Those species with relatively high vacuole content were less sensitive to glyphosate, as vacuolar sequestration removes the herbicide from the translocatable pool, as well as from glyphosate in the plastid.

The shikimate pathway and EPSPS reside in the plant plastid stroma, where the pathway is required for cell maintenance, whether the cell is green or not. As mentioned earlier, glyphosate binds preEPSPS (della-Cioppa and Kishore 1988), so it does not necessarily have to be taken up by the plastid to kill the cell if the EPSPS is poisoned entering the plastid. The relative amount of binding of glyphosate to preEPSPS versus EPSPS in plant cells has not been determined. If glyphosate does enter the plastid, it is probably transported by either a phosphate or an amino acid transporter. Apparently, there is more than one type of glyphosate transporter, as overexpression of one associated with the tonoplast can cause glyphosate resistance, based on sequestration of glyphosate in the plant vacuole (reviewed by Sammons and Gaines 2014). A glutamate/aspartate transporter has recently been reported to also be a glyphosate transporter in the soil bacterium *Bacillus subtilis* (Wicke et al. 2019). The same transporter is also involved in glufosinate transport. Plants also have glutamate/aspartate transporters, but the glutamate transporter of the plastid transports the amino acid from the plastid to the cytosol (Renné et al. 2003), so it may not transport glyphosate into the plastid where EPSPS functions in the shikimate pathway. How much of the glyphosate taken up by the cell that enters the plastid and the mechanism of plastid uptake of glyphosate are still unknown.

3.3 How Inhibition of EPSPS Kills Plants

Only inhibition of EPSPS by glyphosate leads to the processes that eventually kill the plant. Thousands of papers have been published on secondary and tertiary biochemical and physiological effects of glyphosate on plants that provide little insight into its "mode of action." In some cases, people mistake indirect effects for direct effects of glyphosate. For example, many papers describe elevated levels of reactive oxygen species (ROS) in response to glyphosate and insinuate that this effect is somehow unrelated to inhibition of EPSPS (e.g., Gomes et al. 2016). ROS generation is a general effect of stress in plants (Suzuki et al. 2012). Thus, elevation of ROS is a tertiary effect of all herbicides that is not directly related to the target site, except for herbicides that have more direct effects on photosynthesis (photosystem II inhibitors, such as atrazine (2-chloro-4-ethylamino-6-isopropylamino-1,3,5-triazine; CAS # 1912-24-9) and photosystem I energy diverters such as paraquat) (Dayan et al. 2019) and chlorophyll synthesis inhibitors that cause the photodynamic compound protoporphyrin IX (CAS 553-12-8) to accumulate (Dayan and Duke 2003). The latter are all protoporphyrinogen oxidase (PPO, EC 1.3.3.4) inhibitors, such as acifluorfen (5-[2-chloro-4-(trifluoromethyl)phenoxy]-2-nitrobenzoic acid; CAS # 50594-66-6).

Inhibition of EPSPS causes at least three linked effects that can contribute to phytotoxicity. See Fig. 4 for some of the compounds mentioned in this discussion. The most obvious is loss of aromatic amino acids and compounds derived from them, such as IAA and PQ that are essential for plant metabolism. Aromatic amino acids are required for protein synthesis. IAA, derived from tryptophan, is required for many aspects of plant growth and development. PQ, derived from tyrosine, is required for photosynthetic electron transport and is a co-factor phytoene desaturase (EC 1.3.99.31) (PDS) (Breitenbach et al. 2001), an enzyme required for synthesis of carotenoids. PQ is also required for proper functioning of PPO, which is required for chlorophyll synthesis (Brzezowski et al. 2019). At one time, the strong effect of glyphosate on chlorophyll accumulation in plants under some conditions led some to think that it was more than a secondary effect (Kitchen et al. 1981). Many phenylpropanoids (e.g., certain phytoalexins and all flavonoids) and lignin are derived from phenylalanine and tyrosine. Lignin accounts for a major fraction of the carbon passing through the shikimate pathway, especially in woody plants. Blocking production of only these latter products of the shikimate pathway might not kill a plant, at least not quickly enough to be considered as a herbicide. For example, blocking production of phenylalanine-derived secondary compound production in soybean by inhibiting phenylalanine ammonia-lyase (PAL) (EC 4.3.1.24) with the specific inhibitor L-α-aminooxy-β-phenylpropionic acid (AOPP; CAS # - 42990-62-5) does not cause herbicide-like effects (Duke et al. 1980), but, like glyphosate, it reduces production of compounds needed for pathogen resistance, thereby causing greater susceptibility to plant pathogens (e.g., Moerschbacher et al. 1990; Carver et al. 1992) (see Sect. 3.4). An interesting aspect of the paper by Duke et al. (1980) is that both glyphosate and AOPP induced high levels of extractable PAL activity, indicating the phenylalanine pools in glyphosate-treated plants are not only not replenished, but are also probably further reduced by enhanced in vivo PAL activity. The same phenomenon was found in maize treated with glyphosate (Duke and Hoagland 1978). A problem with loss of aromatic amino acids and essential compounds made from them as the only mechanism of action is that, even though glyphosate clearly causes depletion of free pools of aromatic amino acids, providing glyphosate-treated plants with supplementary aromatic amino acids does always not provide strong rescue of glyphosate-treated plant cells or tissues (e.g., Haderlie et al. 1977; Jenson 1985) or bacteria (Fischer et al. 1986). Other, more indirect effects of inhibition of EPSPS may contribute to the herbicidal effects of glyphosate.

There is evidence that prephenate (1-(2-carboxy-2-oxoethyl)-4-hydroxycyclohexa-2,5-dienecarboxylic acid; CAS # 126-49-8) and/or arogenate (1-[(2*S*)-2-azaniumyl-2-carboxylatoethyl]-4-hydroxycyclohexa-2,5-diene-1-carboxylate; CAS # 53078-86-7) may be feedback inhibitors of the shikimate pathway at the level of 3-deoxy-D-arabinoheptulosonate-7-phosphate synthase (EC 2.5.1.54) (DAHPS) (Fischer et al. 1986; Jenson 1985; Herrmann 1995). In a few plant species, one or more aromatic amino acids may act as a feedback inhibitor at the DAHPS level (Maeda and Dudareva 2012; Zulet-González et al. 2020). The relative gene expression of several enzymes of the shikimate pathway in *Amaranthus palmeri* leaf discs was elevated by exposure to glyphosate, and a mixture of the aromatic amino acids reduced this effect (Zulet-González et al. 2020). Arogenate and prephenate levels will still be depleted in glyphosate-treated plants cells provided exogenous aromatic amino acids, so deregulation of the shikimate pathway may not be completely corrected. Reduced products of the shikimate pathway will result in elevated DAHPS activity, subsequently causing consumption of erythrose-4-phosphate (E4P; CAS 585-18-2), PEP and ATP, to produce uncontrolled production of shikimate and other derivatives of intermediates of the shikimate pathway that occur before EPSPS, depleting carbon fixation pathways of key intermediates (e.g., PEP and erythrose-4-phosphate) and ATP. Glyphosate reduces carbon flow to the carotenoid pathway (Corniani et al. 2014), but part of this reduction could be due to reduced PDS activity because of reductions of PQ synthesis from tyrosine. If depletion of carbon fixation intermediates is sufficient, greatly reduced carbon fixation would be relatively rapid (<2 h), as seen in glyphosate-treated sugar beets (Geiger et al. 1986; Servaites et al. 1987). Cessation of carbon fixation in strong sunlight will result in energy dissipation through destructive oxidative processes. The symptoms of glyphosate toxicity in most species are not consistent with this mechanism. However, these symptoms are seen in sugar beets (Geiger and Bestman 1990; Madsen et al. 1986) and in GR *Ambrosia trifida* in which the effects are so rapid that the sprayed foliage dies very rapidly like that of glyphosate-sensitive sugar beet (van Horn et al. 2018; Moretti et al. 2018). In the case of GR *A. trifida*, the foliage dies before glyphosate can be translocated to meristems from which the plant regrows (similar to what is seen with paraquat treatment). In the few plant species like sugar beet and GR *A. trifida*, the drain of intermediates and ATP caused by deregulation of the shikimate pathway by glyphosate may be rapid, causing strong inhibition of carbon fixation, resulting in photosystem energy dissipation via ROS, a rapid process. This process probably occurs to a lesser degree in other plant species under certain environmental situations (e.g., strong sunlight).

The herbicide efficacy of glyphosate on some weeds species diminishes with increases in atmospheric CO_2 concentrations (Ziska and Teasdale 1999, 2000; Ziska et al. 2004; Ziska and Goins 2006). The enzyme responsible for most carbon fixation by plants is ribulose-1,5-bisphosphate carboxylase (4.1.1.39) (RUBISO). RUBISCO is a very inefficient enzyme because of its low affinity for CO_2 and the competition of CO_2 and O_2 for the same binding site. Photorespiration occurs when RUBISCO uses O_2 instead of CO_2, resulting in adding oxygen to ribulose-1,5-bisphosphate to produce 3-phosphoglycerate ((2*R*)-2-hydroxy-3-phosphonooxypropanoic acid; CAS 820-11-1)) (PGA) and 2-phosphoglycolate (2-phosphonatooxyacetate; CAS 13147-57-4) (2PG). 2PG inhibits some enzymes involved in carbon fixation. Thus, photorespiration not only wastes energy produced by the photosystems of photosynthesis, but also inhibits carbohydrate production from fixed CO_2. Plants that rely on RUBISCO for carbon fixation are termed C3 plants because RUBISCO produces PGA, a three-carbon compound, by combining ribulose-1,5-bisphosphate and CO_2. Elevated CO_2 levels increase the enzymatic efficiency of RUBISCO, enhancing photosynthesis in C3 plants. Some plants, such as most grasses (Poaceae), have a more efficient means of carbon fixation, in which CO_2 is first fixed by the enzyme PEP carboxylase (EC 4.1.1.31) to produce oxaloacetate (2-oxobutanoic acid; CAS 328-42-7), a four-carbon compound. Thus, these plants are termed C4 plants. With PEP carboxylase, CO_2 does not compete significantly with O_2, and CO_2 levels are not as limiting for C4 plants as with C3 plants. The anatomy of the leaves of C4 plants is usually characterized by an inner ring of cells (bundle sheath cells) that fix carbon with RUBISCO, surrounded by mesophyll cells that fix carbon with PEP carboxylase. The mesophyll cells provide high concentrations of CO_2 to bundle sheath cells, so that their RUBISCO is more efficient.

The reduction of glyphosate activity by elevated CO_2 (up to 250 ppm above ambient 360 ppm) levels is more pronounced and consistent in C3 than in C4 plants (Ziska and Goins 2006; Fernando et al. 2016) (Table 1), as might be expected because C3 plants do not have a means of concentrating CO_2 to enhance carbon fixation as C4 plants do. In fact, C4 plant growth is saturated at 360 ppm atmospheric

Table 1 Glyphosate efficacy changes with increased ambient CO_2 levels of various C3 and C4 plants

Species	Carbon fixation pathway	Change in efficacy	Reference
Chenopodium album	C3	Reduced	Ziska and Teasdale (1999)
Circium arvense	C3	Reduced	Ziska et al. (2004)
Conyza canadensis	C3	Reduced	Matzrafi et al. (2019)
Elytrigia repens	C3	Reduced	Ziska and Teasdale (2000)
Amaranthus retroflexus	C4	None	Ziska and Teasdale (1999)
Parthenium hysterophorus	C3 & C4	None	Bajwa et al. (2019)
Parthenium hysterophorus	C3 & C4	Reduced	Cowie et al. (2020)
Chloris gayana	C4	Reduced	Manea et al. (2011)
Cyperus esculentus	C4	None	Marble et al. (2015)
Cyperus rotundus	C4	None	Marble et al. (2015)
Eragrostis curvula	C4	Reduced	Manea et al. (2011)
Paspalum dilatatum	C4	Reduced	Manea et al. (2011)
Sporobolus indicus	C4	None	Manea et al. (2011)

Adapted and updated from Ziska (2014)

CO_2 (Leegood 2002), a level slightly lower than current atmospheric CO_2 concentration (410 ppm), making C4 plants less likely to respond positively to CO_2 above current levels. Glyphosate efficacy is compromised in a few C4 plants by elevated CO_2 concentrations and not others (Table 1). In the case of *Parthenium hysterophorus*, different tissues and different developmental stages can be C3 or C4. Accordingly, researchers have reported elevated CO_2 (600 to 800 ppm) to have both no effect (Bajwa et al. 2019) or a reduction (Cowie et al. 2020) on glyphosate efficacy on *P. hysterophorus*, but how much of the tissues were C3 and C4 in the two studies was not reported.

The clear decrease in glyphosate efficacy in C3 plants could be due to two causes. The additional growth of C3 plants at elevated CO_2 concentrations will dilute a glyphosate concentration, reducing the amount per unit of fresh weight. Furthermore, the additional fixation of CO_2 in C3 plants at high CO_2 concentrations should reduce the effect of glyphosate in draining metabolic intermediates from carbon fixation pathways. Thus, the reduced effect of glyphosate on C3 plants at high CO_2 concentrations supports the view that part of the mode of action of glyphosate is deregulation of the shikimate pathway to drain intermediates from metabolic pathways. These findings suggest that future glyphosate use will increasingly favor C3 weeds (e.g., *Chenopodium album*, *A. theophrasti*, and *Convolvulus arvensis*) as atmospheric CO_2 levels increase.

A third part of the mode action of glyphosate may be accumulation of toxic derivatives of the shikimic acid pathway (Dayan and Duke 2020). Most plant species have very low levels of S3P (a substrate of EPSPS) or shikimate (the substrate of shikimate kinase (EC 2.7.1.71), the enzyme just before EPSPS) (Fig. 4), but treatment with glyphosate causes high levels of accumulation of shikimate and to a lesser extent hydroxybenzoic acids (e.g., protocatechuate) and quinate (CAS 77-95-2), another product of a shikimate pathway intermediate (3-dehydroquinate; CAS 10534-44-8) (Fig. 4). Quinate can also be generated from shikimate by quinate hydrolyase (EC 4.2.1.10) (Bentley 1990). Plants treated with acetolactate synthase (ALS; EC 2.2.1.6; also called acetohydroxy acid synthase – AHAS) inhibitor herbicides also accumulate high levels of quinate (Orcaray et al. 2010). The mechanism of this effect of ALS inhibitors is unknown.

The levels of shikimate that accumulate in response to glyphosate treatment generally dwarf those of quinate and hydroxybenzoates. Although no data on the phytotoxicity of shikimate could be found, there are reports that shikimate inhibits PEP carboxylase at high concentrations ($I_{50} = 71$ µM for leaf and ca. 5 mM for nodular PEP carboxylase) (Colombo et al. 1998; de María et al. 2006). There is an additive effect of shikimate and protocatechuate as PEP carboxylase inhibitors (de María et al. 2006), so that the combined concentrations of these inhibitors could be sufficient in some tissues of glyphosate-treated plants to significantly inhibit PEP carboxylase. As mentioned above, this enzyme is a key enzyme in carbon fixation in C4 plants. It also has an important role in C/N metabolism in C3 plants (Chollet et al. 1996). It is amazing that there has been no further research to determine whether shikimate itself is causing metabolic disruption through inhibition of PEP carboxylase. A more indirect contribution of toxicity by glyphosate-caused shikimate accumulation may be through shikimate-caused induction of genes of the shikimate pathway (Zulet-González et al. 2020), further deregulating the pathway to cause metabolic disruption.

Quinate is moderately phytotoxic, causing some of the effects of glyphosate (Orcaray et al. 2010; Zabalza et al. 2017, 2020; Zulet et al. 2013), and, as mentioned above, shikimate can be converted to quinate in vivo. Therefore, at least part of the effects of glyphosate in some plant species may be due to high levels of quinate. A non-phytotoxic application concentration (400 mM) of quinate applied with a mildly phytotoxic application rate (0.21 kg a.e./ha) of glyphosate-killed *Amaranthus palmeri* plants (Zulet-González et al. 2019). Treatment with quinate did not increase the shikimate levels in the plants over that caused by glyphosate alone. Neither glyphosate nor quinate alone caused increases in the extractable activity of the shikimate pathway enzyme anthranilate synthase (EC 4.1.3.27), but glyphosate with quinate caused a four-fold increase in the enzyme. In quinate-sensitive *Papaver rhoeas*, the mode of action of quinate as a herbicide appeared to be related to general perturbations in carbon/nitrogen metabolism, rather than to specific effects on the shikimate pathway (Zabalza et al. 2020).

Shikimate and quinate are both usually found at very low levels (undetectable to a few ppm of dry weight) in plant tissues of the majority of plant species, making their accumulation an excellent biomarker for glyphosate exposure. However, a few plant species accumulate high levels of shikimate (e.g., star anise (*Illicium verum*) and sweetgum (*Liquidambar styraciflua*)) (Enrich et al. 2008; Ghosh et al. 2012) and quinate (e.g., *Cinchona officianalis*) (Eliel and Ramirez 1997) without exposure to glyphosate. In order to avoid autotoxicity, these plants probably have a means of compartmentalizing these compounds away from cells involved in normal growth and development, as is commonly found with many other compounds that can cause autotoxicity to plants (reviewed by Duke et al. 1999). Interestingly, both shikimate and quinate can be starting compounds for synthesis of the anti-influenza pharmaceutical oseltamivir (ethyl (3*R*,4*R*,5*S*)-5-amino-4-acetamido-3-(pentan-3-yloxy)-cyclohex-1-ene-1-carboxylate; CAS # – 196618-13-0) (Ghosh et al. 2012; Federspiel et al. 1999), and glyphosate-treated plants have been proposed as a source of these oseltamivir precursors (Matallo et al. 2014; Hobbie et al. 2017).

In summary, glyphosate probably adversely affects plants by more than just reducing levels of aromatic amino acids and necessary compounds derived from these three amino acids. The role of deregulation of the shikimate pathway with ensuing disruption of carbon assimilation and of phytotoxic quinate accumulation probably varies between species and within a species, depending on the developmental and environmental factors. Variations in the roles of these processes between different tissues and cell types in a plant are also likely. Thus, the mode of action of glyphosate is apparently more complex than might be expected. Nonetheless, all of the effects are ultimately due to the inhibition of EPSPS.

The fact that there are no commercial herbicides that target other enzymes of the shikimate pathway may reflect that other targets may not cause all of the metabolic dysfunctions caused by glyphosate, even though they block the shikimate pathway. For example, the natural cyanobacterial compound 7-deoxy-sedoheptulose, an inhibitor of 3-dehydroquinate synthase (DQS; EC 4.2.3.4), an early step of the shikimate pathway (Fig. 4), is phytotoxic and has been proposed as a herbicide (Brilisauer et al. 2019). Inhibition of this enzyme does not cause quinate or shikimate to accumulate; however, it does cause accumulation of the substrate of DQS, 3-deoxy-D-arabino-heptulosonate-7-phosphate (CAS # 2627-73-8). Thus, it should cause deregulation of the shikimate pathway.

Some have claimed that glyphosate causes either direct effects on plants due to its ability to chelate divalent metal cations, and they have claimed such an effect occurs when glyphosate is applied to GR crops (e.g., Yamada et al. 2009; Zobiole et al. 2010; Martinez et al. 2018; Mertins et al. 2018). This purported effect on plant mineral nutrition was proposed to be linked to greater plant disease in GR crops treated with glyphosate (e.g., Johal and Huber 2009; Kremer and Means 2009). Glyphosate does reduce the ability of non-GR plants to fight plant disease, but this phenomenon is related to reduced levels of defense compounds (see Sect. 3.4) and not to effects on mineral nutrition. The topic of mineral chelation in plants and its potential role in the mode of action of glyphosate was reviewed by Duke et al. (2012), who concluded that the data debunking this hypothesis are much stronger than those supporting it. Since this review was published, additional support has accumulated in support of the view that none of glyphosate's mode of action is associated with effects on plant mineral nutrition (e.g., Costa et al. 2018; Duke et al. 2018b; Kandel et al. 2015; Reddy et al. 2018). These papers found no effects of glyphosate applications on mineral content of GR maize and GR soybean treated with recommended glyphosate application rates in replicated field experiments over more than 1 year at sites in different states of the USA, one Canadian province, and Brazil. The generally steady increase in yields of cotton, maize, and soybean USA, after more than 90% adoption of GR varieties, argues against there being any significant phytotoxicity issues with glyphosate in these crops. A recent short review summarized the state of the current knowledge of this topic (Duke and Reddy 2018).

There have been exceedingly few recent papers that meaningfully probe the mode of action of glyphosate, but there have been many papers describing secondary and tertiary effects. Modern metabolomic, transcriptomic, proteomic, and other methods (e.g., Maroli et al. 2016, 2018a, b; Patterson et al. 2020) provide tools for a better understanding of the more direct effects of inhibiting the shikimate pathway at the EPSPS site. However, interpretation of massive amounts of metabolomic or transcriptomic data to gain insight into a herbicide mode of action can be challenging (Duke et al. 2013, 2018a).

3.4 Role of Microbes in Glyphosate Efficacy

An important part of the mode of action of glyphosate in the field is the role of reduction of plant defenses to plant pathogens. A sublethal application of some, but not all, herbicides can predispose a plant to greater susceptibility of a herbicide (see review by Altman and Campbell (1977)), but this effect is more pronounced with glyphosate (Hammerschmidt 2018). Glyphosate is a more effective herbicide in soil containing microbes than in sterilized soil because of the reduction in plant defenses to soil-borne pathogens by glyphosate (Lévesque and Rahe 1992; Schafer et al. 2012, 2013) (Fig. 8). The reduction in shikimate pathway-derived plant defense compounds (e.g., phytotalexins and lignin) against plant pathogens by glyphosate has been used to enhance and synergize the efficacy of microbial bioherbicides (Christy et al. 1993; Duke et al. 2007; Hoagland et al. 2018; Gressel 2010). For example, Sharon et al. (1992) found a concentration of glyphosate (50 μM) that caused no visible phytotoxicity to the weed *Cassia obtusifolia* (now *Senna obtusifolia*) to almost completely block synthesis of the shikimic pathway-derived phytoalexin 2-(*p*-hydroxyphenyloxy)-5,7-dihydroxychromene. This glyphosate concentration, combined with a dose of the mycoherbicide derived from *Alternaria cassia* conidia that caused only a few necrotic spots on the foliage when used alone, completely killed the weed. This is but one of many examples of the lowering of plant defenses to pathogens by glyphosate. The topic of glyphosate's effects on plant disease via inhibition of shikimate pathway-derived defenses is reviewed in detail by Hammerschmidt (2018) and Duke et al. (2018b).

Although the contribution of soil pathogens to glyphosate efficacy has been clearly demonstrated under controlled conditions, little is known of this effect in the field. The magnitude of this augmentation to glyphosate efficacy by the activity by pathogens would be dependent on several factors, such as both the types and amounts of pathogens in the soil and environmental conditions. We know that glyphosate acts as a fungicide on some plant pathogens (see Sect. 6), so the balance between direct effects of glyphosate on the pathogen and indirect effects from reducing the plant's capacity to produce pathogen defenses could be complicated, depending on many factors.

Fig. 8 *Ambrosia trifida* grown in sterile (**a**) and unsterile (**b**) soil sprayed with different rates (kg ae/ha) of glyphosate at 21 days after spraying. From Schafer et al. (2012) with permission

4 Metabolic Degradation of Glyphosate in Microbes and Plants

There are non-enzymatic means by which glyphosate can degrade by breakage of the C-P bond. For example, Barrett and McBride (2005) reported that both glyphosate and AMPA are degraded by breaking the C-P bond by Mn oxide in aqueous media. Glyphosate degraded faster than AMPA by this mechanism. Because AMPA has a longer half-life than glyphosate in soil (e.g., Simonsen et al. 2008), this mechanism could contribute to glyphosate metabolism in soil. Metal ions in solution have also been implicated in abiotic degradation of glyphosate to AMPA (Yael et al. 2014).

Biological degradation in nature is clearly the predominant mechanism of breakdown of glyphosate, because degradation in sterile soil is nil (e.g, Torstensson and Aamisepp 1977). Soil type can influence the rate of degradation (e.g., Qiao et al. 2020), but how much of this variation is due to differences in bioavailability and microbial differences has not be determined. There are two means of metabolic degradation of glyphosate (Borggaard and Gimsing 2008; Duke 2011; Nandula et al. 2019; Zhan et al. 2018). The predominant route is via a glyphosate oxidoreductase (GOX; EC 1.5.3.23) that converts glyphosate to AMPA and glyoxylate

(CAS # 563-96-2) (see Fig. 3 of Green and Siehl 2020), a common, natural metabolite. AMPA is also a degradation product of some detergents, so some of the AMPA found in the environment is from this source (e.g., Botta et al. 2009). A gene (*goxv247*) that encodes GOX from the soil microbe *Ochrobactrum anthropi* was identified, cloned, and used as a glyphosate resistance transgene in the first commercialized GR canola cultivars (Green 2009). An apparently less common means of glyphosate metabolic degradation is by a C-P lyase that converts the herbicide to the sarcosine (CAS # 107-97-1), a natural product, and inorganic phosphate (e.g., Kishore and Jacob 1987; Jacob et al. 1988) (see Fig. 3 of Green and Siehl 2020). The reviews of Zhan et al. (2018) and Singh et al. (2020) provide similar lists of microbes that degrade glyphosate. Most of those listed degrade it with a GOX enzyme and most are bacteria, although some fungi, such as *Penicillium citrinum* (Zboińska et al. 1992), *Alternaria* sp., and *Trichoderma* spp. (Krzyśko-Łupicka and Orlik 1997) also degrade glyphosate. Sarcosine is seldom found or found in very small amounts in studies on the degradation of glyphosate in soils (e.g., Al-Rajab et al. 2008) and plants (Duke 2011). However, being a natural metabolite, it may have a shorter half-life than AMPA, which might mask the importance of this metabolic degradation route. Also, sarcosine if not always looked for in studies of the degradation of glyphosate (e.g., Arregui et al. 2003).

AMPA, which is more environmentally persistent than glyphosate, requires a C-P lyase enzyme to be degraded. Microbes that break down glyphosate with a C-P lyase can also metabolize AMPA, using both glyphosate and AMPA as a sole source of phosphorus (Selvapanidiyan and Bhatnagar 1994), although some of these microbes apparently have both a GOX and a C-P lyase (Lerbs et al. 1990; Obojska et al. 2002). The greater persistence of AMPA than glyphosate in soils may indicate that microbes that use this degradation pathway are less common than those with GOX. The biochemistry and genetics of C-P lyases that metabolize glyphosate and AMPA are reviewed by Hove-Jensen et al. (2014). The finding that glyphosate is readily broken down by many different microbes has led to numerous papers and patents on use of such microbes to remove glyphosate from soil (e.g., Ermakova et al. 2010) and water (e.g., Hallas et al. 1992). The need for such bioremediation is unlikely in normal use of glyphosate for weed management because of its relatively short half-life in soil and water (Blake and Pallett 2018; Rodríguez-Gil et al. 2020).

A number of publications exist on other microbial enzymes that will transform glyphosate to non-herbicidal compounds. These other glyphosate-degrading enzymes include glyphosate *N*-acetyltransferase (GAT) (Castle et al. 2004), a bacterial glycine oxidase (Nicolia et al. 2014), and a glyphosate decarboxylase (Hammer et al. 2007). There are no data indicating that any of these routes of degradation of glyphosate are significant in the environment. In the case of GAT, the enzymatic activity with glyphosate is so low for the enzyme obtained from *Bacillus licheniformis* that several rounds of gene shuffling with selection for the best enzymatic activity were needed to generate a gene encoding a GAT that could be used to produce a GR crop (Castle et al. 2004). Although most of the genes for these glyphosate-transforming enzymes were proposed for producing GR crops, only a gene for GOX was used in one GR crop (Green 2009), and it is no longer used (see discussion below). GR crops with the highly engineered *GAT* gene reached a high level of development (Green et al. 2008), but were never commercialized.

Glyphosate degradation in the environment can be enhanced by certain animals in soil and water. For example, glyphosate and AMPA degradation in soil containing earthworms (*Eisenia fetida*) is faster than the same soil without earthworms (Lescano et al. 2020). In this study, the earthworms were not harmed by glyphosate. Degradation of glyphosate in water is enhanced by the presence of the golden mussel (*Limnoperna fortunei*) (Gattás et al. 2020). However, there is no evidence in either of these papers that the animals themselves degrade glyphosate.

Plants also degrade glyphosate, predominantly by a GOX-type activity, although evidence of metabolism by a C-P lyase has been reported in a few species of higher plants (Duke 2011). The amount of GOX activity varies considerably, from little or no metabolism in some grasses to much higher levels in some legumes, including soybeans (Reddy et al. 2008; Duke 2011). Vemanna et al. (2017) showed that rice (*Oryza* spp.) has an aldo-keto reductase (AKR) that acts as a GOX and that, when used as an overexpressed transgene, can provide glyphosate resistance to tobacco (*Nicotiana tabacum*). There are many AKRs in plants with a wide spectrum of substrates, and some of these monomeric enzymes are associated with abiotic stress (Sengupta et al. 2015). Pan et al. (2019) reported that the evolved glyphosate resistance of a GR weed (*E. colona*) is due to elevated AKR activity due to two upregulated *AKR* genes. This is the most clearly confirmed case of evolved resistance to glyphosate by enhanced metabolic degradation (Duke 2019), although a few other cases have been reported (summarized by Baek et al. (2020)). However, this GR *E. colona* with enhanced AKR-mediated degradation of glyphosate was later shown to also have a GR EPSPS (McElroy and Hall 2020). The relative contributions of the two resistance mechanisms have not been determined. Whether AKR is the only enzyme responsible for GOX type of glyphosate metabolism to AMPA in plants is unknown. We also do not know if all plants have an AKR with GOX activity. In plants in which low levels of glyphosate metabolism occurs, finding AMPA in glyphosate-treated plants is difficult. In such plants, metabolism could be due in part or wholly to endophyte metabolism of the herbicide, as endophyte-mediated metabolism of other herbicides has been documented (Tétard-Jones and Edwards 2016), and some endophyte-type microbes can metabolize glyphosate (discussed in Sect. 6).

The amount of AMPA found in glyphosate-treated plants will be both a function of the amount of glyphosate applied to the plant and the sensitivity of the plant to that application rate, as a very toxic amount of glyphosate will reduce the ability of the plant to metabolize it. This was considered in the study of Reddy et al. (2008), in which the ratios of glyphosate to AMPA in non-GR plant species treated with an application rate of glyphosate that inhibits growth by 50% were compared at 7 days after treatment in a variety of plant species. This ratio will be affected by many factors, such as time after treatment, species, and degradation of AMPA. AMPA was found in most species, and the ratio of glyphosate to AMPA was less than 10 in three species, indicating strong metabolism of glyphosate. No AMPA was detected in four species, including both GR and non-GR maize. Nevertheless, in a later field study using higher application rates of glyphosate, the same scientists, using the same analytical methods, found low levels of AMPA in glyphosate-treated GR maize leaves in one field site (Mississippi, USA) two years in a row, but not in another site (Illinois, USA) with different GR cultivars (Reddy et al. 2018). However, the harvested seeds at the Mississippi site had no AMPA, whereas a very low AMPA level (ca. 30 ng/g dry wt) was found in seeds in one of 2 years in Illinois. Bernal et al. (2012) also reported levels of AMPA in glyphosate-treated (1.6 kg/ha) GR maize leaves that were 65-fold less than the glyphosate levels at a week after treatment. Both glyphosate and AMPA concentrations decreased with time after spraying, but the ratio of glyphosate to AMPA decreased with time, indicating that AMPA degrades and/or translocates more slowly than glyphosate in vivo. AMPA and glyphosate compete for movement into the vacuole and the cell and perhaps the plastid (Ge et al. 2013), so these processes will also influence the ratio. Hearon et al. (2021) reported AMPA to be readily taken up by GR maize from soil treated with either AMPA or glyphosate, so all AMPA found in GR maize is not necessarily from degradation of glyphosate in the plant. They also claimed conversion of glyphosate to AMPA *in planta*, but this was not rigorously proven. The finding that microbe-free cell cultures of maize can metabolize glyphosate to AMPA (Komoßa et al. 1992) proves that maize has an enzyme that can act as a GOX at relatively low in vivo activity levels compared to some other species. In general, however, members of the Poaceae (Gramineae) like maize have very little capability for degrading glyphosate (Duke 2011).

The question of how much glyphosate and AMPA ends up in harvested GR crops is of great interest because of the current human toxicology controversy. Because glyphosate preferentially translocates from sprayed foliage to metabolic sinks such as developing seeds and storage organs (e.g., sugar beet roots) (Duke et al. 2003a; Gougler and Geiger 1981) and GR crops are not impaired in any significant way by the application rates of glyphosate used for weed management (Nandula et al. 2007), one would expect high levels of glyphosate and/or AMPA in harvested parts of GR crops. The only GR crops for which glyphosate and AMPA residue data from peer-reviewed papers exist are GR soybean, sugar beet, and maize. As mentioned above, neither glyphosate nor AMPA is found in processed sugar from GR sugar beets (Barker and Dayan 2019). In a field study, only trace amounts (<0.1 ppm) of glyphosate were found in GR sugar beet roots 2 weeks after spraying glyphosate either 0.825 kg a.e./ha once or 1.26 kg a.e./ha twice approximately 6 weeks apart. At harvest, the glyphosate concentrations in the fresh sugar beet root from different fields ranged from 1.5 (one glyphosate application) to 32 ppb (two applications). The USEPA MRL for fresh sugar beet is 10 ppm. Glyphosate is exuded from roots of glyphosate-treated non-GR plants (e.g., Coupland and Caseley 1979; Rodrigues et al. 1982; Kremer et al. 2005; Laitenen et al. 2007; Barker and Dayan 2019) provided evidence that rapid loss of glyphosate in sugar beet roots is due to root exudation. There are more reports of glyphosate exudation from plant roots than for any other herbicide (Ghanizadeh and Harrington 2020), but there are no reports that this means of glyphosate loss from the plant contributes to glyphosate resistance (Duke 2019; Baek et al. 2020). Roots of some species can retain glyphosate for long periods, as glyphosate and AMPA were found in the roots of perennial, herbaceous plants where glyphosate was applied at a rate of 2.16 kg a.e./ha a year earlier, with tissue glyphosate concentrations ranging from 77 to 1,050 ppb and AMPA from 16 to 48 ppb (Wood 2019). AMPA was not found in all species and, when found, shoot concentrations of both were much lower than root concentrations.

Both glyphosate and AMPA accumulate in the harvested seed of GM soybean (Arregui et al. 2003; Duke et al. 2003b, Bohm et al. 2014, Bøhn et al. 2014), but reported residues are generally well within the maximum tolerance level. For example, the MRL for glyphosate content for soybean seed in the USA is 20 ppm (United States Electronic Code of Federal Regulations 2020), and concentrations as high as 10 ppm have been reported in a survey of GR soybeans grown in the USA (Bøhn et al. 2014). The yearly analysis of soybean samples in the USA by the U.S. Department of Agriculture, Agricultural Marking Service (USDA AMS 2020) reports no presumptive tolerance violations. Only trace amounts of glyphosate or AMPA are sometimes found in GR maize grain (Reddy et al. 2018; Costa et al. 2018). No glyphosate has been reported in maize in the annual USDA AMS (2020) survey. Very low levels or no glyphosate in seeds of glyphosate-treated GR maize is surprising because, just as developing soybean seeds are metabolic sinks that accumulate glyphosate along with photosynthate, developing maize seeds should also accumulate glyphosate along with sucrose from sprayed leaves. Because at least three labs have found either trace amounts or no glyphosate in seed of glyphosate-treated GR maize from different locations and in multiple years, maize seeds of glyphosate-treated GR maize apparently do not accumulate significant glyphosate or AMPA residues.

The first commercial varieties of GR canola contained transgenes for both a bacterial GOX and a GR form of EPSPS (Green 2009). These two transgenes provide a resistance factor of about 50-fold (Nandula et al. 2007). Only one paper has examined AMPA formation in one of these GR varieties in detail (Corrêa et al. 2016). In a laboratory study, at a very low application rate of radiolabeled glyphosate, virtually all of the glyphosate applied to the plants was converted to AMPA within 7 days, whereas very little AMPA was produced in a conventional, non-GR, isogenic variety. Only AMPA and no glyphosate was found in untreated leaves of GR canola. Whether this AMPA was translocated from treated leaves or was formed by oxidation of translocated glyphosate in the untreated leaves was not determined. In a greenhouse study, plants treated with 3.3 kg a.e./ha converted about a third of the glyphosate taken up to AMPA within 2 weeks. How much the added GOX activity contributed to glyphosate resistance in this GR crop is unknown because there are no publicly available data comparing glyphosate resistance imparted by only the GOX gene, only the GR EPSPS gene, and the two genes together in the same canola germplasm or even different germplasms. However, later varieties of GR canola have only a transgene for GR EPSPS, so the contribution of the GOX was apparently not necessary unless the level of expression of the GR EPSPS gene in these first canola varieties was insufficient for robust resistance. Why the GOX transgene was used with a GR EPSPS in the first GR canola varieties was never disclosed.

AMPA is weakly phytotoxic (Hoagland 1980; Gomes et al. 2014), and GR crops are not resistant to AMPA (Reddy et al. 2004; Ding et al. 2011), indicating that

AMPA has one or more molecular targets other than EPSPS. Amounts of AMPA or glyphosate applied to GR soybeans that result in the same levels of AMPA within the plant tissues result in similar phytotoxicity symptoms (Reddy et al. 2004). Under rare environmental conditions, glyphosate-treated GR soybean accumulates enough AMPA to cause chlorosis (called "yellow flash" by farmers). This effect is not seen in all GR soybean varieties (Cerny et al. 2014). These differences could be due to different AMPA levels accumulating in the different varieties, but differences in glyphosate degradation between varieties have not been determined under the same conditions. The yellow flash effect is temporary and has not been found to affect yield of the crop. Yellow flash is not seen in GR maize, which produces little or no AMPA when treated with glyphosate. Reddy et al. (2004) concluded that yellow flash in glyphosate-treated soybean is due to the phytotoxicity of accumulated AMPA. As noted above, glyphosate-treated canola with the GOX gene accumulates high levels of AMPA, but yellow flash has not been reported in canola. The reason(s) for this difference is unclear, especially since the phytotoxicity effects of treating GR soybean and GR canola with AMPA are similar (Nandula et al. 2007).

5 Non-target Vegetation Effects

Glyphosate is non-selective, so it can be harmful to almost all plant species if the dose is high enough. Non-target vegetation can be exposed to glyphosate by exposure to the root or foliage. Although some have discussed the potential effects of glyphosate on non-target vegetation by root exposure (e.g., Saunders and Pezeshki 2015), this type of exposure is almost irrelevant because, as discussed above, glyphosate is virtually inactive in most soils. Even if glyphosate were significantly bioavailable to plants in soil, glyphosate is not effectively taken up and translocated acropetally from the roots, and the concentrations found in ground and surface waters are generally lower than amounts needed for a significant physiological effect. Glyphosate drift from sprayed fields to foliage of plants outside the field is the main source of exposure of non-target plants. The amount of glyphosate needed to cause phytotoxicity varies between species. Drift levels of glyphosate can also vary considerably, and even the amount of a herbicide reaching a plant within a sprayed field can be highly variable (Velini et al. 2017). Because glyphosate translocates readily from foliage to growing parts of the plant, good coverage of the target weed is not needed for efficacy. Thus, large spray droplets, without good coverage of the weed, can be effective in delivering lethal glyphosate quantities to target plants. The larger the spray droplet, the less the drift problem, especially for an essentially non-volatile compound like glyphosate. Even with aerial spraying of glyphosate, plant injury is usually minimal at distances of >20 m downwind from sprayed fields (Marrs et al. 1993; Reddy et al. 2010). For mature plants of many species, there is minimal damage at distances of less than 20 m. There are reports of significant effects of very high simulated glyphosate drift levels on non-GR crops. For example, a simulated drift level of 100 g/ha was found to adversely affect nitrogen metabolism in non-GR soybeans (Bellaloui et al. 2006). However, there was no effect on yield, seed protein, or seed oil content by this relatively high "drift" level. Wild plant species are generally less sensitive to glyphosate than domesticated plant species (Cederland 2017). An analysis by Cederland (2017) found that drift of 5 g a.e./ha of glyphosate would not result in even minor adverse effects of drift on 95% of plant species, and that drift levels of 1 to 2 g a.e./ha of glyphosate would essentially cause no harm to any vascular plants. However, there can be stimulatory effects of glyphosate on plant growth at such low application rates (hormesis – as discussed in Sect. 2.5). Nevertheless, there has been a report of injury to an endangered plant species (*Pimelea spicata*) from glyphosate drift from a non-agricultural use (Matarczyk et al. 2002), but the "drift" concentration of glyphosate was not provided.

As mentioned above and below, glyphosate can influence plant disease by directly inhibiting the pathogen or reducing plant defenses against plant disease, and this effect could cause effects on plant communities subjected to glyphosate drift. Such potential effects have not been studied, other than the beneficial effects of simulated glyphosate drift in *Eucalyptus grandis*, due to its fungicidal effects on rust (dos Santos et al. 2019) – see Sect. 6.

As with any postemergence herbicide, the effects of glyphosate on non-target vegetation vary with the amount of drift, plant species, environmental conditions, and other factors. Although more plant species might be expected to be affected by glyphosate drift than by drift of any single selective herbicide, in most cases, especially in GR crops, glyphosate replaced several selective herbicides. Thus, the effects of glyphosate on non-target vegetation should be contrasted with the combined effects of the herbicides that it replaced. The relatively short environmental half-life of glyphosate and its lower drift potential than many of the herbicides that it replaced could mean adverse effects on non-target vegetation are likely to be less or, at the most, similar. However, with the increasing evolution and spread of GR weeds (Heap and Duke 2018; Baek et al. 2020), some of the herbicides that glyphosate replaced are now being sprayed again, along with glyphosate (e.g., Gage et al. 2019), reducing the early environmental benefits glyphosate used in GR crops (Cerdeira and Duke 2006; Cerdeira et al. 2007; Duke and Powles 2009).

6 Effects of Glyphosate on Microbes in Agriculture

Fungi and bacteria, as well as members or the phylum Apicomplexa, contain EPSPS that is sensitive to glyphosate (Dill et al. 2010; Roberts et al. 1998). However, there is considerable variation in the EPSPS among microbes, with some having glyphosate-sensitive, class I EPSPS (similar to that in higher plants) and others having relatively insensitive class II enzyme (Funke et al. 2007; Mir et al. 2015). Glyphosate can act as a fungicide and a bactericide on microbes with class I EPSPS. Duke et al. (2018b) recently reviewed much of the literature on this topic. Table 2 provides examples of the effects of glyphosate on variety of microbes as reported in

Table 2 Effects of glyphosate on various microbes in laboratory studies

Microbe species	Dose	Inhibition (%)	Reference
Bacteria			
Aerobacter aerogenes	1 mM	20%	Amrhein et al. (1983)
Bradyrhizobium japonicum	0.5 mM	10–41%[a]	Moorman et al. (1992)
Burkholderia galdioli	20 mM	0–80%	Kuklinsky-Sobral et al. (2005)
Pseudomonas oryzihabitans	20 mM	100%	Kuklinsky-Sobral et al. (2005)
Fungi			
Septoria sp.	<0.6 mM	90%	Dill et al. (2010)
Pseudocercosporella sp.	<0.6 mM	90%	Dill et al. (2010)
Botrytis sp.	<0.6 mM	90%	Dill et al. (2010)
Phytophthora sp.	6 mM	90%	Dill et al. (2010)
Rhizoctonia sp.	6 mM	90%	Dill et al. (2010)
Fusarium sp.	6 mM	90%	Dill et al. (2010)
Gaeumannomyces sp.	6 mM	90%	Dill et al. (2010)
Puccinia sp.	30 mM	90%	Dill et al. (2010)
Pyricularia sp.	30 mM	90%	Dill et al. (2010)
Alternaria sp.	0.6 mM	18–63%	Grossbard (1985)
Aspergillis niger sp.	3 mM	100%	Grossbard (1985)
Cladosporium herbarum	3 mM	100%	Grossbard (1985)
Fusarium lateritium	0.6 mM	33–73%	Grossbard (1985)
Gliocladium roseum	3 mM	100%	Grossbard (1985)
Penicillium sp.	0.6 mM	37–67%	Grossbard (1985)
Stachybotrys chartarum	0.3 mM	100%	Grossbard (1985)
Trichoderm polysporum	0.6 mM	27–68%	Grossbard (1985)
Neurospora crassa	2 mM	0%	Roisch and Lingens (1980)
Pythium ultimum	50 mM	35%	Kawate et al. (1992)

[a]In some studies the effects varied, depending on media and other variables

the literature. The concentrations in the papers listed in Table 2 were given as molarity. The molarity of 1 kg a.e./ha of glyphosate ranges from ca. 2.4 to 15 mM with glyphosate manufacturer recommended spray volumes of 40 to 250 L/ha (Monsanto 2020). Thus, the actual concentration of glyphosate that is applied to plants in the field is often sufficiently high to adversely affect many microbes. However, the concentrations of glyphosate in plants and soils will be lower than the concentration in the spray solution, reducing the possibility of there being an antimicrobial or antifungal effect of glyphosate.

Unfortunately, direct comparisons of effects between species from the data in Table 2 are not possible because of the different methods in the different papers. Furthermore, much of the literature is on the effects of formulated glyphosate, which does not differentiate between effects of formulation ingredients and that of glyphosate (Duke 2018b). Hormesis is common with fungitoxic compounds (Pradhan et al. 2017), so this phenomenon may occur with numerous plant pathogens at low glyphosate concentrations. For example, a sub-millimolar concentration (ca. 0.33 mM) of glyphosate stimulated mycelial dry weight accumulation of the plant pathogens *Fusarium solani* f. sp. *pisi* and *Pythium ultimum* (Kawate et al. 1992). The fact that glyphosate can serve as a source of phosphorus for some fungi (e.g., Adelowo et al. 2014) could contribute to hormesis.

Glyphosate is a relatively weak fungicide on most fungi in in vitro assays (Dill et al. 2010), and, as discussed above, glyphosate in non-GR plants reduces the shikimic acid pathway-based plant defenses, giving microbial plant pathogens an advantage, even though glyphosate could be toxic to the pathogen at the right dose. The bioavailable concentration of glyphosate in soil of sprayed weeds in the field may be insufficient to directly affect these pathogens, although the concentration for inhibition of the growth of some fungal plant pathogens is less than 1 μM in an in vivo assay (*Puccinia* spp. in wheat; Dill et al. 2010). This was a thousand times time less than glyphosate's activity in an in vitro assay (Table 2). Dill et al. (2010) attributed this discrepancy to the fact that *Puccinia* species are obligate pathogens that may not be amenable to in vitro screens.

The high application rates of glyphosate used on GR crops can thus have beneficial effects for the crop by their fungicidal effects on plant pathogens. This is particularly true for rusts. For example, a weed-killing application rate of glyphosate (0.84 kg/ha) applied to GR wheat 1 day before inoculation with wheat leaf rust (*Puccinia triticina*) prevented significant infection compared to plants that were not treated with glyphosate (Feng et al. 2005) (Fig. 9). This application rate in a typical carrier volume of 100 L/ha has a glyphosate concentration of ca. 5 mM, a concentration found to be fungitoxic to several fungi with in vitro assays (Table 2). Anderson and Kolmer (2005) reported similar results with glyphosate on wheat

Fig. 9 The effect of glyphosate treatment on severity of wheat leaf rust (*Puccinia triticina*) in GR wheat leaves 13 days after inoculation with the rust. Treatment A, no spray; treatment B, glyphosate formulation (0.84 kg ae/ha in a commercial formulation) 14 days before inoculation; treatment C, glyphosate formulation at 1 day before inoculation. From Feng et al. (2005). Copyright (2005) National Academy of Sciences, U.S.A

leaf rust and wheat stem rust (*P. graminis* f. sp. *tritici*) in GR wheat, obtaining good disease prevention with applications 22 days before inoculation that was evident 20 days after inoculation. Feng et al. (2005, 2008) also found preventative and curative effects of glyphosate on *P. striiformis* f.sp. *tritici* in GR wheat and suppression of Asian soybean rust (*Phakopsora pachyrhizi*) in GR soybeans. Glyphosate is inhibitory to some other cereal fungal pathogens, including *Septoria nodorum* (leaf blotch) (Harris and Grossbard 1979), *Pyrenophora tritici-repentis* (tan spot) (Sharma et al. 1989), *Gaeumannomyces graminis* (take-all) (Wong et al. 1993), and *Rhizoctonia solani* (Rhizoctonia root rot) (Wong et al. 1993) in studies not involving spraying infected live plants. These results suggest that glyphosate would have a beneficial effect on controlling these diseases in GR crops.

Rust infections in non-cereal GR crops are also reduced by glyphosate. Alfalfa rust (*Uromyces striatus*) was controlled in GR alfalfa by glyphosate (Samac and Foster-Hartnett 2012). It had both preventive and curative effects. Although phytotoxic to glyphosate-susceptible *Eucalyptus grandis,* glyphosate reduced rust infection by *Austropuccinia psidii* at sublethal doses to the tree (dos Santos et al. 2019; Tuffi-Santos et al. 2011). Glyphosate at 0.84 kg a.e./ha has been reported to reduce disease symptoms of *Rhizoctonia solani* in GR cotton (Pankey et al. 2005).

Examples of no effect of glyphosate on a plant disease in a GR crop include a multi-year, multisite study of the influence of glyphosate on Goss's wilt (*Clavibacter michiganensis* ssp. *nebraskensis*) in GR maize (Williams et al. 2015), and a massive, multi-year study in five US states and one Canadian province on the effect of glyphosate on sudden death syndrome (*Fusarium virguliforme*) in GR soybean (Kandel et al. 2015). Earlier work (Njiti et al. 2003; Sanogo et al. 2001) found no influence of glyphosate on sudden death syndrome in GR soybeans. In a two-year field study, Harikrishnan and Yang (2002) found no effect of glyphosate on root rot and damping off caused by *Rhizoctonia solani* in GR soybeans. Likewise, there was no effect of glyphosate on *R. solani* virulence in GR sugar beet (Barnett et al. 2012) and GR cotton (Baird et al. 2004). Another example is the negative findings of Lee et al. (2000, 2003) and Nelson et al. (2002) on the effects of glyphosate on white mold (*Sclerotinia sclerotiorum*) in GR soybean. Baley et al. (2009a) found no effect of glyphosate on virulence of several pathogens (*Gaeumannomyces graminis var. tritici, Pythium ultimum, Rhizoctonia oryzae* and *R. solani*) to GR wheat.

As discussed above, glyphosate makes glyphosate-sensitive plants more susceptible to plant pathogens by reducing synthesis of shikimate pathway-derived defense compounds. However, there is no viable rationale for why GR crops would be more susceptible to plants pathogens, as claimed by some (e.g., Johal and Huber 2009; Yamada et al. 2009). GR crops are about 50-fold more resistant to glyphosate than isogenic lines of the same crops (Nandula et al. 2007), and the lack of shikimate accumulation when these crops are treated with glyphosate (e.g., Velini et al. 2008) indicates that shikimate pathway-based pathogen defenses should not be impaired by glyphosate treatment. As mentioned above, a connection between mineral nutrition of GR crops and disease susceptibility has not been proven (Duke et al. 2012). The preponderance of well-replicated field studies in many geographically diverse locations has found either reduction of or no effect on plant disease in glyphosate-treated GR crops. In his extensive review, Hammerschmidt (2018) concludes that neither the glyphosate resistance gene (discussed in Green and Siehl (2020)) nor glyphosate applied to GR crops makes these crops more susceptible to plant pathogens. His only caveat is that treatment of glyphosate-susceptible plants in the near vicinity of GR crops could cause a temporary increase in inoculum of soil-borne plant pathogens that could increase GR crop disease. However, evidence for this being a significant problem in field situations is lacking.

In summary, glyphosate can act as a fungicide on some plant pathogens in GR crops, and it has been patented for this use (Baley et al. 2009b; Kohn and South 2020). The latter patent claims suppression of the non-rust diseases *Fusarium virguliforme*, *Phialophora gregata*, *Diaporthe phaseolorum*, and *Macrophomina phaseolina* in GR soybeans, generally increasing yields. However, use of glyphosate as a fungicide is not on the glyphosate label, probably partly because it is not as good as most commercial fungicides (e.g., for fungal disease management), having little or no effects on many such microbes. Also, the timing for application of glyphosate for weed management and that for disease control are unlikely to coincide. Peer-reviewed comparisons of glyphosate with commercial fungicides in field settings are not available. Nevertheless, the fungicidal effect of glyphosate on some plant pathogens is an unrecognized benefit of unknown magnitude in GR crops. However, it has little or no effect on many plant pathogens in these crops. Evidence of enhanced plant disease caused by glyphosate in GR crops is weak and, in some cases, may be the result of indirect effects of glyphosate such as increases in pathogen inoculum coming from nearby glyphosate-susceptible plants. However, such an effect must be rare, as the yields of maize, soybean, and cotton in the USA after there was more than 90% of adoption of GR cultivars of these crops has continued to rise at the same rate as before GR crops were introduced (Duke and Reddy 2018).

A virtually unexplored area of research is the effect of glyphosate on diseases of GR weeds. GR hairy fleabane (*Conyza bonariensis*) is more susceptible to powdery mildew caused by *Podosphaera erigerontis-canadensis* than a susceptible biotype (Pazdiora et al. 2019). However, in weeds that have evolved very high levels of glyphosate resistance such as *Amaranthus palmeri* with multiple copies of EPSPS (Gaines et al. 2010, 2011) (more than 20-fold resistant, requiring more than 7 kg a.e./ha to get the level of control that 0.2 kg ae/ha provides with susceptible biotypes), or in *E. indica* with a two codon change (Yu et al. 2015) (threonine to isoleucine at codon 102 and proline to serine at codon 106 – known as the TIPS mutation in EPSPS, requiring more than 30 kg a.e./ha to achieve the effect of 0.3 kg a.e/ha), recommended field rates (0.5–2 kg/ha) of glyphosate could increase their fitness by providing protection from some plant pathogens, in addition to the potential benefits of hormesis as discussed in Sect. 2.5.

Some non-pathogenic microbes interfere with plant pathogens, giving the host plant some protection (e.g., Haidar et al. 2016). For example, some endoyphytic bacteria can suppress plant diseases (Sturtz et al. 2000). If the concentrations of glyphosate reaching these microbes were more toxic to them than to the plant pathogen, glyphosate could enhance the success of the pathogen in GR crops. One study (Kuklinsky-Sobral et al. 2005) found the endophyte species of soybeans grown in soil treated with glyphosate were different than those in soil without a glyphosate treatment. No mention was made of whether the soybean varieties used were GR or not. The total population density of endophytes in the stem and roots (ca. 1,000 and 40,000 CFU (colony-forming units)/g fresh tissue, respectively) was unaffected by growing plants in glyphosate-treated soil, and was reduced from ca. 300 to 100 CFU/g fresh tissue in leaves. A later study found GR soybean cultivars treated with glyphosate to generally have a greater abundance of endophytic bacterial communities (de Almeida Lopes et al. 2016). The endophyte species in GR soybean were different from those of the non-GR cultivars. The experiments were not designed to determine whether the differences were due to glyphosate application or to the genetics of the different cultivars.

Some plant growth-promoting endophytes might be benefitted by glyphosate if they can use it as a source of phosphorous with a C-P lyase, as found in the endophyte *Enterobacter cloacae* (Kryuchkova et al. 2014). Such endophytes might be involved in glyphosate metabolism attributed to the plant, but those metabolizing it with a C-P lase are unlikely to be significantly involved in plant metabolism of glyphosate, because, as previously discussed, sarcosine is rarely reported as a glyphosate metabolite in plants. As far as I can determine, no publications have demonstrated any effects glyphosate on plant disease due to adverse effects on endophytes. Publications that show no effects of glyphosate on endophytes may be rare because "no effect" publications are considered low priority and rejected by the "so what?" rationale of many journals. Thus, unpublished studies such as that by Nolan (2016), who found no effect of glyphosate application on endophytic bacteria associated with roots of GR maize, whereas tillage practices and maize cultivars had effects, are less likely to appear in refereed journals.

Mycorrhizae are much like fungal endophytes, but they form obvious physical interactions with plants which provide benefits to the plant, such as increasing root surface area and enhancing water and nutrient uptake. Arbuscular mycorrhizal fungi (AMF) form structures in cortical roots cells called arbuscles that are involved in exchange of phosphorous, carbon, water, and other nutrients. Glyphosate (2.25 kg a.e./ha) applied to soil reduces root colonization by AMF in glyphosate-susceptible plants, with the effect being influenced by tillage and presence of endophytes (Helander et al. 2018). For example, in *Festuca pratensis*, there was less effect of glyphosate on the number of arbuscles with tillage or endophytes than without. Whether the glyphosate effects were due to effects on the plant, the AMF, or both was not determined. Four treatments of a high glyphosate rate (3 kg a.e./ha X 4, for total of 12 kg/ha) in a single year for 4 years in succession reduced root colonization of meadow grass (*Lolium arundinaceum*) by AMF and certain endophytes (Druille et al. 2016). There was no effect at an application rate that kills most weeds (0.8 kg a.e./ha per treatment X 4, for a total of 3.2 kg a.e./ha). The high rates used to get such an effect are unrealistic, as such high rates (12 kg/ha/year) are not needed to kill almost all unwanted vegetation, and the combined costs of the glyphosate and its application would be economically prohibitive. In a field study such as this, whether the effects are direct effects on the microbes or indirect effects from killing almost all of the plant life is unclear.

Before GR soybeans were commercially introduced, glyphosate was found to be toxic to *Bradyrhizobium japonicum* grown in vitro (Moorman et al. 1992) (Table 2), the microbe responsible for nitrogen fixation in soybean nodules. In this study, there was some variation in the sensitivity of different strains of the microbe to the herbicide. Variation in sensitivity could be due to differences in degradation of glyphosate by *B. japonicum*, as this microbe has the genetics for a C-P lyase (Hove-Jensen et al. 2014). Hydroxybenzoic acids, upstream by-products of shikimic acid (Lydon and Duke 1988), accumulated in the treated microbes (Moorman et al. 1992), indicating that the toxicity is due to inhibition of EPSPS. Glyphosate causes accumulation of shikimate and protocatechuate in *Bradyrhizobium* sp. nodules also (de María et al. 2006). Moorman et al. (1992) reasoned that since soybean nodules are metabolic sinks and because glyphosate preferentially translocates to metabolic sinks, there could be problems with glyphosate translocating to nodules in GR soybeans, where it could adversely affect *B. japonicum*, thereby reducing nitrogen fixation. Reddy et al. (2001) later found no effects of 1.12 kg a.e./ha glyphosate on nodule number or biomass in GR soybean, but 2.24 kg a.e./ha reduced both of these parameters and also reduced leghemoglobin by 6 to 18%. They stated that the adverse effects of the higher rate of glyphosate were of minimal consequence due to the potential of soybean to compensate for short durations of stress. King et al. (2001) found that glyphosate (1.68 kg a.e./ha) applied to twice GR soybeans interfered with nitrogen fixation, but the effect varied with cultivar and location. The effects were not long lived where there was adequate soil moisture throughout the growing season.

Subsequently, Reddy and Zablotowicz (2003) found that glyphosate accumulated in the nodules of glyphosate-treated (0.84 kg/ha) GR soybeans, up to ca. 150 ng/g dry weight. This concentration is similar to that reported in seeds (ca. 200 ng/g dry weight) of glyphosate-treated (0.84 kg/ha) GR soybeans (Duke et al. 2003b). Nodule biomass was reduced ca. 25%, and leghemoglobin was reduced as much as 10%. However, the crop recovered from these effects of glyphosate. A more comprehensive study found nitrogen fixation and/or assimilation in GR soybean to be only slightly affected at glyphosate label use rates (0.84 and 1.68 kg a.e./ha), whereas applications above label use rates (2.52 kg/ha applied twice) consistently reduced nitrogen assimilation, and reduced yield slightly in 1 year out of three (Zablotowicz and Reddy 2007). Bohm et al. (2014) found no effects of glyphosate at 0.96 kg a.e./ha applied twice (1.92 kg a.e./ha total) on nitrogen fixation in field-grown GR soybean. The composition and amounts of both free and protein amino acids of harvested seeds of GR soybean were unaffected by glyphosate treatment (applied at 0.87 kg/ha at both 5 and 7 weeks after planting) (Duke et al. 2018b), indicating no significant effects on nitrogen metabolism of the plant. In summary, glyphosate is unlikely to significantly affect nitrogen metabolism in GR crops when applied at recommended application rates, however, there is recent evidence that some farmers are using significantly higher than recommended rates (Miyazaki et al. 2019) that could adversely affect nitrogen fixation in nodules, thereby affecting yields and quality of harvested seed. Nevertheless, the facts that over 90% of the soybeans grown in the USA are GR (Duke 2018a) and that yields of soybeans have risen in a close to linear fashion since the introduction of GR soybean (Duke and Reddy 2018) support the view that such adverse effects are thus far uncommon.

Some have claimed that glyphosate applied for weed management disrupts soil microflora (e.g., Kremer and Means 2009; Druille et al. 2016; van Bruggen et al. 2018), whereas others have found little effect of single season use (e.g., Hart et al. 2009; Weaver et al. 2007) or repeated use of glyphosate in cropping situations (e.g., Barriuso et al. 2011; Schlatter et al. 2017; Kepler et al. 2020). To put effects of glyphosate on soil microbial communities in perspective, several factors must be considered. As discussed earlier, glyphosate is biologically unavailable to plants in most soils because it binds so tightly to certain soil components and rarely moves farther than a few centimeters into soil. Furthermore, it has a relatively short half-life in most soils in most climates, due to microbial metabolism, mostly to AMPA (Borggaard and Gimsing 2008; Blake and Pallett 2018; Zhan et al. 2018), so even though largely unavailable to plants in soil, it is available to at least some soil microbes. Some microbes can use glyphosate as a sole source of phosphorus, due to a microbial C-P lyase (Selvapanidiyan and Bhatnagar 1994), and, as discussed above, some bacteria and fungi are adversely affected by glyphosate because of glyphosate effects on their EPSPS. Finally, outside of areas sprayed to kill weeds with glyphosate, the concentrations would be expected to be much lower. All of these factors argue against glyphosate having a long-lasting effect on soil microflora, especially outside of sprayed fields. However, glyphosate might be expected to cause soil microflora perturbations in soils of sprayed fields, especially those treated with higher than label rates (>2.0 kg a.e./ha). For example, those microflora that can utilize glyphosate as a phosphorus source might increase, while those adversely affected would decrease in abundance. Nevertheless, as found by Kepler et al. (2020), agronomic practices other than glyphosate use are much more likely to influence soil microbial communities in agricultural fields. Another factor to consider is the effects of glyphosate formulation ingredients. For example, Mendonca et al. (2019) found the polyethoxylated tallow amine used in some glyphosate formulations is toxic to plant-beneficial soil *Pseudomonas* species and that addition of glyphosate did not add toxicity with two of the species, but it did significantly add toxicity with a third species. Unfortunately, the study was not done in soil, which probably would have reduced or eliminated any glyphosate toxicity, and there was no treatment with glyphosate alone, making the results difficult to interpret. Another example of potential effects of glyphosate on microbes being confounded by use of a formulated product is a study in which a glyphosate formulation was applied to soil in which potatoes were later grown (Gómez-Gallego et al. 2020). The plants were then infested with Colorado potato beetle (*Leptinotarsa decemlineata*) larvae, and the microbes found in resulting adult insects was altered, compared to those from plants grown without glyphosate in the soil. There were no treatments with glyphosate alone, making the role of glyphosate impossible to determine.

Most of the one-year studies have found small, but transient effects of glyphosate on soil microflora. For example, Weaver et al. (2007), in a laboratory study, found three-fold a recommended rate of glyphosate (i.e., 0.84 kg a.e./ha X 3) for weed management to cause only a small and brief (<7 days) effect on soil microflora.

Similar results were obtained by Ratcliff et al. (2006), who found that glyphosate applied to forest soils caused few and transient changes in bacterial and fungal communities. They concluded that application of recommended rates of glyphosate to these soils has a benign effect on microbial community structure. Another example is that of Zabaloy et al. (2016), who found in a two-year study that glyphosate had negligible effects on eubacteria and ammonia-oxidizing bacteria and concluded that glyphosate use at recommended rates poses low risk to soil microbiota. Their highest application rate was 1.2 kg a.e./ha. In a three-year study, Bohm et al. (2014) found no effects of the yearly use of two applications of 0.96 kg a.e./ha of glyphosate on soil microbe populations. Two studies have examined the effects of long-term use of glyphosate in field situations on soil microbia. In the first of these studies, Schlatter et al. (2017) compared the effects of 20 years of glyphosate use on bacterial populations in wheat soils with wheat soils where glyphosate had never been used. Glyphosate use was related to only 2 to 5% of the variation in bacterial communities, whereas most of the variation was associated with cropping history, year, location, and proximity to roots. Less than 1% of the taxa were affected by glyphosate use, and most of these were increased. In a well-replicated study repeated in 2 years in two widely separated states (Maryland and Mississippi) of the USA and with two GR crops (maize and soybean), glyphosate application had almost no effects on soil fungal and prokaryote communities, whereas geography, farming systems, and seasons had profound effects (Kepler et al. 2020). Glyphosate had been used with GR crops for 15 previous concurrent years in one of the study sites.

There are many generations for a microbial species during a crop growing season. Furthermore, the number of individual microbes of a microbial species in a field is many orders of magnitude greater than that of weeds. Thus, if glyphosate is toxic to a microbe, the probability of evolved resistance to it is theoretically very high, although I am unaware of documentation of this in an agricultural field. However, in the laboratory, GR microbes can easily be selected for in glyphosate-containing growth media. For example, Amrhein et al. (1983) produced GR *Aerobacter aerogenes* by repeated (9X) transfer of a cultures to 5 mM glyphosate. The mechanism of resistance was a 10- to 30-fold increase in EPSPS activity, a mechanism that has also evolved in some plant species (Baek et al. 2020). Microbes with glyphosate-insensitive class II EPSPS (Funke et al. 2007; Mir et al. 2015) isolated from fields with a history of extensive glyphosate use have been reported (Firdous et al. 2018), but there was no determination as to whether the microbe was simply enriched in the field or evolved a highly GR EPSPS from a less resistant EPSPS.

7 Conclusions

Glyphosate is a remarkably successful herbicide that has dominated the herbicide market for decades. Its many attributes include its non-selectivity, high level of efficacy on most species, excellent translocation, relatively slow action on most weed species, and its relative safety to non-target organisms. It was a major herbicide before the introduction of GR crops in 1996, and became the clearly dominant herbicide worldwide after their introduction. Glyphosate has been an important tool in managing cover crops and has promoted reduced and no-tillage agriculture in both non-GR and GR crops, thereby reducing soil erosion, moisture loss, and use of fossil fuels. It is used extensively in non-GR crops for preplant and postharvest weed management, as well as in orchards, vineyards, and silviculture. It is widely used for weed management in non-agricultural settings such as turf, roadsides, and aquatic weed management. In sugarcane it is used at low application rates to enhance sucrose yields, and in some agronomic crops it is used as a harvest aid to quickly kill the crop to facilitate mechanized harvesting. It is the only commercial herbicide that acts by inhibition of EPSPS or any other enzyme of the shikimate pathway. It has no other molecular target as a herbicide, however, inhibition of EPSPS causes several effects that contribute to its adverse effects on plants. These include: 1) depletion of aromatic amino acids needed for synthesis of proteins, IAA, PQ, and secondary products required for plant defense; 2) deregulation of the shikimate pathway, leading to loss of intermediates in carbon fixation and other biosynthetic pathways; and 3) accumulation of the phytotoxic shikimate pathway intermediate, quinic acid. The importance of each of these aspects of glyphosate's mode of action varies between species and within a species with biotic and abiotic environmental factors and plant growth stage. Plants and microbes can metabolically degrade glyphosate by converting it to AMPA and glyoxylate, and to a lesser extent by breaking the C-P bond, creating sarcosine and inorganic phosphate. Many fungi and bacteria are sensitive to glyphosate, and glyphosate can act as a fungicide against some plant pathogens in GR crops. *Bradyrhizobium* spp. are sensitive to glyphosate, and glyphosate can inhibit nitrogen fixation in GR soybeans, although this effect has not been found to significantly influence soybean yields at recommended glyphosate application rates. The effects of glyphosate on microflora of crop soils is generally low and transient, with weather and other agricultural practices such as tillage having much stronger effects. Glyphosate has been a valuable tool in economically managing weeds in many settings for the past 45 years.

Acknowledgments I thank Drs. Ana Zabalza and Douglas Sammons for their advice about certain aspects of the review.

The author has no conflicts of interest.

References

Abbas T, Nadeem MA, Tanveer A, Zohaib A, Rasool T (2015) Glyphosate hormesis increases growth and yield of chickpea (*Cicer arietinum* L.). Pak J Weed Sci Res 21:533–542

Abbas T, Nadeem MA, Tanveer A, Maqbool R, Shehzad A, Farooq N (2016) Glyphosate herbicide causes hormesis in wheat. Pak J Weed Sci Res 22:575–586

Adams CR, Weise C, Cobb LC (2014) Effect of season and number of glyphosate applications on control of invasive Mexican petunia (*Ruellia simplex*). Ecol Restorations 32:133–137

Adelowo FE, Olu-Arotiowa OA, Amuda OS (2014) Biodegradation of glyphosate by fungal species. Adv Biosci Bioeng 2:104–118

Al-Rajab AJ, Amellal S, Shiavon M (2008) Sorption and leaching of ^{14}glyphosate in agricultural soils. Agron Sustain Dev 28:419–428

Altman J, Campbell CL (1977) Effects of herbicides on plant diseases. Annu Rev Phytopathol 15:361–385

Amrhein N, Deus B, Gehrke P, Steinrücken HC (1980) The site of the inhibition of the shikimate pathway by glyphosate. II. Interference of glyphosate with chorismate formation *in vivo* and *in vitro*. Plant Physiol 66:830–844

Amrhein N, Johänning D, Schab J, Schulz A (1983) Biochemical basis for glyphosate-tolerance in a bacterium and a plant tissue culture. FEBS Lett 157:191–196

Anderson KS, Johnson KA (1990a) Kinetic and structural analysis of enzyme intermediates: lessons from EPSP synthase. Chem Rev 90:1131–1149

Anderson KS, Johnson KA (1990b) "Kinetic competence" of the 5-enolpyruvylshikikimate-3-phosphate synthase tetrahedral intermediate. J Biol Chem 265:5567–5572

Anderson JA, Kolmer JA (2005) Rust control in glyphosate tolerant wheat following application of the herbicide glyphosate. Plant Dis 89:1136–1142

Anderson KS, Sikorski JA, Johnson KA (1988) A tetrahedral intermediate in the EPSP synthase reaction observed by rapid quench kinetics. Biochemistry 27:7395–7406

Antier C, Kudsk P, Reboud X, Ulber L, Baret PV, Messéan A (2020a) Glyphosate use in the European sector and framework for its further monitoring. Sustainability 12:5682. https://doi.org/10.3390/su12145682

Antier C, Andersson R, Auskalniene O, Barić K, Baret P, Bessenhofer G, Calha I, Carrola Dos Santos S, De Cauwer B, Chachalis D, Dorner Z, Follak S, Forristal D, Gaskov S, Gonzalez Andujar JL, Hull R, Jalli H, Kierzek R et al (2020b) A survey on the uses of glyphosate in European countries. INRAE. https://doi.org/10.15454/A30K-D531

Arisue N, Hashimoto T (2015) Phylogeny and evolution of apicoplasts and apicomplexan parasites. Parasitol Int 64:254–259

Arregui MC, Lenardón A, Sanchez D, Maitre MI, Scotta R, Enrique S (2003) Monitoring glyphosate residues in transgenic glyphosate-resistant soybean. Pest Manag Sci 60:163–166

Askew SD, Askew WB, Goatley JM (2019) Fineleaf fescue species and variety tolerance to glyphosate. Weed Technol 33:185–191

Baek, Y, Bobadilla LK, Giacomini DA, Montgomery JS, Murphy BP, Tranel PJ (2020) Evolution of glyphosate-resistant weeds. Rev Environ Contam Toxicol. (This volume)

Baird R, Batson W, Watson C, Hightower P (2004) Evaluation of transgenic cotton varieties and a glyphosate application of seedling disease incidence. Mycopathologia 158:363–368

Bajwa AA, Wang H, Chouhan BS, Adkins SW (2019) Effect of elevated carbon dioxide concentration on growth, productivity and glyphosate response of parthenium weed (*Parthenium hysterophorus* L.). Pest Manag Sci 75:2934–2941

Baley GJ, Campbell KG, Yenish J, Kidwell KK, Paulitz TC (2009a) Influence of glyphosate, crop volunteer and root pathogens on glyphosate-resistant wheat under controlled environmental conditions. Pest Manag Sci 65:288–299

Baley GJ, Kohen FC, Pitkin JW, Rinehart J, Taylor JH (2009b) Method for disease control in MON89788. US7608761 B1, 2009

Barker AL, Dayan FE (2019) Fate of glyphosate during production and processing of glyphosate-resistant sugar beet (*Beta vulgaris*). J Agric Food Chem 67:2061–2065

Barnett KA, Sprague CL, Kirk WW, Hanson LE (2012) Influence of glyphosate on Rhizoctonia crown and root rot (*Rhizoctonia solani*) in glyphosate-resistant sugar beet. Weed Sci 60:113–120

Barrett P (1985) Efficacy of glyphosate in the control of aquatic weeds. In: Grossbard E, Atkinson D (eds) The herbicide glyphosate. Butterworths, London, pp 365–374

Barrett KA, McBride MB (2005) Oxidative degradation of glyphosate and aminomethylphosphonate by manganese oxide. Environ Sci Technol 39:9223–9228

Barriuso J, Marin S, Mellado RP (2011) Potential accumulative effects of the herbicide glyphosate on glyphosate-tolerant maize rhizobacterial communities over a three-year cultivation period. PLoS One 6:e27558

Bataillard D, Christe P, Pigeault R (2020) Impact of field-realistic doses of glyphosate and nutritional stress on mosquito life history traits and susceptibility to malaria parasite infection. Ecol Evol 10:5079–5088

Baylis AD (2000) Why glyphosate is a global herbicide: strengths, weaknesses and prospects. Pest Manag Sci 56:299–308

Becerril JM, Duke SO, Lydon J (1989) Glyphosate effects on shikimate pathway products in leaves and flowers of velvetleaf. Phytochemistry 28:695–699

Bellaloui N, Reddy KN, Zablotowicz RM, Mengistu A (2006) Simulated glyphosate drift influence nitrate assimulation and nitgrogen fixation in non-glyphosate-resistant soybean. J Agric Food Chem 54:3357–3364

Belz RG (2014) Is hormesis an underestimated factor in the development of herbicide resistance. Julius-Kühn-Arch 443:81–91

Belz RG, Duke SO (2014) Herbicides and hormesis. Pest Manag Sci 70:698–707

Belz RG, Duke SO (2017) Herbicide-mediated hormesis. Amer Chem Soc Symp Ser 1249:135–148

Belz RG, Sinkkonen A (2019) Low toxin doses change plant size distribution in dense populations – glyphosate exposed *Hordeum vulgare* in a greenhouse study. Envron Internat 132. https://doi.org/10.1016/j.envint.2019.105072

Benbrook CM (2016) Trends in glyphosate herbicide use in the United States and globally. Environ Sci Eur 28:1–15

Bentley R (1990) The shikimate pathway – a metabolic tree with many branches. Crit Rev Biochem Mol Biol 24:307–384

Bernal L, Dussán J (2020) Synergistic effect of *Lysinibacillus sphaericus* and glyphosate on temephos-resistant larvae of *Aedes aegypti*. Parasites Vectors 13:Article no. 13

Bernal J, Martin MT, Soto ME, Nozal MJ, Marotti I, Dinelli G, Bernal JL (2012) Development and application of a liquid chromatography-mass spectrometry method to evaluate the glyphosate and aminomethylphosphonic acid dissipation in maize plants after foliar treatment. J Agric Food Chem 60:4017–4025

Binkholder KM, Fresenburg BS, Teuton TC, Xiong X, Smeda RJ (2011) Selection of glyphosate-resistant annual bluegrass (*Poa annua*) on a golf course. Weed Sci 59:286–289

Blackshaw RE, Harker KN (2016) Wheat, field pea, and canola response to glyphosate and AMPA soil residues. Weed Technol 30:985–991

Blake R, Pallett K (2018) The environmental fate and ecotoxicology of glyphosate. Outlooks Pest Manag 29:266–269

Blevins RL, Cook D, Phillips SH, Phillips RE (1971) Influence of no-tillage on soil moisture. Agron J 63:593–596

Blume CJ, Fei S-Z, Christians NE (2010) Field evaluation of reduced-growth, glyphosate-resistant Kentucky bluegrass in a noncompetitive setting. Crop Sci 50:1048–1056

Bohm GMB, Rombaldi CV, Genovese MI, Castilhos D, Alves BJR, Rumjanek NG (2014) Glyphosate effects on yield, nitrogen fixation, and seed quality in glyphosate-resistant soybean. Crop Sci 54:1737–1743

Bøhn T, Cuhra M, Traavik T, Sanden M, Fagan J, Primicerio R (2014) Compositional differences in soybeans on the market: glyphosate accumulates roundup ready GM soybeans. Food Chem 153:207–215

Boocock MR, Coggins JR (1983) Kinetics of 5-enolpyruvylshikimate-3-phosphate synthase inhibition by glyphosate. FEBS Lett 154:127–133

Borggaard OK, Gimsing AL (2008) Fate of glyphosate in soil and the possibility of leaching to ground and surface waters: a review. Pest Manag Sci 64:441–456

Botta F, Lavison G, Couturier G, Alliot F, Moreau-Guigon E, Fauchon N, Guery B, Chevreuil M, Blanchoud H (2009) Transfer of glyphosate and its degradate AMPA to surface waters through urban sewerage systems. Chemosphere 77:133–139

Bowmer KH, Eberbach PL, McCorkkelle G (1993) Uptake and translocation of ^{14}C-glyphosate in *Alternanthera philoxeeroides* (Mart.) Griseb. (alligator weed) I. Rhizome concentrations required for inhibition. Weed Res 33:53–57

Boyle JH, Dalgleish HJ, Puzey JR (2019) Monarch butterfly and milkweed declines substantially predate the use of genetically modified crops. Proc Natl Acad Sci U S A 116:3006–3011

Breitenbach J, Zhu C, Sandmann G (2001) Bleaching herbicide norflurazon inhibits phytoene desaturase by competition with cofactors. J Agric Food Chem 49:5270–5272

Brilisauer K, Rapp J, Rath P, Schöllhorn A, Bleul L, Weiss E, Stahl M, Grond S, Frochhammer K (2019) Cyanobacterial antimetabolite 7-deoxy-sedoheptulose blocks the shikimate pathway to inhibit the growth of phototrophic organisms. Nat Commun 10:545. https://doi.org/10.1038/s41467-019-08476-8

Brito IPFS, Tropaldi L, Carbonari CA, Velini ED (2018) Hormetic effects of glyphosate on plants. Pest Manag Sci 74:1064–1070

Brookes G, Barfoot P (2018) Environmental impacts of genetically modified (GM) crop use 1996-2016: impacts on pesticide use and carbon emission. GM Crops Food 8:216–228

Brzezowski P, Ksas B, Havaux M, Grimm B, Chazaux M, Peltier G, Johnson X, Alric J (2019) The function of protoporphyrinogen IX oxidase in chlorophyll biosynthesis requires oxidized plastoquinone in *Chlamydomonas reinhardtii*. Commun Biol 2:159. https://doi.org/10.1038/s42003-019-039505

Burt EO (1980) Glyphosate for torpedograss and bermudagrass control. In: Proc 3rd Int Turfgrass Res Conf, vol 3. pp. 257–262

CaJacob CA, Feng PCC, Heck GR, Murtaza FA, Sammons RD, Padgett SR (2003) Engineering resistance to herbicides. In: Christou P, Klee H (eds) Handbook of biotechnology. Wiley, Chichester, pp 353–372

Calabrese EJ, Bachmann KA, Bailer AJ, Bolger PM, Borak J, Cai L Cedergreen N, Cherian MG, Chiueh CC, Clarkson TW et al (2007) Biological stress terminology: integrating the concepts of adaptive response and preconditioning stress within a hormetic dose-response framework. Toxicol Appl Pharmacol 222:122–128

Carbonari CA, Gomes GLGC, Velini ED, Machado RF, Simões PS, Macedo GD (2014) Glyphosate effects on sugarcane metabolism and growth. Am J Plant Sci 5:3585–3593

Carver TLW, Robbins MP, Zeyen RJ, Dearne GA (1992) Effects of PAL-specific inhibition on suppression of activated defence and quantitative susceptibility of oats to *Erysiphe graminis*. Physiol Molec Plant Pathol 41:149–163

Casale J, Lydon J (2007) Apparent effects of glyphosate on alkaloid production in coca plants grown in Columbia. J Forensic Sci 52:573–578

Caseley JC, Coupland D (1985) Environmental and plant factors affecting glyphosate uptake, movement and activity. In: Grossbard E, Atkinson D (eds) The herbicide glyphosate. Butterworths, London, pp 92–123

Castle LA, Siehl DL, Gorton R, Patten PA, Chen YH, Bertain S, Cho H-J, Duck N, Wong J, Liu D, Lassner MW (2004) Discovery and directed evolution of a glyphosate tolerance gene. Science 304:1151–1154

Cedergreen N (2008) Is the growth simulation by low doses of glyphosate sustained over time? Envir Pollut 156:1099–1104

Cederland H (2017) Effects of spray drift of glyphosate on nontarget terrestrial plants – a critical review. Environ Toxicol Chem 36:2879–2886

Cerdeira AL, Duke SO (2006) The current status and environmental impacts of glyphosate-resistant crops: a review. J Environ Qual 35:1633–1658

Cerdeira AL, Gazziero DLP, Duke SI, Matallo MB, Spadoto CA (2007) Review of potential environmental impacts of transgenic glyphosate-resistant soybean in Brazil. J Environ Sci Health Part B 42:539–549

Cerny L, Wu K, Floyd T (2014) Molecular markers associated with yellow flash in glyphosate tolerant soybeans. US Patent 9,617,605 B2. April 11, 2017

Cessna AJ, Darwent AL, Kirkland KJ, Townley-Smith L, Harker KN, Lefkovitch LP (1994) Residues of glyphosate and its metabolite AMPA in wheat seed and foliage following preharvest applications. Can J Plant Sci 74:653–661

Cessna AJ, Darwent AL, Townley-Smith L, Harker KN, Kirkland KJ (2002) Residues of glyphosate and its metabolite AMPA in field pea, barley and flax seed following preharvest applications. Can J Plant Sci 82:485–489

Chahal GS, Jordan DL, Burton JD, Danehower D, York AC, Eure PM, Clewis B (2010) Influence of water quality and coapplied agrochemicals on efficacy of glyphosate. Weed Technol 26:167–176

Chollet R, Vidal J, O'Leary MH (1996) Phosphoenolpyruvate carboxylase: a ubiquitous, highly regulated enzyme in plants. Annu Rev Plant Physiol Plant Mol Biol 47:273–298

Christy AL, Herbst KA, Kostka SJ, Mullen JP, Carlson PS (1993) Synergizing weed biocontrol agents with chemical herbicides. Am Chem Soc Symp Ser 524:87–100

Clay PA, Griffin JL (2000) Weed seed production and seedling emergence responses to late-season glyphosate applications. Weed Sci 48:481–486

Colombo SL, Andreo CS, Chollet R (1998) The interaction of shikimic acid and protein phosphorylation with PEP carboxylase from the C4 dicot *Amaranthus viridis*. Phytochemistry 48:55–59

Corniani N, Velini ED, Silva FML, Nanayakkara NPD, Witschel M, Dayan FE (2014) Novel bioassay for the discover of inhibitors of the 2-*C*-methyl-*D*-erythritol 4-phosphate (MEP) and terpenoid pathways leading to carotenoid biosynthesis. PLoS One 9(7):e103704

Cornish PS (1992) Glyphosate residues in a sandy soil affect tomato transplants. Aust J Exp Agric 32:395–399

Corrêa EL, Dayan FE, Owen DL, Rimando AM, Duke SO (2016) Glyphosate-resistant and conventional canola (*Brassica napus* L.) responses to glyphosate and aminomethylphosphonic acid (AMPA) treatment. J Agric Food Chem 64:3508–3513

Costa FR, Rech R, Duke SO, Carvalho LB (2018) Lack of effects of glyphosate and glufosinate on growth, mineral content, and yield of glyphosate- and glufosinate-resistant maize. GM Crops Food 9. https://doi.org/10.1080/21645698.2018.1511204

Coupland D, Caseley JC (1979) Presence of ^{14}C activity in root exudates and guttation fluid from *Agropyron repens* treated with ^{14}C-labelled glyphosate. New Phytol 83:17–22

Cowie BW, Venter N, Witkowski ETF, Byrne MJ (2020) Implications of elevated carbon dioxide on the susceptibility of the globally invasive weed, *Parthenium hysterophorus* to glyphosate herbicide. Pest Manag Sci 76:2324–2332

Dalley CD, Richard EP (2010) Herbicides as ripeners for sugarcane. Weed Sci 58:329–333

Daouk S, De Alencastro LF, Pfeifer H-R (2013) The herbicide glyphosate and its metabolite AMPA in the Lavaux vineyard area, western Switzerland: proof of widespread export to surface waters. Part II: the role of infiltration and surface runoff. J Environ Sci Health Part B 48:725–736

Dayan FE (2018) Is there a natural route to the next generation of herbicides? Outlooks Pest Manag 29:54–57

Dayan FE, Duke SO (2003) Herbicides: protoporphyrinogen oxidase inhibitors. In: Plimmer JR, Gammon DW, Ragsdale NN (eds) Encyclopedia of agrochemicals, vol 2. Wiley, New York, pp 850–863

Dayan FE, Duke SO (2020) Discovery for new herbicide sites of action by quantification of plant primary metabolite and enzyme pools. Engineering 6:509–514

Dayan FE, Netherland MD (2005) Hydrilla, the perfect aquatic weed, become more noxious than ever. Outlooks Pest Manag 16:277–282

Dayan FE, Barker A, Bough R, Ortiz M, Takano H, Duke SO (2019) Herbicide mechanisms of action and resistance. In: Moo-Young M (ed) Comprehensive biotechnology, vol 4, 3nd edn. Elsevier, Permagon, pp 36–48. https://doi.org/10.1016/B978-0-444-64046-8.00211-1

de Almeida Lopes KB, Carpentieri-Pipolo V, Oro TH, Pagliosa ES, Degrassi G (2016) Culturable endophytic bacterial communities associated with field-grown soybean. J Appl Microbiol 120:740–755

de María N, Becerril JM, García-Plazaola HA, de Felipe MR, Fernández M (2006) New insights on glyphosate mode of action in nodular metabolism: role of shikimate accumulation. J Agric Food Chem 54:2621–2628

de Moraes CP, de Brito IPFS, Tropaldi L, Carbonari CA, Velini ED (2020) Hormetic effect of glyphosate on *Urochloa decumbens* plants. J Envron Sci Health Part B 55:376–381
della-Cioppa G, Kishore GM (1988) Import of a precursor protein into chloroplasts is inhibited by the herbicide glyphosate. The EMBO J 7:1299–1305
della-Cioppa G, Bauer SC, Klein BK, Shah DM, Fraley RT, Kishore G (1986) Translocation of the precursor of 5-enolpyruvylshikimate-3-phosphate synthase into chloroplasts of higher plants *in vitro*. Proc Natl Acad Sci U S A 83:6873–6877
Derting CW (1987) Wiper application. In: McWhorter CG, Gebhardt MR (eds) Methods of applying herbicides, Weed Science Society of America Monograph 4, Champaign. pp. 207–229
Devine MD, Duke SO, Fedtke C (1993) Physiology of herbicide action. Prentice Hall, Englewood Cliffs, p 441
Dias RC, Dadazio TS, Tropaldi L, Carbonari CA, Velini ED (2019) Glyphosate as growth regulator for bahiagrass and broadleaf carpetgrass. Planta Daninha 37:e019213829. https://doi.org/10.1590/S0100-83582019370100144
Dill GM, Sammons RD, Feng PCC, Kohn F, Kretzmer K, Mehrsheikh A, Bleeke M, Honegger JL, Farmer D, Wright D, Haupfear (2010) Glyphosate: discovery development, applications, and properties. In: Nandula VK (ed) Glyphosate resistance in crops and weeds, Wiley, Hoboken, pp. 1–33
Ding W, Reddy KN, Zablotowicz RM, Bellaloui N, Bruns HA (2011) Physiological responses of glyphosate-resistant and glyphosate-sensitive soybean to aminomethylphosphonic acid, a metabolite of glyphosate. Chemosphere 83:593–598
dos Santos SA, Tuffi-Santos LD, Tanaka FAO, Sant'Anna-Santos BF, Rodrigues FA, Alfenas AC (2019) Carfentrazone-ethyl and glyphosate drift inhibits uredinial formation of *Austropuccinia psidii* on *Eucalyptus grandis* leaves. Pest Manag Sci 75:53–62
dos Santos JCC, da Silva DMR, Amorim DJ, Sab MPV, Silva MA (2021) Glyphosate hormesis mitigates the effect of water deficit in safflower (*Carthamus tintoris* L.). Pest Manag Sci 77. https://doi.org/10.1002/ps.6231
Druille M, Garćia-Parisi PA, Golluscio RA, Cavagnaro FP, Omacini M (2016) Repeated annual glyphosate applications may impair beneficial soil microorganisms in temperate grassland. Agric Ecosyst Environ 230:184–190
Duke SO (1988) Glyphosate. In: Kearney PC, Kaufman DD (eds) Herbicides: chemistry degradation, and mode of action, vol 3. Marcel Dekker, New York, pp 1–70
Duke SO (2011) Glyphosate degradation in glyphosate-resistant and -susceptible crops and weeds. J Agric Food Chem 59:5835–5841
Duke SO (2014) Biotechnology: herbicide-resistant crops. In: van Alfen M (ed) Encyclopedia of agriculture and food systems, vol 2. Elsevier, San Diego, pp 94–116
Duke SO (2015) Perspectives on transgenic, herbicide-resistant crops in the USA almost 20 years after introduction. Pest Manag Sci 71:652–657
Duke SO (2018a) The history and current status of glyphosate. Pest Manag Sci 74:1027–1034
Duke SO (2018b) Interaction of chemical pesticides and their formulation ingredients with microbes associated with plants and plant pests. J Agric Food Chem 66:7753–7561
Duke SO (2019) Enhanced metabolic degradation: the last evolved glyphosate resistance mechanism of weeds? Plant Physiol. https://doi.org/10.1104/pp.19.01245
Duke SO, Hoagland RE (1978) Effects of glyphosate on metabolism of phenolic compounds. I. Induction of phenylalanine ammonia-lyase activity in dark-grown maize roots. Plant Sci 11:185–190
Duke SO, Powles SB (2009) Glyphosate-resistant crops and weeds: now and in the future. AgBioforum 12:346–357
Duke SO, Reddy KN (2018) Is mineral nutrition of glyphosate-resistant crops altered by glyphosate treatment? Outlooks Pest Manag 29:206–209
Duke SO, Hoagland RE, Elmore CD (1980) Effects of glyphosate on metabolism of phenolic compounds. V. L-α-aminooxy-β-phenylpropionic acid and glyphosate effects on phenylalanine ammonia-lyase in soybean seedlings. Plant Physiol 65:17–21

Duke SO, Rimando AM, Duke MV, Paul RN, Ferreira JFS, Smeda RJ (1999) Sequestration of phytotoxins by plants: implications for biosynthetic production. In: Cutler HA, Cutler SJ (eds) Natural products: agrochemicals and pharmaceuticals. CRC Press, Boca Raton, pp 127–136

Duke SO, Baerson SR, Rimando AM (2003a) Herbicides: glyphosate. In: Plimmer JR, Gammon DW, Ragsdale NN (eds) Encyclopedia of agrochemicals. Wiley, New York. https://doi.org/10.1002/047126363X.agr119

Duke SO, Rimando AM, Pace PF, Reddy KN, Smeda RJ (2003b) Isoflavone, glyphosate, and aminomethylphosphonic acid levels in seeds of glyphosate-treated, glyphosate-resistant soybean. J Agric Food Chem 51:340–344

Duke SO, Wedge DE, Cerdeira AL, Mattalo MB (2007) Interactions of synthetic herbicides with plant disease and microbial herbicides. In: Vurro M, Gressel J (eds) Novel biotechnologies for biocontrol agent enhancement and management. Springer, Dordrecht, pp 277–296

Duke SO, Lydon J, Koskinen WC, Moorman TB, Chaney RL, Hammerschmidt R (2012) Glyphosate effects on plant mineral nutrition, crop rhizophere microbiota, and plant disease in glyphosate-resistant crops. J Agric Food Chem 60:10375–10397

Duke SO, Bajsa J, Pan Z (2013) Omics methods for probing the mode of action of natural and synthetic phytotoxins. J Chem Ecol 39:333–347

Duke SO, Pan Z, Bajsa-Hirschel J, Sánchez-Moreiras AM, Vaughn JN (2018a) Use of omics methods to determine the mode of action of natural phytotoxins. Am Chem Soc Symp Ser 1294:33–46

Duke SO, Rimando AM, Reddy KN, Cizdziel JV, Bellaloui N, Shaw DR, Williams MM, Maul JE (2018b) Lack of transgene and glyphosate effects on yield and mineral and amino acid content of glyphosate-resistant soybean. Pest Manag Sci 74:1166–1173

Dusky JA, Kang MS, Glaz B, Miller JD (1986) Response of eight sugarcane cultivars to glyphosine and glyphosate ripeners. J Plant Growth Regul 4:225–235

Eliel EL, Ramirez MB (1997) (-)-Quinic acid: configurational (stereochemical) descriptors. Tetrahedron Asymmetry 8:2551–3554

Enrich LB, Scheuermann ML, Mohadjer A, Matthias KR, Eller CF, Newman MS, Fujinaka M, Poon T (2008) *Liquidamber styraciflua*: a renewable source of shikimic acid. Tetrahedron Lett 49:2503–2505

Ermakova IT, Kiseleva NI, Shushkova T, Zharikov M, Zharikov GA, Leontievsky AA (2010) Bioremediation of glyphosate-contaminated soils. Appl Microbiol Biotechnol 88:585–594

Eschenburg S, Kabsch W, Healy ML, Schonbrunn E (2003) A new view of the mechanisms of UDP-N-acetylglucosamine enolpyruvyl transferase (MurA) and 5-enolpyruvylshikimate-3-phosphate synthase (AroA) derived from X-ray structures of their tetrahedral reaction intermediate states. J Biol Chem 278:49215–49222

Federspiel M, Fischer R, Hennig M, Mair H-J, Oberhouser T, Rimmler G, Ableiz T, Bruhin J, Estermann H, Gandert C, Göckel V, Götzö S, Hoffmann U, Huber G, Janatsch G, Lauper S, Rõckel-Stäbler O, Trussardi R, Zwahlen AG (1999) Industrial synthesis of the key precursor in the synthesis of the anti-influenza drug oseltamivir phosphate (Ro 64-0796/002, GS-4104-02): ethyl (3*R*,4*S*, 5S)-4,5-epoxy-3-(1-ethyl-propoxy)-cyclohex-1-ene-1-carboxylate. Org Proc Res Dev 3:266–274

Feng PCC, Sandbrink JJ, Sammons RD (2000) Retention, uptake, and translocation of ^{14}C-glyphosate from track-spray applications and correlation to rainfastness in velvetleaf (*Abutilon theophrasti*). Weed Tech 14:127–132

Feng PCC, Baley GJ, Clinton WP, Bunkers GJ, Alibhai MF, Paulitz T, Kidwell KK (2005) Glyphosate inhibits rust diseases in glyphosate-resistant wheat and soybean. Proc Natl Acad Sci U S A 102:17290–17295

Feng PCC, Clark C, Andrade G, Balbi MC, Caldwell P (2008) The control of Asian rust by glyphosate in glyphosate-resistant soybeans. Pest Manag Sci 64:353–359

Fernando N, Manalil S, Florentine SK, Chauhan BS, Seneweera S (2016) Glyphosate resistance of C_3 and C_4 weeds under rising atmospheric CO_2. Front Plant Sci 7:7. https://doi.org/10.3389/fpls.2016.00910

Firdous S, Iqbal S, Anwar S, Jabeen H (2018) Identification and analysis of 5-enolpyruvylshikimate-3-phosphate synthase EPSPS gene from glyphosate-resistant *Ochrobacterium intermedium* Sq20. Pest Manag Sci 74:1184–1196

Fischer RS, Berry A, Gaines CG, Jensen RA (1986) Comparative action of glyphosate as a trigger of energy drain in eubacteria. J Bacteriol 168:1147–1154

Franz JE, Mao MK, Sikorski JA (1997) Glyphosate: a unique global herbicide. Amer Chem Soc Monograph 189. American Chemical Society, Washington

Freedman B (1990) Controversy over the use of herbicides in forestry, with particular reference to glyphosate usage. J Environ Sci Health Part C 8:277–286

Fry DJ (1991) Centipedegrass response to plant growth regulators. HortSci 26:40–42

Fu S, Chen HYH, Bell FW, Sharma M, Delaney JR, Peterson G (2008) Effects of timing of glyphosate application on jack pine, black spruce, and white spruce plantations in northern Manitoba. For Chron 84:37–45

Funke T, Healy-Fried ML, Han H, Alberg DG, Bartlett PA, Schönbrunn E (2007) Differential inhibition of class I and class II 5-enolpyruvylshikimate-3-phosphate synthases by tetrahedral reaction intermediate analogues. Biochemistry 46:13344–13351

Gage KL, Krausz RF, Walters SA (2019) Emerging challenges for weed management in herbicide-resistant crops. Agriculture 9. https://doi.org/10.3390/agriculture9080180

Gaines TA (2018) The importance of glyphosate in non-GM settings. Outlooks Pest Manag 29:255–257

Gaines TA, Zhang W, Wang D, Bukun B, Chisholm ST, Shaner DL, Nissen SJ, Patzoldt WL, Tranel PJ, Culpepper AS, Grey TL, Webster TM, Vencill WK, Sammons RD, Jiang J, Preston C, Leach JE, Westra P (2010) Gene amplification confers glyphosate resistance in *Amaranthus palmeri*. Proc Natl Acad Sci U S A 107:1029–1034

Gaines TA, Shaner DL, Ward SM, Leach JE, Preston C, Westra P (2011) Mechanism of resistance of evolved glyphosate-resistant Palmer Amaranth (*Amaranthus palmeri*). J Agric Food Chem 59:5886–5889

Gattás F, Espinosa M, Babay P, Pizzaro H, Cataldo D (2020) Invasive species versus pollutants: potential of *Limnoperna fortunei* to degrade glyphosate-based commercial formulations. Ecotox Environ Saf 201. https://doi.org/10.1016/j.ecoenv.2020.110794

Ge X, d'Avignon A, Ackerman JJH, Ostrander E, Sammons RD (2013) Application of ^{31}P-NMR spectroscopy to glyphosate studies in plants: insights into cellular uptake and vacuole sequestration correlated to herbicide resistance. In: Handbook on herbicides: biological activity, classification and health and environmental implications. Nova Science Publishers, pp 55–83

Ge X, d'Avignon A, Ackerman JJH, Sammons RD (2014) Glyphosate uptake, vacuolar sequestration, and tonoplast pump activity in glyphosate-resistant horseweed. Plant Physiol 166:1256–1268

Geiger DR, Bestman HD (1990) Self-limitation of herbicide mobility by phototoxic action. Weed Sci 38:324–329

Geiger DR, Kapitan SW, Tucci MA (1986) Glyphosate inhibits photosynthesis and allocation of carbon to starch in sugar beet leaves. Plant Physiol 82:468–472

Geisy JP, Dobson S, Solomon KR (2000) Ecotoxicological risk assessment for roundup® herbicide. Rev Environ Contam Toxicol 167:35–120

Gélinas P, Gagnon F, McKinnon C (2018) Wheat preharvest herbicide application, whole-grain flour properties, yeast activity and the degradation of glyphosate in bread. Int J Food Sci Technol 53:1597–1602

Ghanizadeh H, Harrington KC (2020) Perpective: root exudation of herbicides as a novel mode of herbicide resistance in weeds. Pest Manag Sci 76:2543–2547

Ghosh S, Chisti Y, Banerjee UC (2012) Production of shikimic acid. Biotech Adv (6):1425–1431

Gimsing AL, Borggaard OK (2001) Effect of KCl and $CaCl_2$ as background electrolytes on the competitive adsorption of glyphosate and phosphate on goethite. Clays Clay Minerals 49:270–275

Givens WA, Shaw DR, Kruger GR, Johnson WG, Weller SC, Young BG, Wilson RG, Owen MDK, Jordan D (2009) Survey of tillage trends following the adoption of glyphosate-resistant crops. Weed Technol 23:15–155
Gomes MP, Smedbol E, Chalifour A, Hénault-Ethier L, Labrecque M, Lepage L, Lucotte M, Juneau P (2014) Alteration of plant physiology by glyphosate and its by-product aminomethylphosphonic acid: an overview. J Exp Bot 65:4691–4703
Gomes MP, Le Manac'h SG, Maccario S, Labreque M, Lucotte M, Juneau P (2016) Differential effects of glyphosate and aminomethylphosphonic acid (AMPA) on photosynthesis and chlorophyll metabolism in willow plants. Pestic Biochem Physiol 130:65–70
Gómez-Gallego C, Rainio M, Collado MC, Mantziari A, Salminen S, Saikkonen K, Helander M (2020) Glyphosate-based herbicide affects the composition of microbes associated with Colorado potato beetle (*Leptinotarsa decemlineata*). FEMS Microbiol Lett 367. https://doi.org/10.1093/femsle/fnaa050
Gougler JA, Geiger DR (1981) Uptake and distribution of *N*-phosphonomethylglycine in sugar beet plants. Plant Physiol 68:668–672
Green JM (2009) Evolution of glyphosate-resistant crop technology. Weed Sci 57:109–117
Green JM, Siehl DL (2020) History and outlook for glyphosate-resistant crops. Rev Environ Contam Toxicol. (this volume)
Green JM, Hazel CB, Forney DR, Pugh LM (2008) New multiple-herbicide crop resistance and formulation technology to augment the utility of glyphosate. Pest Manag Sci 64:332–339
Gressel J (2010) Herbicides as synergists for mycoherbicides, and vice versa. Weed Sci 58:324–328
Griffin JL, Boudreaux JM, Miller DK (2010) Herbicides as harvest aids. Weed Sci 58:355–358
Grossbard E (1985) Effects of glyphosate on the microflora: with reference to the decomposition of treated vegetation and interaction with some plant pathogens. In: Grossbard E, Atkinson D (eds) The herbicide glyphosate. Butterworths, London, pp 159–185
Grossbard E, Atkinson D (1985) The herbicide glyphosate. Butterworths, London, p 490
Haderlie LC, Widholm JM, Slife FW (1977) Effect of glyphosate on carrot and tobacco cells. Plant Physiol 60:40–43
Haidar R, Fermaud M, Calvo-Garrido C, Roudet J, Deshcamps A (2016) Modes of action for biocontrol of *Botryis cinera* by antagonistic bacteria. Phytopathol Mediterr 55:301–322
Hallas LE, Adams WJ, Heitkamp MA (1992) Glyphosate degradation by immobilized bacteria: field studies with industrial wastewater effluent. Appl Envir Microbiol 58:1215–1219
Hammer PE, Hinson TK, Duck NB, Koziel (2007) Methods to confer herbicide resistance. US patent application 20070107078 A1, pp 1-53
Hammerschmidt R (2018) How glyphosate affects plant disease development: it is more than enhanced susceptibility. Pest Manag Sci 74:1054–1063
Harikrishnan R, Yang XB (2002) Effects of herbicides on root rot and damping-off caused by *Rhizoctonia solani* in glyphosate-tolerant soybean. Plant Dis 86:1369–1373
Harring T, Streibig JC, Husted S (1998) Accumulation of shikimic acid: a technique for screening glyphosate efficacy. J Agric Food Chem 46:4406–4412
Harrington KC, Ghanizadeh H (2017) Herbicide application using wiper applicators – a review. Crop Prot 102:56–62
Harris CA, Gaston CP (2004) Effects of refining predicted dietary intakes of pesticide residues: a case study using glyphosate. Food Addit Contam 21:857–864
Harris D, Grossbard E (1979) Effects of the herbicides gramoxone W and roundup on *Septoria nodorum*. Trans Br Mycol Soc 73:27–34
Hart MM, Powell JR, Gulden RH, Dunfield KE, Pauls KP, Swanton CJ, Klironomos JN, Antune PM, Koch AM, Trevors JT (2009) Separating the effect of crop from herbicide on soil microbial communities in glyphosate-resistant corn. Pedobiologia 52:253–262
Hartzler RG (2010) Reduction in common milkweed (*Asclepias syriaca*) occurrence in Iowa cropland from 1999 to 2009. Crop Protect 29:1542–1544
Haughton AJ, Bell JR, Boatman ND, Wilcox A (2001) The effect of the herbicide glyphosate on non-target spiders: part II. Indirect effects on *Lepthyphantes tenuis* in field margins. Pest Manag Sci 57:1037–1042

Heap I, Duke SO (2018) Overview of glyphosate-resistant weeds worldwide. Pest Manag Sci 74:1040–1049

Hearon SE, Wang M, McDonald TJ, Phillips TD (2021) Decreased bioavailability of aminomethylphosphonic acid (AMPA) in genetically modified corn with activated carbon or calcium montmorillonite clay inclusion in soil. J Environ Sci 100:131–143

Helander M, Saloniemi I, Omacini M, Druille M, Salminen J-P, Saikkonen K (2018) Glyphosate decreases mycorrhizal colonization and affects plant-soil feedback. Sci Total Environ 642:285–291

Herbicide Resistance Action Committee (2020) HRAC mode of action classification 2020. https://hracglobal.com/files/HRAC_Revised_MOA_Classification_Herbicides_Poster.png. Accessed 8 Aug 2020

Herrmann KM (1995) The shikimate pathway as an entry to aromatic secondary metabolism. Plant Physiol 107:7

Hoagland RE (1980) Effects of glyphosate on metabolism of phenolic compounds. VI. Effects of glyphosine and glyphosate metabolites on phenylalanine ammonia-lyase activity, growth, and protein, chlorophyll, and anthocyanin levels in soybean (*Glycine max*) seedlings. Weed Sci 28:393–400

Hoagland RE, Boyette CD, Jordan RH, Stetina KC (2018) Interaction of the bioherbicide *Myrothecium verrucaria* with technical-grade glyphosate on glyphosate-susceptible and –resistant palmer amaranth. Am J Plant Sci 9:2306–2319

Hobbie KA, Rooney N, Scott RP, Anderson KA (2017) An alternative method to produce shikimic acid chemical feedstock by applying glyphosate to forage crops. Crop Sci 57:945–950

Hove-Jensen B, Zechel DL, Jochimsen B (2014) Utilization of glyphosate as phosphate source: biochemistry and genetics of bacterial carbon-phosphorous lyase. Microbiol Mol Biol Rev 78:176–197

Itou T, Hayama K, Sakai A, Tanouchi H, Okuda S, Kushima H, Kajimoto T (2015) Developing an effective glyphosate application technique to control *Bischofia javanica* Blume, an invasive alien tree species in the Ogasawara Islands. J Forest Res 20:248–253

Jacob GS, Garbow JR, Hallas LE, Kimack NM, Kishore GM, Schaefer J (1988) Metabolism of glyphosate in *Pseudomonas* sp. strain LBr. Appl Environ Microbiol 54:2953–2958

Jang SJ, Mallory-Smith C, Kuk YI (2020) Inhibition of wheat growth planted after glyphosate application to weeds. Weed Sci 68:373–381

Jansons M, Pugajeva I, Bartkevičs V (2018) Occurrence of glyphosate in beer from the Latvian market. Food Additives Contamin 35A:1767–1775

Jaworski EG (1972) Mode of action of *N*-phosphonomethylglycine: inhibition of aromatic amino acid biosynthesis. J Agric Food Chem 20:1195–1198

Jeffery LS, English JR, Connell J (1981) The effects of fall application of glyphosate on corn (*Zea mays*), soybeans (*Glycine max*), and johnsongrass (*Sorghum halepense*). Weed Sci 29:190–195

Jenson RA (1985) The shikimate/arogenate pathway: link between carbohydrate metabolism and secondary metabolism. Physiol Plant 66:164–168

Johal GS, Huber DM (2009) Glyphosate effects on diseases of plants. Eur J Agron 31:144–152

Johnson BJ (1976) Glyphosate for weed control in dormant bermudagrass. Weed Sci 24:140–143

Johnson BJ (1990) Response of bahiagrass (*Paspalium notatum*) to plant growth regulators. Weed Technol 4:895–899

Kandel YR, Bradley CA, Wise KA, Chilvers MI, Tenuta AU, Davis VM, Esker PD, Smith DL, Licht MA, Mueller DS (2015) Effect of glyphosate application on sudden death syndrome of glyphosate-resistant soybean under field conditions. Plant Dis 99:347–354

Kawate MK, Kawate SC, Ogg AG, Kraft JM (1992) Response of *Fusarium solani* f. sp. *pisi* and *Phythium ultimum* to glyphosate. Weed Sci 40:497–502

Kay SH (1995) Efficacy of wipe-on applications of glyphosate and imazapyr on common reed in aquatic sites. J Aquat Plant Manag 33:25–26

Kepler RM, Epp-Schmidt DJ, Yarwood SA, Cavigelli MA, Reddy KN, Duke SO, Bradley CA, Williams MM, Buyer JS, Maul JE (2020) Soil microbial communities in diverse

agroecosystems exposed to the herbicide glyphosate. Appl Environ Microbiol 86:e01744–e01719. https://doi.org/10.1128/AEM.01744-19
Khan S, Zhou JL, Ren L, Mojiri A (2020) Effects of glyphosate on germination, photosynthesis and chloroplast morphology in tomato. Chemosphere 258. https://doi.org/10.1016/j.chemosphere.2020.127350
King CA, Purcell LC, Vories ED (2001) Plant growth and nitrogenase activity of glyphosate-tolerant soybean in response to foliar glyphosate applications. Agron J 93:179–186
Kishore GM, Jacob GS (1987) Degradation of glyphosate by *Pseudomonas* sp. PG2982 via a sarcosine intermediate. J Biol Chem 262:12164–12168
Kishore GM, Shah DM (1988) Amino acid biosynthesis inhibitors as herbicides. Annu Rev Biochem 57:627–663
Kitchen LM, Witt WW, Rieck CE (1981) Inhibition of chlorophyll accumulation by glyphosate. Weed Sci 29:513–516
Knowles WS, Anderson KS, Andrew SS, Phillion DP, Ream JE, Johnson KA, Sikorski (1993) Synthesis and characterization of N-amino-glyphosate as a potent analog inhibitor of *E.Coli* EPSP synthase. Bioorgan Med Chem Lett 3:2863–2868
Kohn DC, South MS (2020) Use of glyphosate for disease suppression and yield enhancement in sobean. US10555527B2
Kolakowski BM, Miller L, Murray A, Leclair A, Bietlot H, van de Riet JM (2020) Analysis of glyphosate residues in foods from the Canadian retail markets between 2015 and 2017. J Agric Food Chem 68:5201–5211
Komoßa D, Gennity I, Sandermann H (1992) Plant metabolism of herbicides with C-P bonds: glyphosate. Pestic Biochem Physiol 43:85–94
Koskinen WC, Marek LJ, Hall KE (2016) Analysis of glyphosate and aminomethylphosphonic acid in water, plant materials and soil. Pest Manag Sci 72:423–432
Kremer RJ, Means NE (2009) Glyphosate and glyphosate-resistant crop interactions with rhizosphere microorganisms. Eur J Agron 31:153–161
Kremer RJ, Means NE, Kim KS (2005) Glyphosate affects soybean root exudation and rhizosphere microorganisms. Int J Environ Anal Chem 85:1165–1174
Krutz LJ, Locke MA, Steinrade RW Jr (2009) Interactions of tillage and cover crop on water, sediment, and pre-emergence loss in glyphosate-resistant cotton: implications for the control of glyphosate-resistant weed biotypes. J Environ Qual 38:1240–1247
Kryuchkova YV, Burygin GL, Gogoleva NE, Gogolev YV, Chernyshova MP, Makarov OE, Fedorov EE, Turkovskaya OV (2014) Isolation and characterization of glyphosate-degrading rhizosphere strain, *Enterobacter cloacae* K7. Microbiol Res 169:99–105
Krzyśko-Łupicka T, Orlik A (1997) Use of glyphosate as the sole source of phosphorus or carbon for the selection of soil-borne fungal strains capable to degrade this herbicide. Chemosphere 34:2601–2605
Kudsk P, Mathiasson SK (2020) Pesticide regulation in the European Union and the glyphosate controversy. Weed Res 68:214–222
Kuklinsky-Sobral J, Araújo WL, Mendes R, Pizzirani-Kleiner AA, Azevedo L (2005) Isolation and characterization of endophytic bacteria from soybean (*Glycine max*) grown in soil treated with glyphosate herbicide. Plant Soil 273:91–99
Laitenen P, Rämö S, Siimes K (2007) Glyphosate translocation from plants to soil – does it constitute a significant proportion of residues in soil? Plant Soil 300:51–60
Lanaro R, Costa JL, Cazenave SOS, Zanolli-Filho LA, Tavares MFM, Chasin AAM (2015) Determination of herbicides paraquat, glyphosate, and aminomethylphosphonic acid in marijuana samples by capillary electrophoresis. J Foren Sci 60:S241–S247
Lange AH, Fischer BB, Elmore CL, Kempen HM, Schlesselman J (1975) Roundup – the end of perennial weeds in tree and vine crops? Calif Agric:6–7
Lazzaro L, Feretti G, Bianchi E, Benesperi R (2019) Treatment by glyphosate-based herbicide allowed recovering native species after *Oxalis pes-caprae* L. invasion: indications from a Mediterranean island. Plant Biosyst 153:651–659

Lee LJ, Ngim J (2000) A first report of glyphosate-resisant goosegrass (*Eleusine indica* (L) Gaertn) in Malaysia. Pest Manag Sci 56:336–339
Lee CD, Penner D, Hammerschmidt R (2000) Influence of formulated glyphosate and activator adjuvants on *Sclerotinia sclerotiorumin* glyphosate-resistant and -susceptible *Glycine max*. Weed Sci 48:710–715
Lee CD, Penner D, Hammerschmidt R (2003) Glyphosate and shade effects on glyphosate-resistant soybean defense response to *Sclerotinia sclerotiorum*. Weed Sci 51:294–298
Leegood RC (2002) C_4 photosynthesis: principles of CO_2 concentration and prospects for introduction into C_3 plants. J Exp Bot 53:581–590
Legendre BL, Finger CK (1987) Response of sugarcane varietis to the chemical ripener glyphosate. Proc. Plant Growth Reg Soc Am 14:479–484
Lerbs W, Stock M, Parthier B (1990) Physiological aspects of glyphosate degradation in *Alcaligenes* spec. strain GL. Arch Microbiol 153:146–150
Lescano MR, Masin CE, Rodriguez AR, Godoy JL, Zalazar CS (2020) Earthworms to improve glyphosate degradation in biobeds. Environ Sci Pollut Res 27:27023–27031
Lévesque CA, Rahe JE (1992) Herbicide interactions with fungal root pathogens, with special reference to glyphosate. Annu Rev Phytopathol 30:579–602
Linz GM, Homan HJ (2011) Use of glyphosate for managing invasive cattail (*Typha* spp.) to disperse blackbird (*Icteridae*) roosts. Crop Protect 30:98–104
Locke MA, Bryson CT (1997) Herbicide-soil interaction in reduced tillage and plant residue management systems. Weed Sci 45:307–320
Lockhart WL, Billeck BN, Baron CL (1989) Bioassays with a floating aquatic plant (*Lemna minor*) for effects of sprayed and dissolved glyphosate. Hyrobiologia 188:353–359
Lopez EG (1993) Effect of glyphosate on different densities of water hyacinth. J Aquat Plant Manag 31:255–257
Lundgren JG, Gassmann AJ, Bernal J, Duan JJ, Ruberson J (2009) Ecological compatibility of GM crops and biological control. Crop Prot 28:1017–1030
Lydon J, Duke SO (1988) Glyphosate induction of elevated levels of hydroxybenzoic acids in higher plants. J Agric Food Chem 36:813–818
Madsen KH, Heitholt JJ, Duke SO, Smeda RJ, Streibig JC (1986) Photosynthetic parameters in glyphosate-treated sugarbeet (*Beta vulgaris* L.). Weed Res 35:81–88
Maeda H, Dudareva N (2012) The shikimate pathway and aromatic amino acid biosynthesis in plants. Annu Rev Plant Biol 63:73–105
Malalgoda M, Ohm J-B, Howatt KA, Green A, Simsek S (2020) Effects of pre-harvest glyphosate use on protein composition and shikimic acid accumulation in wheat. Food Chem 332:127422
Manea A, Leishman MR, Downey PO (2011) Exotic C4 grasses have increased tolerance to glyphosate under elevated carbon dioxide. Weed Sci 59:28–36
Marble SC, Prior SA, Runion GB, Torbert HA (2015) Control of yellow and purple nutsedge in elevated CO_2 environment with glyphosate and halosulfuoron. Front Plant Sci 6:Article 1. https://doi.org/10.3389/fpls.2015.00001
Maroli AS, Nandula VK, Duke SO, Tharayil N (2016) Stable isotope resolved metabolomics reveals the role of anabolic and catabolic processes in glyphosate-induced amino acid accumulation in *Amaranthus palmeri* biotypes. J Agric Food Chem 64:7040–7048
Maroli AS, Gaines TA, Foley ME, Duke SO, Doğramici M, Anderson JV, Horvath DP, Chao WS, Tharayil N (2018a) Omics in weed science: a perspective from genomics, transcriptomics, and metabolomics approaches. Weed Sci 66:681–695
Maroli AS, Nandula VK, Duke SO, Gerard P, Tharayil N (2018b) Comparative metabolomic analysis of two *Ipomoea lacunosa* biotypes with contrasting glyphosate tolerance captures glyphosate-induced differential perturbations in cellular physiology. J Agric Food Chem 66:2027–2039
Marrs RH, Frost AJ, Plant RA, Lunnis P (1993) Determination of buffer zones to protect seedlings of non-target plants from the effects of glyphosate spray drift. Agric Ecosyst Environ 45:283–293

Marshall EJP, Solomon KR, Carrasquilla G (2009) Coca (*Erythroxylum coca*) control is affected by glyphosate formulations and adjuvants. J Toxicol Environ Heath 72A:930–936

Martinez DA, Loening UE, Graham MC (2018) Impacts of glyphosate-based herbicides on disease resistance and health of crops: a review. Environ Sci Eur 30:2. https://doi.org/10.1186/s12302-018-0131-7

Marzabadi MR, Gruys KJ, Pansegrau PD, Walker MC, Yuen HK, Sikorski JA (1996) An EPSP synthase inhibitor joining shikimate 3-phosphate with glyphosate: synthesis and ligand bonding studies. Biochemistry 35:4199–4210

Matallo MB, Almeida SDB, Franco DAS, Cerdeira AL, Gazziero DLP (2014) Glyphosate as a tool to produce shikimic acid in plants. Planta Daninha 32:601–608

Matarczyk JA, Willis AJ, Vranjic JA, Ash JE (2002) Herbicides, weeds and endangered species: management of Bitou bush (*Chrysanthemoides monilifera* spp. *rotunda*) with glyphosate and impacts on the endangered shrub, *Pimelea spicata*. Biol Conserv 108:133–141

Mateos-Naranjo E, Redondo-Gómez S, Cox L, Cornejo J, Figueroa ME (2009) Effectiveness of glyphosate and imazamox on the control of the invasive cordgrass *Spartina densiflora*. Ecotoxicol Environ Saf 72:1694–1700

Matzrafi M, Brunharo C, Tehranchian P, Hanson BD, Jasieniuk M (2019) Increased temperatures and elevated CO_2 levels reduce the sensitivity of *Conyza canadensis* and *Chenopodium album* to glyphosate. Sci Rep 9:2228. https://doi.org/10.1038/s41598-019-38729-x

McAllister RS, Haderlie LC (1985) Translocation of ^{14}C-glyphosate and $^{14}CO_2$-labeled photoassimilates in Canada thistle (*Cirsium arvense*). Weed Sci 33:153–159

McConkey GA, Pinney JW, Westhead DR, Plueckhahn K, Fitzpatrick TB, Mcheroux P, Kappes B (2004) Annotating the *Plasmodium* genome and the enigma of the shikimate pathway. Trends Parasitol 20:60–65

McElroy JS, Hall ND (2020) *Echinochloa colona* with reported resistance to glyphosate conferred by aldo-keto reductase also contains a Pro-106-Thr EPSPS target site mutation. Plant Physiol. https://doi.org/10.1104/pp.20.00064

McWhorter DB, Derting CW (1985) Methods of application of glyphosate. In: Grossbard E, Atkinson D (eds) The herbicide glyphosate. Butterworths, London, pp 241–259

Melander B, Munier-Jolain N, Charles R, Wirth J, Schwarz J, van der Weide R, Bonin L, Jensen PK, Kudsk P (2013) European perspectives on the adoption of nonchemical weed management in reduced-tillage systems for arable crops. Weed Technol 27:231–240

Mendonca CM, Reed ML, Kukurugya MA, Aristilde L (2019) Adverse metabolic outcomes in soil *Pseudomonas* species exposed to polyethoxylated tallow amine and glyphosate. Envion Sci Technol Lett 6:448–455

Mertins M, Hoss S, Neumann G, Afzal J, Reichenbecher W (2018) Glyphosate, a chelating agent-relevant for eccological risk assessment? Environ Sci Pollut Res Internat 25:5298–5317

Mesnage R, Antoniou MN (2017) Facts and fallacies in the debate on glyphosate toxicity. Front Public Health. https://doi.org/10.3389/fpubh.2017.00316

Mills PJ, Kania-Korwel I, Fagan J, McEvoy LK, Laughlin GA, Barrett-Conner E (2017) Excretion of the herbicide glyphosate in older adults between 1993 and 2016. JAMA 318:1610–1611

Mir R, Jallu S, Singh TP (2015) The shikimate pathway: review of amino acid sequence, function and three-dimensional structures of the enzymes. Crit Rev Microbiol 41:172–189

Miyazaki J, Bauer-Panskus A, Bøhn T, Reichenbecher W, Then C (2019) Insufficient risk assessment of herbicide-tolerant genetically engineered soybeans intended for import into the EU. Environ Sci Eur 31:92. https://doi.org/10.1186/s12302-19-0274-1

Moerschbacher BM, Noll U, Gorrichon L, Reisener H-J (1990) Specific inhibition of lignification breaks hypersensitive resistance of wheat to stem rust. Plant Physiol 93:465–470

Mohseni-Moghadam M, Wolfe S, Dami I, Doohan D (2016) Response of wine grape cultivars to simulated drift rates of 2,4-D, dicamba, and glyphosate, and 2,4-D or dicamba plus glyphosate. Weed Technol 30:807–814

Mollaee M, Matloob A, Mobli A, Thompson M, Chauhan BS (2020) Response of glyphosate-resistant and susceptible biotypes of *Echinochloa colona* to low doses of glyphosate in different soil moisture conditions. PLoS One 15:e0233428. https://doi.org/10.1371/journal.pone.0233428

Monsanto (2020) Sprayers and water volumes. https://www.monsanto-ag.co.uk/roundup/roundup-amenity/application-information/sprayers-and-water-volumes/#:~:text=Knapsack%20Spraying,and%20250%20litres%20per%20hectare

Moorman TB, Becerril JM, Lydon J, Duke SO (1992) Production of hydroxybenzoic acids by *Bradyrhizobium japonicum* strains after treatment with glyphosate. J Agric Food Chem 40:289–293

Moretti ML, van Horn CR, Robertson R, Segobye K, Weller SC, Young BG, Johnson WG, Sammons RD, Wang D, Ge X, d'Avignon A, Gaines TA, Westra P, Green AC, Jeffery T, Lespérance MA, Tardif FJ, Sikkema PH, Hall JC, McLean MD, Lawton MB, Schulz B (2018) Glyphosate resistance in *Ambrosia trifida*: part 2. Rapid response physiology and non-target-site resistance. Pest Manag Sci 74:1079–1088

Morillo E, Undabeytia T, Maqueda C, Ramos A (2000) Glyphosate adsorption on soils of different characteristics. Influence of copper addition. Chemosphere 40:103–107

Morin F, Vera V, Nurit F, Tissut M, Marigo G (1997) Glyphosate uptake in *Catharanthus roseus* cells: role of a phosphate transporter. Pestic Biochem Physiol 58:13–22

Morishita DW (2018) Impact of glyphosate-resistant sugar beet. Pest Manag Sci 74:1050–1053

Nandula VK, Reddy KN, Rimando AM, Duke SO, Poston DH (2007) Glyphosate-resistant and -susceptible soybean (*Glycine max*) and canola (*Brassica napus*) dose response and metabolism relationships with glyphosate. J Agric Food Chem 55:3540–3545

Nandula VK, Riechers DE, Ferhatoglu Y, Marrett M, Duke SO, Dayan FE, Goldberg-Cavellari A, Tétard-Jones C, Wortley DU, Onkokesung N, Brazier-Hicks M, Edwards R, Gaines T, Iwakami S, Jugulam M, Ma R (2019) Herbicide metabolism: crop selectivity, bioactivation, resistance mechanisms, and regulation. Weed Sci 67:149–175

Nascente AS, Crusciol CAC, Stone LF, Cobucci T (2013) Upland rice yield as affected by previous summer crop rotation (soybean or upland rice) and glyphosate management of cover crops. Planta Daninha 31:147–155

Nelson KA, Renner KA, Hammerschmidt R (2002) Cultivar and herbicide selection affects soybean development and the incidence of Sclerotinia stem rot. Agron J 94:1270–1281

Nguyen CT, Dang LH, Nguyen DT, Tran KP, Giang BL, Tran NQ (2019) Effect of GA3 and gly plant growth regulators on productivity and sugar content of sugarcane. Agriculture 9:136. https://doi.org/10.3390/agriculture9070136

Nicolia A, Ferradini N, Molla G, Biagetti E, Pollegioni L, Veronesi F, Rosellini D (2014) Expression of an evolved engineered variant of a bacterial glycine oxidase leads to glyphosate resistance in alfalfa. J Biotechnol 184:201–208

Njiti VN, Myers O Jr, Schroeder D, Lightfoot D (2003) Roundup ready soybean: glyphosate effects on *Fusarium solani* root colonization and sudden death syndrome. Agron J 95:1140–1145

Nolan BL (2016) The effects of tillage, glyphosate, and genetic modification on bacterial root endophyte composition in Zea mays. Honors Thesis 138. https://egrove.olemiss.edu/hon_thesis/138

Norsworthy JK, Burgos NR, Oliver LR (2001) Differences in weed tolerance to glyphosate involve different mechanisms. Weed Technol 15:725–731

Obojska A, Ternan NG, Lejczak B, Kafarski P, McMullan G (2002) Organophosphonate utilizatjon by the thermophile *Geobacillus caldoxyylosilyticus* T20. Appl Environ Microbiol 68:2081–2084

Onyon LJ, Volans GN (1987) The epidemiology and prevention of paraquat poisoning. Human Toxicol 6:19–29

Orcaray L, Igal M, Marino D, Royuela M (2010) The possible role of quinate in the mode of action of glyphosate and acetolactate synthase inhbitors. Pest Manag Sci 66:262–269

Pan L, Yu Q, Han H, Mao L, Nyporko A, Fan L, Bai L, Powles S (2019) Aldo-keto reductase metabolizes glyphosate and confers glyphosate resistance in *Echinocloa colona*. Plant Physiol. https://doi.org/10.1104/pp.19.00979

Pankey JH, Griffin JL, Colyer PD, Schneider RW, Miller DK (2005) Preemergence herbicide and glyphosate effect on seedling disease in glyphosate-resistant cotton. Weed Technol 19:312–318

Patterson EL, Saski C, Küpper A, Beffa R, Gaines TA (2020) Omics potential in herbicide-resistant weed management. Plan Theory 8:607. https://doi.org/10.3390/plants8120607

Pazdiora PC, Piasecki C, Dorneles KR, Agostinetto D, Vargas L, Dallagnol LJ (2019) Glyphosate-resistant *Conyza bonariensis* is susceptible to powdery mildew caused by *Podosphaera erigerontis-canadensis*. Plant Dis 103:365

Pehl CE, Shelnutt HE (1990) Hexazinone influences on *Pinus taeda* seedlings. Forest Ecol Manage 35:271–276

Pereira FCM, Tayengwa R, Alves PLDCA, Peer WA (2019) Phosphate status affects phosphate transporter expression and glyphosate uptake and transport in grand eucalyptus (*Eucalyptus grandis*). Weed Sci 67:29–40

Pincelli-Souza RP, Bortolheiro FPAP, Carbonari CA, Velini ED, Silva MA (2020) Hormetic effect of glyphosate persists during the entire growth period and increase sugarcane yield. Pest Manag Sci 76:2388–2394

Pleasants JM, Oberhauser KS (2013) Milkweed loss in agricultural fields because of herbicide use: effect on the monarch butterfly population. Insect Conserv Diver 6:135–144

Pollegioni L, Schonbrunn E, Siehl D (2011) Molecular basis of glyphosate resistance – different approaches through protein engineering. FEBS J 278:2753–2766

Pombo G, Orzolek MD, Tukey LD, Pyzik TP (1985) The effect of paclobutrazol, diaminozide, glyphosate and 2,4-D in gel on the emergence and growth of germinated tomato seeds. J Hort Sci 60:353–367

Pradhan S, Flores FJ, Melouk H, Walker NR, Molineros JE, Garzon CD (2017) Chemical hormesis on plant pathogenic fungi and oomycetes. Amer Chem Soc Symp Ser 1249:121–133

Qiao C, Wang C, Pang R, Tian F, Han L, Guo L, Luo J, Li J, Pang T, Xie H, Fang J (2020) Environmental behavior and influencing factors in peach orchard ecosystem. Ecotoxicol Environ Saf 206:111209. https://doi.org/10.1016/j.ecoenv.2020.111209

Rajaa R, Nguyen TT, Slaughter DC, Fennimore SA (2020) Real-time weed-crop classification and localization technique for robotic weed control in lettuce. Biosyst Eng 192:257–274

Ramirez-Villacis DX, Finkel OM, Salas-González I, Fitzpatrick CR, Dangl JL, Jones CD, Leon-Reyes A (2020) Root microbiome modulates plant growth promotion induced by low doses of glyphosate. mSphere 5(4):e00484–e00420

Ratcliff AW, Busse MD, Shestak CJ (2006) Changes in microbial community structure following herbicide (glyphosate) additions to forest soils. App Soil Ecol 34:114–124

Reddy KN, Koger CH (2004) Live and killed hairy vetch cover crop effects on weeds and yield in glyphosate-resistant corn. Weed Technol 18:835–840

Reddy KN, Zablotowicz RM (2003) Glyphosate-resistant soybean response to various salts of glyphosate and glyphosate accumulation in soybean nodules. Weed Sci 51:496–502

Reddy KN, Hoagland RE, Zablotowicz RM (2001) Effect of glyphosate on growth, chlorophyll, and nodulation in glyphosate-resistant and susceptible soybean (*Glycine max*) varieties. J New Seeds 2:37–52

Reddy KN, Duke SO, Rimando AM (2004) Aminomethlphosphonic acid, a metabolie of glyphosate, causes injury in glyphosate-treated, glyphosate-resistant soybean. J Agric Food Chem 52:5139–5143

Reddy KN, Rimando AM, Duke SO, Nandula VK (2008) Aminomethylphosphonic acid accumulation in plant species treated with glyphosate. J Agric Food Chem 56:2125–2130

Reddy KN, Ding W, Zablotowicz RM, Thomson SJ, Huang Y, Krutz LJ (2010) Biological responses to glyphosate drift from aerial application in non-glyphosate-resistant corn. Pest Manag Sci 66:1148–1154

Reddy KN, Cizdziel JV, Williams MM, Maul JE, Rimando A, Duke SO (2018) Glyphosate resistance technology has minimal effect on maize mineral nutrition and yield. J Agric Food Chem 66:10139–10146

Renné P, Dressen U, Hebbeker U, Hille D, Flügge UI, Westhoff P, Weber APM (2003) The Arabidopsis mutant dct is deficient in the plastidic glutamate/malate translocator DiT2. Plant J 35:316–331

Roberts F, Roberts CW, Johnson JJ, Kyle DE, Krell T, Coggins JR, Coombs GH, Milhous WK, Tzipori S, Ferguson DJ, Chakrabarti D, McLeod R (1998) Evidence for the shikimate pathway in apicomplexan parasites. Nature 393:801–805

Roberts CW, Roberts F, Lyons RE, Kirisits MJ, Mui EJ, Finnerty J, Johnson JJ, Ferguson DJP, Coggins JR, Krell T et al (2002) The shikimate pathway and its branches in apicomplexan parasites. J Infect Dis 185(Suppl):S25–S36

Rodrigues JJ, Worsham AD, Corbin FT (1982) Exudation of glyphosate from wheat (*Triticum aestivum*) plants and its effects on interplanted corn (*Zea mays*) and soybeans (*Glycine max*). Weed Sci 30:316–320

Rodríguez-Gil JL, Prosser RS, Duke SO, Solomon KR (2020) Ecotoxicology of glyphosate, its formulants, and environmental degradation products. Rev Environ Contam Toxicol. (this volume)

Roisch U, Lingens F (1980) The mode of action of the herbicide *N*-(phosphonomethyl)glycine: its effect on the growth and enzymes of aromatic amino biosynthesis of *Escherichia coli*. Hoppe Seylers Z Physiol Chem 361:1049–1058

Rose-Fricker C (2002) Glyphosate tolerant fescue grass variety. US Patent 6,423,887 B1, July, 23, 2002

Samac D, Foster-Hartnett D (2012) Effect of glyphosate application on foliar diseases of glyphosate-tolerant alfalfa. Plant Dis 96:1104–1110

Sammons RD, Gaines TA (2014) Glyphosate resistance: state of knowledge. Pest Manag Sci 70:1367–1377

Sammons RD, Gruys KJ, Anderson KS, Johnson KA, Sikorski JA (1995) Reevaluating glyphosate as a transition-state inhibitor of EPSP synthase: identification of an EPSP synthase•EPSP•glyphosate ternary complex. Biochemistry 34:6433–6440

Sammons RD, You J, Qi Y, Flasinski S, Kavanaugh C, Washam J, Ostrander E, Wang D, Heck G (2018) Evaluation of glyphosate resistance in *Arabidopsis thaliana* expressing an altered target site EPSPS. Pest Manag Sci 74:1174–1183

Samsel A, Seneff S (2013) Glyphosate's suppression of cytochrome P450 enzymes and amino acid biosynthesis by the gut microbiome: pathways to modern diseases. Entropy 15:1416–1463

Sanogo S, Yang XB, Lundeen P (2001) Field response of glyphosate-tolerant soybean to herbicides and sudden death syndrome. Plant Dis 85:773–779

Saunders LE, Pezeshki R (2015) Glyphosate in runoff waters and in the root-zone: a review. Toxics 3:462–480

Schafer JR, Hallett SG, Johnson WG (2012) Response of giant ragweed (*Ambrosia trifida*), horseweed (*Conyza canadensis*), and common lambsquaters (*Chenopodium album*) biotypes to glyphosate in the presence of absence of soil microorganisms. Weed Sci 60:641–649

Schafer JR, Hallett SG, Johnson WG (2013) Soil microbial root colonization of glyphosate-treated giant ragweed (*Ambrosia trifida*), horseweed (*Conyza canadensis*), and common lambsquarters (*Chenopodium album*) biotypes. Weed Sci 61:289–295

Schlatter DC, Yin C, Hulbert S, Burke I, Paulitz T (2017) Impact of repeated use on wheat-associated bacterial are small and depend on glyphosate use history. Appl Environ Microbiol 83:e0134–e0117. https://doi.org/10.1128/AEM.0134-17

Schönherr J (2002) A mechanistic analysis of penetration of glyphosate salts across astomatous cuticular membranes. Pest Manag Sci 58:342–351

Schrübbers LC, Valverde BE, Sørensen JC, Cedergreen N (2014) Glyphosate spray drift in *Coffea arabica* – sensitivity of coffee plants and possible use of shikimic acid as a biomarker for glyphosate exposure. Pest Biochem Physiol 115:15–22

Sebastian DJ, Nissen SJ, Sebastian JR, Beck KG (2017) Seed bank depletion: the key to long-term downy brome (*Bromus tectorum* L.) management. Rangeland Ecol Manag 70:477–483

Selvapanidiyan A, Bhatnagar RK (1994) Isolation of a glyphosate-metabolizing *Pseudomonas*: detection, partial purification and localization of carbon-phosphorous lyase. Appl Microbiol Biotechnol 40:876–882

Sengupta D, Naik D, Reddy AR (2015) Plant aldo-keto reductases (AKR) as multi-tasking soldiers involved in diverse plant metabolic processes and stress defense: a structure-function update. J Plant Physiol 179:40–55

Servaites JC, Tucc MA, Geiger DR (1987) Glyphosate effects on carbon assimilation, ribulose bisphosphate carboxylase activity, and metabolite levels in sugar beet leaves. Plant Physiol 85:370–374

Shaner DL, Nadler-Hassar T, Henry WB, Koger CH (2005) A rapid *in vivo* shikimate accumulation assay with excised leaf discs. Weed Sci 53:769–774

Sharma U, Adee EA, Pfender WF (1989) Effect of glyphosate herbicide on pseudothecia formation by *Pyrenophora triticirepentis* in infested wheat straw. Plant Dis 73:647–650

Sharon A, Amsellem Z, Gressel J (1992) Glyphosate suppression of an elicited response. Increased susceptibility of *Cassia obtusifolia* to a mycoherbicide. Plant Physiol 98:654–659

Shieh W-J, Geiger DR, Buczynski SR (1993) Distribution of imported glyphosate in quackgrass (*Elytrigia repens*) rhizomes in relation to assimilate accumulation. Weed Sci 41:7–11

Simonsen L, Fomsgaard IS, Svensmark B, Spliid NH (2008) Fate and availability of glyphosate and AMPA in agricultural soil. J Environ Sci Health 43B:365–375

Singh BK, Shaner DL (1998) Rapid determination of glyphosate injury to plants and identification of glyphosate-resistant plants. Weed Technol 12:527–530

Singh S, Kumar V, Datta S, Wani AB, Dhanjal D, Romero R, Singh J (2020) Glyphosate uptake, translocation, resistance emergence in crops, analytical monitoring, toxicity and degradation: a review. Environ Chem Lett 18:663–702

Solomon KR (2020) Estimated exposure to glyphosate in humans via environmental, occupational, and dietary pathways: an updated review of the scientific literature. Pest Manag Sci 76:2878–2885

Solomon KR, Anadón A, Carrasquilla G, Cerdeira AL, Marshall J, Sanin L-H (2007) Coca and poppy eradication in Colombia: environmental and human health assessment of aerially applied glyphosate. Rev Environ Contam Toxicol 190:43–125

Sørensen FW, Gregersen M (1999) Rapid lethal intoxication caused by the herbicide glyphosate-trimesium (touchdown). Human Exp Toxicol 18:735–737

Steinrücken HC, Amrhein N (1980) The herbicide glyphosate is a potent inhibitor of 5-enolpyruvyl-shikimic acid-3-phosphate synthase. Biochem Biophys Res Commun 94:1207–1212

Steinrücken HC, Amrhein N (1984) 5-Enolpyruvylshikimate-3-phosphate synthase of *Klebsiella pneumoniae*. 2. Inhibition by glyphosate [N-(phosphonomethyl)glycine]. Eur J Biochem 143:351–357

Sturtz AV, Christie BR, Nowak J (2000) Bacterial endophytes: potential role in developing sustainable systems of crop production. Crit Rev Plant Sci 19:1–30

Suzuki N, Koussevitsky S, Mittler R, Miller G (2012) ROS and redox signaling in the response of plants to abiotic stress. Plant Cell Environ 35:259–270

Takano HK, Dayan FE (2020) Glufosinate-ammonium: a review of the current state of knowledge. Pest Manag Sci 76:3911–3925

Takano HK, Patterson EL, Nissen SJ, Dayan FE, Gaines TA (2019) Predicting herbicide movement across semi-permeable membranes using three phase partitioning. Pestic Biocchem Physiol 159:22–26

Tétard-Jones C, Edwards R (2016) Potential roles for microbial endophytes in herbicide tolerance in plants. Pest Manag Sci 72:202–209

Thogmartin WE, Wiederholt R, Oberhauser K, Drum RG, Diffendorfer JE, Altizer S, Taylor OR, Pleasants J, Semmens D, Semmens B, Erickson R, Libby K, Lopez-Hoffman L (2017) Monarch butterfly population decline in North America: identifying the threatening processes. R Soc Open Sci 4:170760. https://doi.org/10.1098/rsos.170760

Tohge T, Watanabe M, Hoefgen R, Fernie AR (2013) Shikimate and phenylalanine biosynthesis in green linage. Front Plant Sci 4:64

Torstensson NTL, Aamisepp A (1977) Detoxification of glyphosate in soil. Weed Res 17:209–212

Tuffi-Santos LD, Graça RN, Alfenas AC, Ferreira FA, Melo CA, Machado MS (2011) Glyphosate reduces urediniospore development and *Puccinia psidii* disease severity on *Eucalyptus grandis*. Pest Manag Sci 67:876–880

United States Electronic Code of Federal Regulations, §180.364 Glyphosate; tolerances for residues. https://www.ecfr.gov/cgi-bin/retrieveECFR?gp=1&SID=725587cc6c286ff9ca686aca984ab681&ty=HTML&h=L&mc=true&r=SECTION&n=se40.26.180_1364. Accessed 23 Jul 2020

United States Geological Survey (2020) Estimated annual agricultural pesticide use, Pesticide use maps – glyphosate. https://water.usgs.gov/nawqa/pnsp/usage/maps/show_map.php?year=2016&map=GLYPHOSATE&hilo=L&disp=Glyphosate. Accessed 28 Jul 2019

USDA AMS (2020) United States Department of Agriculture Agricultural Marketing Service, Pesticide Data Program. https://www.ams.usda.gov/datasets/pdp. Accessed 26 Jan 2020

Van Bruggen AHC, He MM, Shin K, Mai V, Jeong KC, Finckh MR, Morris JG (2018) Environmental and health effects of the herbicide glyphosate. Sci Total Environ 616–617:255–268

van Horn CR, Moretti ML, Robertson RR, Segobye K, Weller SC, Young BG, Johnson WG, Schulz B, Green AC, Jeffery T, Lespérance MA, Tardiff FJ, Sikkema PH, Hall JC, McLean MD, Lawton MB, Sammons RD, Wang D, Westra P, Gaines TA (2018) Glyphosate resistance in *Ambrosia trifida*: part 1. Novel rapid cell death response to glyphosate. Pest Manag Sci 74:107–1078

Velini ED, Alves E, Godoy MC, Meschede DK, Souza RT, Duke SO (2008) Glyphosate applied at low doses can stimulate plant growth. Pest Manag Sci 64:489–496

Velini ED, Trindade MLB, Barberis LRM, Duke SO (2010) Growth regulation and other secondary effects of herbicides. Weed Sci 58:351–354

Velini ED, Carbonari CA, Trindade MLB, Gomes GLGC, Antuniassi UR (2017) Variations in pesticide doses under field conditions. Am Chem Soc Symp Ser 1249:47–60

Vemanna RS, Vennapusa AR, Easwaran M, Chandrashekar BK, Rao H, Ghanti K, Sudhakar C, Mysore KS, Makarla U (2017) Aldo-keto reductase enzymes detoxify glyphosate and improve herbicide resistance in plants. Plant Biotechnol J 15:794–804

Viana RS, Lisboa LAM, Figueredo PAM, Ramos SB, Ferrari S, May A, Prado EP, Miasaki CT, Ferreira IS, Brenha JAM (2019) Productivity and biochemical characteristics of sugarcane when submitted to the action of chemical ripeners. Aust J Basic Appl Sci 13:64–71

Wagner R, Kogan M, Parada AM (2003) Phytotoxic activity of root absorbed glyphosate in corn seedlings (*Zea mays* L.). Weed Biol Manag 3:228–232

Wang Z-Y, Brummer EC (2012) Is genetic engineering ever going to take off in forage, turf and bioenergy crop breeding? Ann Bot 110:1317–1325

Wang WZ, Yang BP, Feng XY, Cao ZY, Feng CL, Wang JG, Xiong GR, Shen LB, Zeng J, Zhao TT, Zhang SZ (2017) Development and characterization of transgenic sugarcane with insect resistance and herbicide tolerance. Front Plant Sci 8:1535. https://doi.org/10.3389/fpls.2017.01535

Wauchope RD (1976) Acid dissociation constants or arsenic acid, methylarsonic acid (MAA), dimethylarsinic acid (cacodylic acid), and *N*-(phosphonomethyl) glycine. J Agric Food Chem 24:717–721

Weaver MA, Krutz LJ, Zablotowicz RM, Reddy KN (2007) Effects of glyphosate on soil microbial communities and its mineralization in a Mississippi soil. Pest Manag Sci 63:388–393

Weise A, Schulte M, Theuvsen L, Steinmann H-H (2018) Interactions of glyphosate use with farm characteristics and cropping systems in Central Europe. Pest Manag Sci 74:1155–1165

Westerman RB, Murray DS (1994) Silverleaf nightshade (*Solanum elaeagnifolium*) control in cotton (*Gossipium hirsutum*) with glyphosate. Weed Technol 8:720–727

Weston LA (1990) Cover crop and herbicide influence on row crop seedling establishment. Weed Sci 38:166–171

Whigham DK, Stoller EW (1979) Soybean desiccation by paraquat, glyphosate, and ametryn to accelerate harvest. Agron J 71:630–633

Wicke D, Schulz LM, Lentes S, Scholz P, Poehlein A, Gibhardt J, Daniel R, Ischebeck T, Commichau FM (2019) Identification of the first glyphosate transporter by genomic adaptation. Envion Microbiol 21:1287–1305

Wiese A, Schulte M, Theuvsen L, Steinmann H-H (2018) Interactions of glyphosate use with farm characteristics and cropping patterns in Central Europe. Pest Manag Sci 74:1155–1165

Williams MM, Bradley CA, Duke SO, Maul JE, Reddy KM (2015) Goss's wilt incidence in sweet corn is independent of transgenic traits and glyphosate. HortScience 50:1791–1794

Wong PTW, Dowling PM, Tesoriero LA, Nicol HI (1993) Influence of preseason weed management and in-crop treatments on two successive wheat crops. 2. Take-all severity and incidence of Rhizoctonia root rot. Aust J Exp Agric 33:173–177

Wood LJ (2019) The presence of glyphosate in forest plants with different life strategies one year after application. Can J For Res 49:586–594

Xiang W-S, Wang X-J, Ren T-R, Ju X-L (2005) Expression of a wheat cytochrome P450 monooxygenase in yeast and its inhibition by glyphosate. Pest Manag Sci 61:402–406

Xu J, Smith S, Smith G, Wang W, Li Y (2019) Glyphosate contamination in grains and foods: an overview. Food Control 106:106710

Yael JA, Fuhr JD, Bocan GA, Millone AD, Tognalli N, dos Santos Afonso M, Martiarena ML (2014) Abiotic degradation of glyphosate into aminomethylphosphonic in the presence of metals. J Agric Food Chem 62:9651–9656

Yamada T, Kremer RJ, Castro PRC, Wood BW (2009) Glyphosate interactions with physiology, nutrition, and disease of plants: threat to agricultural sustainability. Eur J Agron 31:111–113

Yu Q, Jalaludin A, Han H, Chen M, Sammons RD, Powles SB (2015) Evolution of a double amino acid substitution in the 5-enolpyruvylshikimate-3-phosphate synthase in *Eleusine indica* conferring high-level glyphosate resistance. Plant Physiol 167:1440–1447

Zabaloy MC, Carné I, Viassolo R, Gómez MA, Gornez E (2016) Soil ecotoxicity assessment of glyphosate use under field conditions: microbial activity and community structure of eubacteria and ammonia-oxidizing bacteria. Pest Manag Sci 72:684–691

Zabalza A, Orcaray L, Fernández-Escalada M, Zulet-González A, Royuela M (2017) The pattern of shikimate pathway and phenylpropanoids after inhibition by glyphosate or quinate feeding in pea roots. Pestic Biochem Physiol 141:96–102

Zabalza A, Zulet-González A, Barco-Antoñanzas EMV, Gil-Monreal M, Royuela M (2020) Physiological approach to the use of the natural compound quinate in the control of sensitive and resistant *Papaver rhoeas*. Plan Theory 9:1215. https://doi.org/10.3390/plants9091215

Zablotowicz RM, Reddy KN (2007) Nitrogenase activity, nitrogen content, and yield responses to glyphosate in glyphosate-resistant soybean. Crop Prot 26:370–376

Zapiola ML, Mallory-Smith CA (2012) Crossing the divide: gene flow produces intergenic hybrids in feral transgenic creeping bentgrass population. Mol Ecol 21:4672–4680

Zboińska E, Maliszewska I, Lejczak B, Kafarski P (1992) Degradation of organophosphonates by *Penicillium citrinum*. Lett App Micobiol 15:269–272

Zhan H, Feng Y, Fan X, Chen S (2018) Recent advances in glyphosate biodegradation. Appl Microbiol Biotechnol 102:5033–5043

Ziska LH (2014) Climate, CO_2 and invasive weed management. In: Dukes JS (ed) Ziska LH. CAB International, Invasive species and global climate change, pp 293–304

Ziska LH, Goins EW (2006) Elevated atmospheric carbon dioxide and weed populations in glyphosate treated soybean. Crop Sci 46:1354–1359

Ziska LH, Teasdale BJA (1999) Future atmospheric carbon dioxide may increase tolerance to glyphosate. Weed Sci 47:608–615

Ziska LH, Teasdale JR (2000) Sustained growth and increased tolerance to glyphosate observed in a C3 perrenial weed, quackgrass (*Elytrigia repens*), grown at elevated carbon dioxide. Funct Plant Biol 27:159–166

Ziska LH, Faulkner S, Lydon J (2004) Changes in biomass and root:shoot ration of field-grown Canada thistle (*Cirsium arvense*), as noxious, invasive weed, with elevated CO_2: implications for control with glyphosate. Weed Sci 52:584–588

Zobiole LHS, Oliveira RS, Huber DM, Constantin J, Castro C, Oliviera FA, Oliveira A (2010) Glyphosate reduces shoot concentrations of mineral nutrients in glyphosate-resistant soybeans. Plant Soil 328:57–79

Zoller O, Rhyn P, Rupp H, Zarn JA, Geiser C (2018) Glyphosate residues in Swiss market foods: monitoring and risk evaluation. Food Additives Contam 11B:83–91

Zulet A, Zabalza A, Royuela M (2013) Phytotoxic and metabolic effects of exogenous quinate on *Pisum sativum* L. J Plant Growth Regul 32:779–788

Zulet-González A, Fernández-Escalada M, Zabalza A, Royuela M (2019) Enhancement of glyphosate efficiency on *Amaranthus palmeri* by exogenous quinate application. Pestic Biochem Physiol 158:1–11

Zulet-González A, Barco-Antoñanzas M, Gil-Monreal M, Royuela M, Zabalza A (2020) Increased glyphosate-induced gene expression in the shikimate pathway is abolished in the presence of aromatic amino acids and mimicked by shikimate. Front Plant Sci. https://doi.org/10.3389/fpls.2020.00459

History and Outlook for Glyphosate-Resistant Crops

Jerry M. Green and Daniel L. Siehl

Contents

Abstract Glyphosate-resistant (GR) crops, commercially referred to as glyphosate-tolerant (GT), started the revolution in crop biotechnology in 1996. Growers rapidly accepted GR crops whenever they became available and made them the most rapidly adopted technology in agriculture history. Adoption usually meant sole reliance on glyphosate [*N*-(phosphonomethyl)glycine, CAS No. 1071-83-6] for weed control. Not surprisingly, weeds eventually evolved resistance and are forcing growers to change their weed management practices. Today, the widespread dissemination of GR weeds that are also resistant to other herbicide modes-of-action (MoA) has greatly reduced the value of the GR crop weed management systems. However, growers continue to use the technology widely in six major crops throughout North and South America. Integrated chemistry and seed providers seek to sustain glyphosate efficacy by promoting glyphosate combinations with other herbicides and stacking the traits necessary to enable the use of partner herbicides. These include glufosinate {4-[hydroxy(methyl)phosphinoyl]-DL-homoalanine, CAS No. 51276-47-2}, dicamba (3,6-dichloro-2-methoxybenzoic acid, CAS No. 1918-00-9), 2,4-D

J. M. Green (✉)
Green Ways Consulting LLC, Landenberg, PA, USA
e-mail: jerry.m.green@greenwaysconsulting.com

D. L. Siehl
Sr. Scientist (ret.), Corteva Agriscience, Wilmington, DE, USA
e-mail: dsiehl.biochem@gmail.com

J. B. Knaak (ed.), *Reviews of Environmental Contamination and Toxicology Volume 255*,
Reviews of Environmental Contamination and Toxicology 255,
https://doi.org/10.1007/398_2020_54

[2-(2,4-dichlorophenoxy)acetic acid, CAS No. 94-75-7], 4-hydroxyphenyl pyruvate dioxygenase inhibitors, acetyl coenzyme A carboxylase (ACCase) inhibitors, and other herbicides. Unfortunately, herbicide companies have not commercialized a new MoA for over 30 years and have nearly exhausted the useful herbicide trait possibilities. Today, glyphosate-based crop systems are still mainstays of weed management, but they cannot keep up with the capacity of weeds to evolve resistance. Growers desperately need new technologies, but no technology with the impact of glyphosate and GR crops is on the horizon. Although the expansion of GR crop traits is possible into new geographic areas and crops such as wheat and sugarcane and could have high value, the Roundup Ready® revolution is over. Its future is at a nexus and dependent on a variety of issues.

Keywords Biotechnology · Formulation · Genetically modified crops · Herbicide-resistant · Herbicide-tolerant · Resistance · Tolerance · Traits · Weed · Weed management

Abbreviations

ACCase	Acetyl coenzyme A carboxylase
ALS	Acetolactate synthase
EFSA	European Food Safety Authority
EPA	Environmental Protection Agency
GM	Genetically modified
GST	Glutathione-S-transferase
HPPD	4-Hydroxyphenyl pyruvate dioxygenase
HR	Herbicide-resistant
HT	Herbicide-tolerant
IP	Intellectual property
ISAAA	International Service for the Acquisition of Agri-Biotech Applications
NTO	Nontarget organism
NTSR	Non-target site resistance
PDS	Phytoene desaturase
PPO	Protoporphyrinogen oxidase
PSII	Photosystem II

1 Introduction

Enabling the use of glyphosate as a selective crop herbicide in 1996 was one of the most important innovations of the twentieth century. It started the plant biotech crop revolution. Growers made glyphosate-resistant (GR) crops, generally known commercially as glyphosate-tolerant (GT), the most rapidly adopted technology in the history of agriculture because it was cheaper, more effective, and more convenient than the selective herbicides they were using. Today, six main crops have transgenes that confer glyphosate resistance: soybeans [*Glycine max* (L.) Merr.], corn (*Zea mays* L.), cotton (*Gossypium hirsutum* L.), canola (*Brassica napa* L.), alfalfa (*Medicago sativa* L.), and sugarbeets (*Beta vulgaris* L.). In 2018, 26 countries (21 developing and 5 industrialized countries) planted 191.7 million hectares of biotech crops, which added 1.9 million hectares to the 2017 record. Most genetically modified (GM) crops are resistant to glyphosate (ISAAA 2020).

Glyphosate was the ideal herbicide for developing herbicide-resistant (HR) crops. Its low-cost, high efficacy on nearly all weeds, low environmental impact, and low toxicity made it a "Once-in-a-Century Herbicide" (Duke and Powles 2008). Glyphosate is readily absorbed and translocated throughout weeds, where it inhibits 5-enolpyruvyl-shikimate-3-phosphate synthase (EPSPS; EC 2.5.1.19), an enzyme of the aromatic biosynthesis pathway in autotrophic organisms (Siehl 1997). When the pathway is blocked, the plant cannot synthesize essential metabolites such as aromatic amino acids, auxin hormones, and quinones including tocochromanols and plastoquinones.

Synthetic chemical herbicides are still the first option for weed control even after 70 years of widespread use. We discuss the impact of GR crops and the resulting evolution of GR weeds on chemical weed control. Despite the prevalence of GR weeds, glyphosate and GR crop systems will continue to have value when used in combination with herbicides with different modes-of-action (MoA) and other weed management tactics. New GR crops could have value through expansion into new geographies and crops, depending on public and regulatory acceptance and the success with weed management practices that sustain glyphosate utility (Bøhn and Millstone 2019; Green 2018).

2 Development of Glyphosate-Resistant Crops

Glyphosate was already widely used for nonselective vegetation control when Monsanto introduced GR crops in 1996. Monsanto began the long process of developing GR crops in 1983 when plant biotechnology was in its infancy. It saw the potential for GR crops when few others did (Kishore et al. 1992). However, achieving commercially acceptable tolerance to glyphosate was more difficult than expected due to the difficulty in finding a form of EPSPS with sufficient insensitivity to glyphosate and the requisite catalytic performance. Eventually, Monsanto scientists discovered an EPSPS with a high degree of insensitivity ($K_i = 1970$ μM) in an *Agrobacterium* strain called CP4, surviving in the manufacturing waste stream at Luling, LA (Barry et al. 1992).

Scientists generally consider two options for creating a herbicide trait; a target enzyme desensitized to inhibition by the herbicide or an enzyme that metabolizes the herbicide into an inactive molecule. For glyphosate, metabolic inactivation is feasible, but desensitization of the target has been the commercially successful approach.

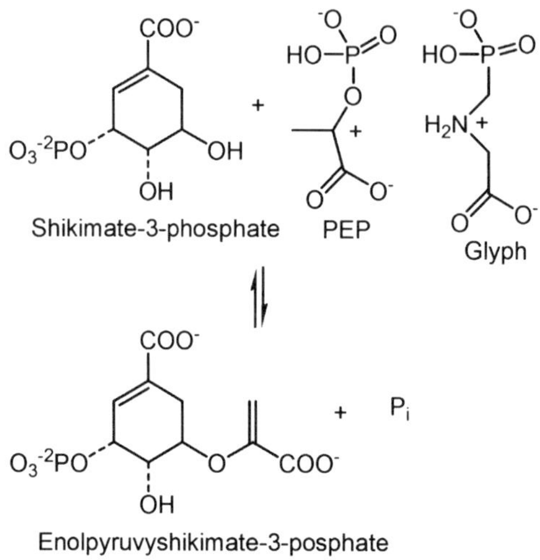

Fig. 1 EPSPS reaction. The reaction is an addition/elimination in which an enzymic base deprotonates the 5-hydroxyl of S3P, allowing the electron pair to attack the oxocarbenium ion of PEP (shown to suggest the species mimicked by glyphosate), generated by the enzyme. Originally published in the Journal of Biological Chemistry. Dong et al. (2019), Desensitizing plant EPSP synthase to glyphosate: Optimized global sequence context accommodates a glycine-to-alanine change in the active site. J. Biol. Chem. 2019; 294: 716–725 © the Author(s)

Search for desensitized EPSPS. EPSPS catalyzes the transfer of a carboxyvinyl group from phosphoenolpyruvate (PEP) to shikimate-3-phosphate (S3P) (Fig. 1). The crystal structure of the *E. coli* enzyme shows glyphosate bound adjacent to S3P in the PEP binding site (Schönbrunn et al. 2001), accounting for the consistent observation that inhibition is competitive with PEP (Boocock and Coggins 1983; Steinrucken and Amrhein 1984; Dong et al. 2019). The reaction proceeds through an oxocarbenium ion of PEP, generated by the enzyme. Glyphosate has a charge distribution and steric configuration like that of the carbenium resonance structure of PEP (Fig. 1). Tight binding (values for K_i for plant EPSPS in the range of 50 nM (Baerson et al. 2002) to 70 nM (Dong et al. 2019)) and observably slow release of glyphosate from an E:S3P:glyph complex (Dong et al. 2019) support the concept that glyphosate is a reaction intermediate analog.

The following is an evaluation of naturally occurring and mutant variants of EPSPS for their ability to confer commercial-level glyphosate resistance in crop plants, either by cis- or transgenic expression. The key parameters are k_{cat} (reactions per unit time at Vmax), $K_{m\ \mathrm{PEP}}$ (enzyme-substrate binding affinity; lower value = higher affinity), and K_i (binding affinity for glyphosate; lower value = higher affinity). Derivatives of those parameters provide an expression of catalytic efficiency (k_{cat}/K_m), selectivity for PEP vs glyphosate (K_i/K_m), and an expression, [$(k_{cat}/K_m)*K_i$] that captures both catalytic efficiency and selectivity. The values in Table 1 are useful for comparing the variants described because they were all obtained with nearly pure enzymes with mutations constructed in the same backbone (maize EPSPS) and analyzed in the same lab by the same procedure (Dong et al. 2019).

Table 1 Kinetic parameters of EPSPS variants found in GR weeds and crops, constructed in the context of *Z. mays* EPSPS (see Dong et al. 2019 for details)

Variant	Origin or application	k_{cat} min^{-1}	$K_{m\ PEP}$ µM	$k_{cat}/K_{m\ PEP}$ min^{-1} μM^{-1}	K_i µM	$(k_{cat}/K_{m\ PEP})$ x K_i, min^{-1}	k_{gly}[a], min^{-1}	$K_{m\ S3P}$ µM
Zm native		1,630	9.5	172	0.066	11	<lod	13.2
Zm-P106S	Many GR weeds	1,540	11.5	134	0.33	44	2.3	15.4
Zm-P106L	Few GR weeds	1,760	47.0	37.5	3.94	148	5.7	27.6
Zm-TIPS	GA21 maize	105	16.2	6.5	731	4,740	25.5	27.5
Zm-T102S	GR *Tridax*[b]	1,600	30.9	51.8	0.69	35.7	2.0	28.8
Zm-G101A	Constructed	1,000	333.0	3.0	1,930	5,780	25.7	84.0
Zm-G101A optimized[c]	Constructed	741	18.1	40.9	839	34,350	189	12.6
CP4	Glyph prod site; most GR crops	411	15.5	26.5	1,970	52,240	176	5.2

Data originally published in the Journal of Biological Chemistry. Dong et al. (2019), Desensitizing plant EPSP synthase to glyphosate: Optimized global sequence context accommodates a glycine-to-alanine change in the active site. J. Biol. Chem. 2019; 294: 716–725 © the Authors

[a]Enzyme turnover (min-1) at 30 µM PEP and S3P, 1 mM glyphosate (see text for rational as a fitness parameter

[b]GR *Tridax procumbens* reported by Li et al. (2018)

[c]Variant generated by random mutagenesis and shuffling of Zm-G101A, contains 20 substitutions relative to Zm-G101A (Dong et al. 2019)

All constructs had an N-terminal 10x-Histidine tag which, coupled with very high expression levels in *E. coli*, facilitate the purification of larger numbers of purified EPSPS variants for kinetic analysis. We calculated the concentration of each variant using a custom extinction coefficient calculated by vNTI, based on its amino acid sequence. We used a highly sensitive continuous spectrophotometric assay wherein the phosphate released from PEP was detected by reacting it with 2-amino-6-mercapto-7-methylpurine ribonucleoside (MESG, CAS No: 55727-10-1), catalyzed by purine nucleoside phosphorylase (EC 2.4.2.1), yielding the highly absorbent 2-amino-6-mercapto-7-methylpurine. Our kinetic parameters are similar to those of Baerson et al. (2002), who use highly sensitive detection of ^{14}C-EPSP produced from ^{14}C-PEP, and Yu et al. (2015), who detect phosphate with the MESG reagent.

Singly mutagenized EPSPS from plants, *E. coli,* or *Salmonella* yielded no variant with properties adequate for conferring commercial tolerance to glyphosate. The known plant mutations and close homologs, e.g., *E. coli* (termed Class I EPSPS), exert their effect by modulating the position of Gly101 (numbering according to mature maize EPSPS [CAA44974.1]; 96 in *E. coli*) in a way that creates interference with the binding of glyphosate through one of its phosphonate oxygens (Schönbrunn et al. 2001). The longer length of glyphosate relative to PEP allows for fine-tuning the differential affinity for the two ligands. In a crystallographic study of the *E. coli* enzyme, changing proline 101 (*E. coli* numbering) to serine or leucine had the effect of moving the alpha carbon of Gly96 closer to glyphosate, consequently reducing affinity for glyphosate (K_i with P101S, 14-fold increase; with P101L, 165-fold increase; Healy-Fried et al. 2007). The P101S substitution did not significantly affect affinity for PEP, while the P101L substitution reduced affinity (increased $K_{m\ \mathrm{PEP}}$) 2.5-fold. The P101S (*E. coli* numbering) mutation was discovered by mutagenesis of bacterial genes (Stalker et al. 1985). Since then, eight species of weeds have emerged as resistant to glyphosate by virtue of substitutions at the equivalent position (proline 106, mature maize EPSPS numbering; Baerson et al. 2002; Huffman et al. 2016; Ngo et al. 2018; Sammons and Gaines, 2014). Depending on the substitution and where no other resistance mechanisms are suspected, the dose required for 50% mortality is twofold to sevenfold greater in resistant plants relative to sensitive ones. The available kinetic data reflect a similar degree of desensitization (increased K_i) to glyphosate. In the *Zea mays* backbone, P106S elevated K_i by fivefold, but $K_{m\ \mathrm{PEP}}$ by only 20% (Table 1). The same mutation in GR goosegrass (*Eleusine indica*) raised $K_{m\ \mathrm{PEP}}$ by 2.3-fold (8.9 μM for P106S vs 3.8 μM in the native EPSPS), but the effect on K_i was much greater (0.048 vs 1.04 μM, Baerson et al. 2002).

Just as the proline to leucine substitution perturbed affinity for PEP while very significantly desensitizing the *E. coli* enzyme to glyphosate (Healy-Fried et al. 2007, vide supra), the same mutation had similar effects in maize EPSPS, where $K_{m\ \mathrm{PEP}}$ was elevated fivefold and K_i, 60-fold (Table 1). P106L has been identified in at least three GR weed species (Kaundun et al. 2011; Chen et al. 2015; Ngo et al. 2018). Ngo and colleagues isolated populations of GR Rhodes grass (*Chloris virgata*) with either the P106S or P106L mutation and noted that the lines containing P106L were 2.9-fold to 4.9-fold more resistant than the lines with P106S. Despite greater desensitization to glyphosate compared with P106S, the fivefold elevated $K_{m\ \mathrm{PEP}}$ with P106L limits the fitness of the enzyme, perhaps accounting for its lower occurrence in GR weeds relative to P106S.

A second glyphosate-desensitized EPSPS variant is a double mutant maize enzyme in which threonine at position 102 is changed to isoleucine in concert with the P106S mutation. The enzyme, termed TIPS, is highly desensitized to glyphosate (10,000-fold increased K_i), whereas its K_m for phosphoenolpyruvate is nearly normal (16.2 μM vs 9.5 for the native; Table 1). However, it has only 6% of the k_{cat} of the native enzyme (Table 1; also, Funke et al. 2009; Yu et al. 2015). Like the Pro106 mutations, the TIPS mutations exert their effect by shifting Gly101 closer to the glyphosate binding site (Funke et al. 2009). The catalytic efficiency ($k_{cat}/K_{m\ \mathrm{PEP}}$) for TIPS EPSPS is only 4% of that for native maize EPSPS (6.5 $\mathrm{min}^{-1}\ \mu\mathrm{M}^{-1}$ for TIPS vs 172 for native, Table 1), insufficient if a tolerance trait is to be created by natural mutagenesis or gene editing. However, given its excellent discrimination between PEP and glyphosate ($K_i/K_{m\ \mathrm{PEP}} = 45$ vs 0.0069 for native maize, calculated from Table 1), it can perform well given sufficiently high transgenic expression, as in GA21 maize (Spencer et al. 2000). While P106X mutations in resistant weeds have been known for years, it was recently shown in a tropical weed that a mutation at the

TIPS partner position, 102, can also confer resistance (Li et al. 2018). In this case, the T102 change was serine, not isoleucine. Kinetic characterization of the variant showed that it was no fitter than P106S and thus unsuitable as a commercial tolerance trait (Table 1).

Stepwise acquisition of both T102I and P106S mutations was documented in *Eleusine indica* (Yu et al. 2015). However, out of a population of 193 individuals, only 1.6% were homozygous for TIPS. The highest frequency allelic combination was TIPS/P106S, suggesting that the normal catalytic efficiency contributed from the P106S allele was more important for fitness than having the second allele encode a highly insensitive but catalytically deficient enzyme.

Moehs et al. (2020) recently used chemical mutagenesis, DNA-based screening, and conventional crossing to create the TIPS mutations in two of the three subgenome homoeologous copies of the EPSPS gene in wheat. The third homoeologous copy had either wild type EPSPS or was homozygous for the T101L mutation. The impaired catalytic capacity of the TIPS mutations in two of the three EPSPS isozymes appeared to impair growth in the absence of glyphosate treatment, despite the presence of a third, wild type, enzyme. The presence of other chemically induced mutations throughout the genome could also have contributed to impaired growth. The plants exhibited a "substantial" tolerance to glyphosate at spray rates of 630 and 870 g/ha.

The most direct way to influence the binding affinity of glyphosate through position 101 (maize numbering) is to substitute alanine for Gly101, which places an additional methyl group in the active site near the phosphate end of PEP or phosphonate end of glyphosate. This mutation was first reported with the enzyme from a GR strain of *Klebsiella* (Sost and Amrhein 1990). The first naturally occurring EPSPS known to have alanine in place of glycine at position 101 was that from *Agrobacterium* strain CP4, the organism found surviving at the glyphosate manufacturing plant and used to develop Roundup Ready® crops (Padgette et al. 1995). It is highly insensitive to glyphosate (K_i = 1970 μM) while maintaining a high affinity for PEP (K_m = 15 μM). However, it has only 15% of the catalytic efficiency (k_{cat}/K_m) of the plant enzyme due mainly to a much lower k_{cat} (Table 1), necessitating tissue-specific, high expression transformation cassettes. Plasmid pPV-GMGT04, for example, has one copy of the CP4 EPSPS gene driven by the Cauliflower Mosaic Virus 35S promoter and a second copy driven by the Figwort Mosaic Virus 35S promoter. Both copies were fused to the petunia EPSPS chloroplast transit peptide for targeting to the organelle with the entire aromatic biosynthesis pathway. An improved expression cassette was used for introducing CP4 EPSPS into Roundup Ready2 Yield soybeans and Roundup Ready Flex cotton (Meyer 2006). The plasmid, designated PV-GMGOX20, contains a chimeric promoter consisting of enhancer sequences from the 35S promoter of the Figwort Mosaic virus and the promoter from the *Tsf1* gene of *Arabidopsis thaliana* encoding elongation factor EF-1 alpha. Grain yield from commercial lines derived from the initial transformation event is 5% greater than that obtained from the original Roundup Ready soybeans (Meyer 2006).

Schönbrunn's group investigated the molecular basis for CP4's exquisite discrimination between glyphosate and PEP using X-ray crystallography. Though CP4 and *E. coli* EPSPS share only 26% amino acid sequence identity, they share the same fold and topology (Pollegioni et al. 2011; Duke 2021), allowing direct comparisons of CP4 with a representative Class I EPSPS. Funke et al. (2006) compared the crystal structure of CP4 ligated with S3P and glyphosate (PDB 2GGA) with a structural model of *E. coli* EPSPS where the contextually equivalent glycine (position 96, *E. coli* numbering) was changed to alanine, also ligated with S3P and glyphosate (Eschenburg et al. 2002). Funke et al. observed that the alanine methyl group in CP4 is 0.3 Å further away from the phosphonate group of glyphosate than the same alanine in the *E. coli* modeled structure. Presumably, the sequence context of CP4 places the methyl group of alanine 96 in an ideal position to interfere with glyphosate binding but not PEP.

Plant EPSPS with the G101A mutation has similar insensitivity to glyphosate as CP4 but only 1.4% of the catalytic efficiency of the native plant enzyme, mainly due to the 40-fold increase in K_m for PEP imposed by the additional methyl group (Table 1; also, Padgette et al. 1991). Its low affinity for PEP precludes the G101A mutation from being found in a GR weed as a single mutation. The divergent amino acid sequence of CP4 versus Class I EPSPSs was thought to provide the structural context for an optimal spatial location of the alanine methyl group. However, scientists at Corteva Agriscience showed that with 17 or more additional mutations (discovered by an iterative process of random mutagenesis, combinatorial gene shuffling, and selection), the enzyme from maize could be adapted to accommodate the G101A mutation, resulting in kinetic parameters equal to or better than those of CP4 (Dong et al. 2019). The maize variants are no closer in homology to CP4 than is the native maize enzyme, showing that the amino acid sequence context provided by CP4 that positions alanine for optimal discrimination between glyphosate and PEP is not unique but can be arrived at by modern methods of protein engineering. In theory, the substitutions defined by in vitro optimization could be created by CRISPR/Cas9-enabled gene editing.

Questions that emerge from the preceding review are (1) which kinetic parameters are most important for enabling glyphosate resistance in crops and weeds, (2) what are the ideal values for them, and (3) do they differ for crops and weeds? The ideal EPSPS for either crops or weeds would exhibit the normal ability to maintain flux through the EPSPS reaction in the presence of glyphosate concentrations up to 1 mM, a concentration attainable in tissues, especially meristems, receiving metabolite flow from treated leaves (Kirkwood et al. 2000). The term $(k_{cat}/K_m)*K_i$ combines an expression of catalytic efficiency (k_{cat}/K_m) with one of affinity for inhibitor compared to the substrate (K_i/K_m) (Lu et al. 2017). However, while the term is useful for assessing the intrinsic capacity for activity in the presence of a competitive inhibitor, it can be misleading for predicting in vivo fitness (reaction velocity under application conditions, i.e., plants sprayed with glyphosate). It omits concentrations of substrates and inhibitor, factors that are not intrinsic to the enzyme, but on which the reaction rate depends, as seen in the Michaelis–Menten equation for reaction velocity (v) in the presence of a competitive inhibitor (I):

$$v = \frac{k_{\text{cat}}\,[\text{E}]\,[\text{S}]}{K_m\left(1 + \frac{[\text{I}]}{\text{K}_i}\right) + [\text{S}]}$$

Reaction velocity is directly proportional to k_{cat} and nearly so to $1/K_m$. A very low value for K_i will greatly increase the denominator, thereby reducing v. Higher values for K_i effectively improve fitness, but only until K_i reaches the approximate inhibitor concentration, after which further increases will proportionately increase (k_{cat}/K_m) $*K_i$, but can only effect an additional twofold increase in v. The ideal gauge of enzyme fitness would be a single rate measurement made under the conditions of the application (pH, ionic strength, substrate, and inhibitor concentrations) if known. For optimizing maize EPSPS-G101A, we used a rate measurement under conditions designed to mimic intracellular conditions (pH 7, 100 mM KCl, 5% ethylene glycol; Dong et al. 2019). Ideally, concentrations of PEP and S3P would have been set at 10 or 15 μM, which we assume approximate in vivo concentrations based on their values for K_m, but the sensitivity of our assay limited us to 30 μM each. Glyphosate was set at 1 mM. The reaction velocity (μM min^{-1}) expressed as a function of enzyme concentration (μM) yields units of min^{-1}, which we termed "k_{gly}". Figure 2 is a graphic comparison of $(k_{\text{cat}}/K_m)*K_i$ with k_{gly}. The two measures of fitness correlated rather well except for CP4. With its very high K_i, CP4 displayed a disproportionately high $(k_{\text{cat}}/K_m)*K_i$. (Note: Values of $(k_{\text{cat}}/K_m)*K_i$ for CP4 and G101A-optimized are much farther apart than they appear on the log scale (See Table 1). This is due to the greater impact of K_i on that parameter compared with its

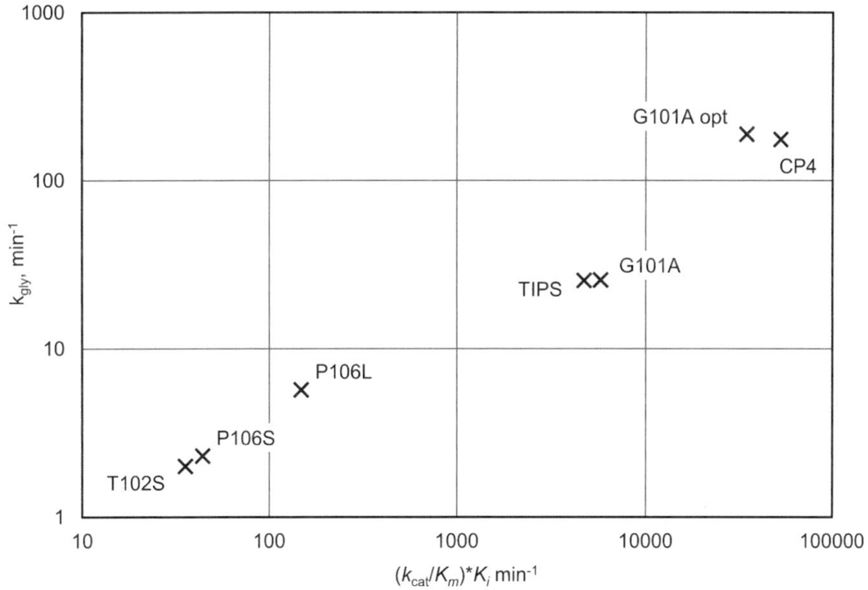

Fig. 2 Fitness ($\text{k}_{\text{gly}}{}^{1}$) of EPSPS variants as a function of values $(k_{\text{cat}}/K_m)*K_i$ $^{1}\text{k}_{\text{gly}}$; reaction velocity ($\text{min}^{-1}$) of the EPSPS variant in the presence of 30 μM each of PEP and S3P, 1 mM glyphosate. For rational as a fitness parameter, see Text, Sect. 2

impact on the velocity equation for competitive inhibition, which our k_{gly} parameter seeks to represent. The relatively low k_{gly} for CP4 given its outstanding selectivity (K_i/K_m) is attributable to its low value for k_{cat}. We conclude that the answer to question 1 is that all three raw kinetic parameters contribute to fitness for glyphosate resistance, and the composite term $(k_{cat}/K_m)*K_i$ is a good surrogate for fitness except when K_i is very much higher than [I]. The ideal GR-enabling EPSPS (Question 2) would have the insensitivity of CP4 ($K_i > \sim 1{,}500$ μM) and a catalytic efficiency approaching that of the native plant enzyme ($k_{cat}/K_m > 150\ min^{-1} \mu M^{-1}$). It would be instructive to learn whether such an enzyme would meet commercial requirements for glyphosate resistance if endowed by gene editing.

Regarding Question 3, from the data presented here, weeds appear to require a far less fit EPSPS variant than crops. The P106S mutation was shown conclusively to solely account for the resistance seen in a Tennessee isolate of GR *Eleusine indica* (Huffman et al. 2016). Yet the mutation has not been exploited as a GR trait in a crop, probably due to its modest insensitivity to glyphosate. In contrast, the TIPS mutations, while shown to carry a severe fitness penalty in weeds (Han et al. 2017), doubtless due to the impaired kcat (Table 1), can enable glyphosate resistance in maize, given sufficiently high transgenic expression (GA21 maize; Spencer et al. 2000). Crop resistance must be sufficient to withstand a double dose of herbicide due to overlapping spray. Further, crops place a high demand for the products of biosynthetic pathways, and any impaired flux will cause yield loss. In contrast, weeds need only to be fit enough to produce viable seeds. Also, a desensitized EPSPS variant encoded on the native gene may have an advantage over a transgene in that it is optimized by nature for appropriate expression in all tissues and growth stages.

Glyphosate Resistance Through Derivatization or Degradation of Glyphosate An alternative way to confer herbicide resistance in crops is to express an enzyme that degrades or derivatizes the herbicide. In one such approach, N-acetylation of glyphosate was discovered in a soil bacterium, *Bacillus licheniformis* (Castle et al. 2004). The activity was far too weak to confer tolerance but was increased 9,000-fold by gene shuffling. Although the native substrate is not known, robust activity ($k_{cat}/K_m = 1{,}500\ min^{-1}\ mM^{-1}$ versus $4\ min^{-1}\ mM^{-1}$ for glyphosate) was found with D-2-amino-3-phosphonopropionate (D-AP3) an isomer of glyphosate (Siehl et al. 2007). Though no antibiotic activity has been ascribed to D-AP3, the existence of an N-acetyltransferase with activity toward it is reminiscent of the mechanism for detoxifying glufosinate.

In microorganisms, glyphosate is degraded by two distinct pathways, as shown in Fig. 3. Glyphosate is not metabolically degraded in most plant species. However, appreciable oxidation to glyoxylate and aminomethylphosphonate (AMPA) was observed in soybean (Komossa et al. 1992; Duke 2011). The enzyme responsible was not identified, but recently, an isolate of *Echinochloa colona* with low-level resistance to glyphosate showed elevated expression of an aldol-keto reductase capable of cleaving glyphosate to AMPA and glyoxylate (Pan et al. 2019). Elsewhere, a bacterial glycine oxidase (GO) was engineered to accept glyphosate as a

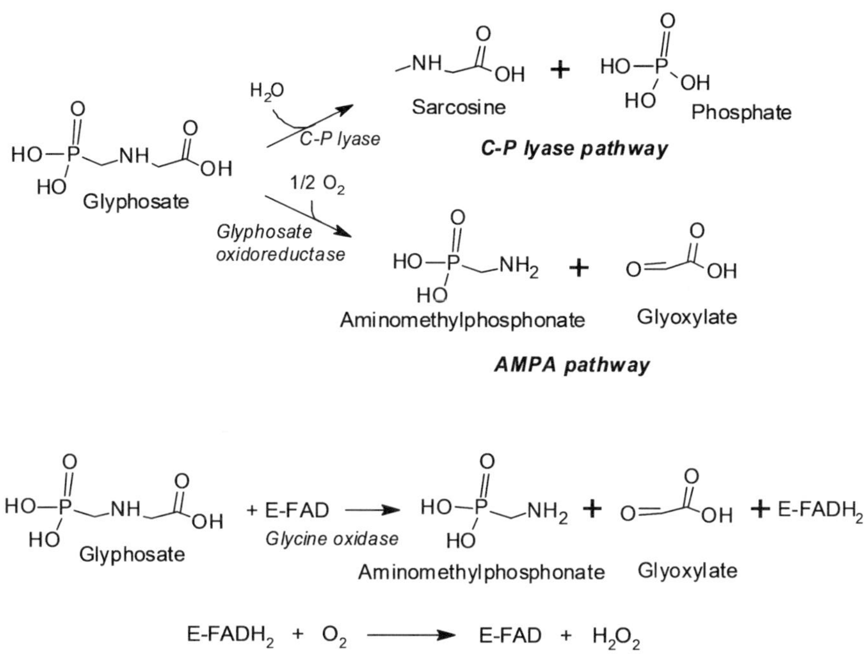

Fig. 3 Microbial pathways of glyphosate degradation. *Top*: Routes of microbial degradation of glyphosate. *Bottom*: Mechanism of the glycine oxidase modified for activity with glyphosate (Pedotti et al. 2009). Redrawn from Pollegioni et al. (2011); used with permission

substrate, yielding glyoxylate and AMPA (Pedotti et al. 2009). They engineered a 15,000-fold shift in the ratio of $k_{cat}/K_{m\ glyph}/k_{cat}/K_{m\ glycine}$ in their improved variants relative to the native enzyme, mainly by raising K_m for glycine 150-fold (0.7–105 mM) while reducing K_m for glyphosate by a similar magnitude (87–0.5 mM). Values for kcat were about 1 s^{-1} for wild type and improved variant with either substrate. Alfalfa plants expressing the improved glyphosate oxidase linked to a chloroplast targeting sequence exhibited "moderate" resistance (Nicolia et al. 2014).

Another glyphosate oxidase termed GOX was identified in *Ochrobactrum anthropi* strain LBAA (Barry and Kishore 1995). Like GO, GOX is a flavoenzyme but catalyzes oxidative cleavage of the C2-N bond of glyphosate, yielding glyoxylate and AMPA, by a different mechanism. GOX acts in concert with CP4 EPSPS to confer glyphosate resistance in the first GR canola.

In addition to cleavage of the C2-N bond catalyzed by glyphosate oxidases, many soil microbes cleave the C-P bond, yielding N-methylglycine (sarcosine) and phosphate (Hove-Jensen et al. 2014). N-methylglycine occurs naturally and can be further metabolized to glycine by several routes. Thus, a tolerance mechanism by which the C-P lyase pathway metabolizes glyphosate would reduce if not eliminate the synthetic pesticide residues. Hove-Jensen and colleagues also elucidated the genetic and mechanistic details of the C-P lyase pathway. The multiple enzymes and transporters required for the pathway are encoded by 14 genes (more or less), usually on a single operon. Seven are considered the "core complex", with the phnJ gene product known to catalyze the key reaction, the S-adenosyl-l-methionine-dependent radical cleavage of 5-phosphoribosyl-1-phosphonate to produce 5-phosphoribosyl 1,2-cyclic phosphate and the corresponding alkane. Though tantalizing as a mechanism for glyphosate tolerance, it is a daunting prospect to express a multigenic trait coding for a multi-enzyme complex that normally functions anaerobically.

There are microbial enzymes that cleave C-P bonds by a hydrolytic mechanism, each specific for a particular phosphonate compound (Villareal-Chiu et al. 2012). It is tempting to use directed evolution to make these single-gene hydrolases accept glyphosate as a substrate. However, all the native substrates have a carbonyl group at the position beta to the phosphorus atom. The carbonyl oxygen can accommodate the electron pair that must be displaced from the phosphorus atom, precluding the C-P bond of glyphosate from being hydrolyzed by this mechanism.

3 Rapid Adoption of Glyphosate-Resistant Crops

Most of the economic impact of GR crops has been due to one gene, CP4 EPSPS (Fig. 4). Today, GR traits are widely available in breeding germplasm. Breeders can easily maintain glyphosate resistance as a background trait in their germplasm and satisfy the expectations of many growers that it is in the seed they purchase (Green 2014). GR soybeans sales started in 1996 with cotton, canola, and corn ensuing soon after. Sales of GR alfalfa and sugarbeets began a decade later. GR crops made the widely used nonselective glyphosate into a selective herbicide. Growers urgently needed the technology when it became available. Weeds were evolving resistance to selective herbicides such as triazines and inhibitors of acetolactate synthase and acetyl-CoA carboxylase, which required growers to use complicated and expensive mixtures to manage. Managing resistant weeds was taking too much time as farms

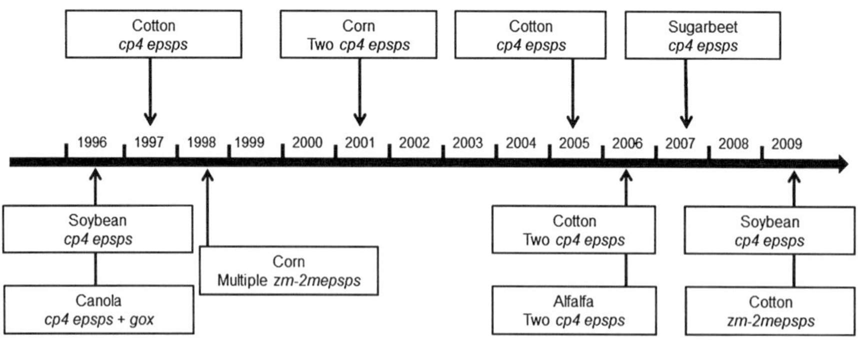

Fig. 4 Timeline for the introduction of commercial transgenic glyphosate traits by crop and resistance gene(s)

were getting bigger and employing fewer people. Glyphosate was initially the ideal solution to control resistant weeds.

GR crops gave the seed industry a new way to create intellectual property (IP) capture value. The first GR crop systems were not perfect (Elmore et al. 2001; Green 2009). Crop yields were low, safety margins narrow, and some application timings were tightly restricted. Monsanto also required growers to pay a technology fee and sign a contract. The contract required growers to agree to not replant seed, which was essential to maintain the trait value in soybeans. Growers strongly objected to the contract but still signed. Control of the glyphosate trait through this agreement was more valuable than a patent because it gave control of the technology indefinitely.

GR crops gave a range of benefits to growers. In addition to enabling the cost savings of using glyphosate instead of more expensive selective herbicides and realizing increased yields due to more effective weed control, GR crops enabled growers to reduce or even eliminate tilling. A pre-plant spray of glyphosate requires less fossil fuel than turning the soil, reducing fuel costs. Incidental but very welcome benefits to both growers and the environment were less soil erosion, reduced carbon dioxide emissions from tractors, and increased carbon sequestration in the soil. The net result of these benefits is that from 1996 to 2015 in North and South America was a cost-saving totaling $70 billion (Brookes et al. 2017).

The success of GR crops had an unfortunate unintended consequence. Although profits and research budgets generally increased, companies shifted funding away from herbicide discovery, which they perceived to have been largely rendered obsolete, to biotechnology and crop genetics (Charles 2001). Furthermore, generic manufacturers took advantage of the expiration of key glyphosate patents, which occurred not long after GR crops became available, and sold glyphosate at low prices. The continued decline in the cost of glyphosate reduced the demand for selective herbicides despite price reductions. In 2002, after 6 years of glyphosate sales in GR crops, the number of herbicides used on 10% or more of the US soybeans had decreased from 11 to just one, glyphosate (Duke and Powles 2009). Thus, GR crops were causing a problem (focusing selection pressure for weed resistance on one herbicide) and inhibiting the solution (discovering new herbicides with new MoA to partner with glyphosate).

4 Evolution and Consequences of Resistant Weeds

When growers adopted GR crops, they usually adopted the practice of using only glyphosate to control weeds (Baek et al. 2021). The crucial question always was whether a glyphosate-only system would be sustainable. Before GR crops, glyphosate was a widely used nonselective herbicide with very few suspected cases of resistance. Some believed that weeds would not evolve resistance because mutations in plant EPSPS were only modestly insensitivity to glyphosate or caused catalytic impairment (see above). Further, plants seemed to have minimal ability to degrade glyphosate to nontoxic metabolites (Duke 2011). However, applying glyphosate alone over vast areas of GR crops put tremendous selection pressure on weeds to evolve resistance, and they eventually did (Powles 2008). GR weeds are common now and reached the tipping point where many growers can no longer rely on glyphosate alone to provide commercially acceptable weed control. The epidemic of GR weeds has significantly reduced the value of the GR crop weed management system. Still, many growers continue to use GR crops throughout North and South America because competing systems are not any better, more expensive, or difficult to use.

In retrospect, growers should have used glyphosate in combination with existing selective herbicides to diversify their weed management practices. Photosystem II (PSII) inhibitors such as triazine and urea herbicides, lipid synthesis inhibitors such as S-metolachlor (2-amino-4-(hydroxymethylphosphinyl)butanoic acid, CAS No. 87392-12-9), and inhibitors of phytoene desaturase (PDS) or protoporphyrinogen oxidase (PPO) could have provided soil residual to control and delayed the evolution and spread of GR weeds (Green and Owen 2011). However, growers *en masse* used only glyphosate year after year. Today, 48 weed species have evolved resistance to glyphosate with at least ten different mechanisms (Sammons et al. 2016, Heap 2020). Unfortunately, many GR weeds in GR crop systems are also resistant to other herbicides. Currently, 20 weed species are known to be resistant to glyphosate and at least one other herbicide type (Heap 2020). These multiple HR weeds complicate weed management and threaten current crop production practices. Many growers are now almost out of options and must use large volumes of old and partially effective selective herbicide mixtures (Perry et al. 2016). The use of diverse herbicide systems in GR crops is now imperative.

The most difficult to control GR weeds in GR crops include *Amaranthus palmeri* S. Wats., *A. tuberculatus* (Moquin-Tandon) J. D. Sauer, *Ambrosia artemisiifolia* L., *A. trifida* L., and *Conyza canadensis* (L.) (Heap 2020). The first response of many growers when they see weed escapes is to reapply glyphosate at higher rates and then use glyphosate mixtures with other herbicides. Using a plethora of combinations of old, imperfect herbicides that growers had stopped using is a temporary solution and not a technological step forward (Green and Owen 2011).

The recent mergers of some of the largest pesticide companies, such as Bayer and Monsanto, Dow and DuPont, and Syngenta with ChemChina and Sinochem, ensure the continuance of herbicide discovery programs with critical mass (Mulvany and Decker 2019). Ironically, the slowdown in herbicide discovery has not meant a decline in the chemical herbicide business. Growers now must spend more money on more herbicides to combat HR weeds, creating a resurgence of the crop protection herbicide business (Sfiligoj 2014). In the same way, seed companies also benefit when growers buy higher-priced seeds with more herbicide traits that enable new options to control HR weeds.

Table 2 Commercial and publicly announced genetically modified transgenic multiple herbicide-resistant (HR) crops

Herbicide types	Crops
Glyphosate and glufosinate	Soybeans, corn, and cotton
Glyphosate, glufosinate, and 2,4-D	Soybeans and cotton
Glyphosate, glufosinate, and dicamba	Soybeans, corn, cotton, and wheat
Glyphosate, glufosinate, and HPPD-inhibitors	Soybeans and cotton
Glyphosate, glufosinate, 2,4-D, and ACCase inhibitors	Corn

Multiple companies are developing new PPO-inhibiting HR crop systems to combine with other traits
Multiple companies are developing HPPD crop traits
ACCase non-transgenic HR crops are also commercial

5 Next Generation of Glyphosate-Resistant Crops

When the patents for the "first generation" GR crops were about to expire (Shah et al. 1990), there was no process established on how to handle generic GM crops as there was for generic pesticides. Improved GR crops overcame some of the deficiencies of the first GR crops and created new intellectual property protection. The improved crops made retaining total control of the first GR crops less important for Monsanto. The new GR soybeans claimed a yield advantage over the original; the new GR cotton claimed improved crop safety with a wider application window, and the new GR canola claimed improved crop safety with a wider application window even without the glyphosate oxidase (*gox*) gene. New GR crops are still being introduced. In February 2019, Argentina approved a GM soybean coded DBN-09004-6 with the CP4 EPSPS and *pat* genes developed by Beijing Dabeinong Biotechnology Co., Ltd., becoming the first GM crop developed by a Chinese company approved for planting outside of China.

The days when crops are only resistant to glyphosate have ended (Green and Castle 2010; Que et al. 2010, Nandula 2019). A new generation of GR crops is well underway with combinations of glyphosate and other herbicide traits (Table 2). Today, most new HR crops are stacks with resistance to glyphosate, glufosinate, and one of four different herbicide types. In 2020, US soybean growers can now choose new varieties with various combinations of five HR traits (Ungelesbee 2019). These varieties have glyphosate and dicamba traits; glyphosate, dicamba, and glufosinate traits; glyphosate, glufosinate, and HPPD-inhibitor traits; and glyphosate, 2,4-D, and glufosinate traits, so growers can apply mixtures of glyphosate with other herbicides. Unfortunately, herbicide companies have not commercialized a new MoA for over 30 years and have nearly exhausted the useful herbicide trait combination. Growers desperately need new herbicide technology (Han et al. 2016; Dayan 2019), but the chance of finding another herbicide with a similar impact to glyphosate is small.

The newest HR crop technologies have resistance to a synthetic auxin herbicide, ACCase-inhibitor, or one of two HPPD-inhibitors (Behrens et al. 2007; Wright et al. 2010). Glyphosate resistance stacked with dicamba resistance is getting the most attention. The dicamba trait is conferred by a monooxygenase from *Pseudomonas maltophilia* (strain DI-6) that converts dicamba to 3,6-dichlorosalicylic acid (DCSA) and formaldehyde (Behrens et al. 2007). The oxygenase reaction requires two electrons and two protons, which in the bacterium originate from NADH and are shuttled via a reductase to ferredoxin. Interestingly, robust tolerance is conferred in plants by transformation only with the oxygenase. The plant has orthologs of the reductase and ferredoxin that are fully adequate to complete the electron transfer.

After three years of strong growth, experts expect plantings of dicamba-resistant crops to plateau this year at about 20 million hectares. Four companies are promoting the technology and hold registrations for foliarly applied dicamba. Dicamba-resistant soybeans and cotton enable new uses of an old herbicide with a long history of off-target drift problems. The first seasons of dicamba use in dicamba-resistant soybeans caused millions of hectares of damage to nontarget sensitive soybeans and other plants (Bradley 2018; Hager 2019). Opinions differ sharply on what caused the problem (Li et al. 2013; Egan et al. 2014). Despite the difficulties, the EPA extended dicamba product registration for two and five more years (EPA 2018, 2020).

The analogous use of 2,4-D in resistant crops is expanding greatly this year, making it about 2 years behind dicamba use and similarly depends on new directions for use and a new salt and formulation. Corteva expects its 2,4-D resistant seed to capture about 20% of the US crop in 2020, the first year it has been widely available. The 2,4-D trait is conferred in soybeans by AAD-12, an Fe(II)/α-ketoglutarate-dependent dioxygenases from *Delftia acidovorans* that degrades the acetic acid side chain of 2,4-D, yielding non-phytotoxic dichlorophenol and glyoxylate (Wright et al. 2010).

Crops with resistance to HPPD-inhibiting herbicides with some soil residual could also help control key GR weeds. The evolution of HPPD-resistant Palmer amaranth and waterhemp before market introduction reduces the value of this technology and requires its use with other herbicides (Green 2012). Two HPPD traits with different characteristics are under development, both of which involve an HPPD with reduced sensitivity (Allen et al. 2012; Miller et al. 2013). As with auxin herbicides, corn generally has natural tolerance to most HPPD herbicides, so the technology has more utility in soybeans and cotton.

BASF and Bayer, in cooperation with Sumitomo, are developing crops resistant to PPO-inhibiting herbicides (Green 2018). BASF and Sumitomo Chemical may have a new generation of broad-spectrum PPO herbicides that could be commercially available early next decade. The concept of matching broad-spectrum resistance-busting PPO-inhibiting herbicides with PPO-resistant crops could be very beneficial if researchers can identify the right herbicides and traits. However, a PPO-resistant crop system is not a new idea. Syngenta had a similar effort with the trade name of Accuron™ more than a decade ago (Li and Nicholl 2005).

Transgenic and non-transgenic crops are also commercially available with resistance to ACCase- and acetolactate synthase (ALS)-inhibiting herbicides (Green and Owen 2011). To get broad-spectrum crop resistance to a range of herbicides, researchers are investigating metabolic degradation by cytochrome P450 monooxygenase and glutathione-S-transferase (GST). Such metabolic mechanisms giving crop safety to a wide range of herbicides would be highly valuable until weeds evolve similar resistance mechanisms (Han et al. 2016; Délye 2013).

Growers need the new HR crop technologies to use with new nonselective and selective herbicides. Unfortunately, the future pipeline of herbicide options with commercial utility for the HR crop stacking is mostly exhausted (Green 2018). Current options already have resistant weed problems and other limitations. Most do not meet the standard of overlapping weed spectrum with an effective and different MoA (Vencill et al. 2012; Young 2015). Currently, three-way herbicide stacks are commercially available in cotton and will soon be in soybeans. Plans are for four-way herbicide stacks in the mid-2020s and a five-way by 2028. These multiple HR crops will help growers manage resistant weeds.

6 Outlook for Glyphosate-Resistant Crops

The agrochemical industry has encountered a downturn in the agricultural economy, as US farm income declined by 40% between 2013 and 2016 (https://www.cobank.com/knowledge-exchange/general/recessions-us-agriculture). Simultaneously, increasing regulations made the introduction of new herbicides and herbicide traits more expensive (Phillips 2020). Together, these trends slowed the introduction of new technology. Some politicians and regulators want to ban these technologies, so growers must contemplate a future without them. Weed scientists in Australia recently did just that, modeling five agronomic settings where glyphosate use would be restricted or banned (Beckie et al. 2020). The participants outlined alternative methods of weed control using nonchemical weed management practices combined with preemergence herbicides. The study is a model for formulating appropriate strategies in regions currently relying on glyphosate and GR crops.

Human safety of GR crops – Glyphosate has been widely used for more than five decades and GR crops for more than two decades. Still, GR crop systems remain controversial and the target of activists. Today, questions about safety dominate the news (Kabat 2019; WSSA 2019). In considering the potential toxicity of the proteins introduced into GR crops, we first point out that there have been no reports of direct carcinogenic, teratogenic, or mutagenic effects associated with the ingestion of proteins in general (Hammond et al. 2013). That is not surprising given that proteins are not taken up intact by the intestine, but are denatured by low pH in the stomach, then hydrolyzed into amino acids and di- and tri-peptides by intestinal proteases. Many pseudo-scientific reports claim adverse effects from consuming food derived from crops containing GR and other traits created by genetic technology. A report claimed that Cry insect control proteins from *Bacillus thuringiensis* caused hematotoxicity in mice when *Bt* spores containing various Cry proteins were administered by stomach tube (Mezzomo et al. 2013). Besides the irrelevant mode of administration, the control was water instead of spores lacking the Cry genes, thereby failing to account for the many substances in the spores that may have caused the observed effects. Other studies use physiologically unattainable doses of up to 1,000-fold typical exposure levels, attempting to demonstrate a hazard, as opposed to an actual risk (Hammond et al. 2013).

In GR crop-derived food, the protein introduced is either CP4 EPSPS or TIPS doubly mutated maize EPSPS. In the latter case, the same mutations are present in naturally occurring GR weeds (see above). Regarding CP4, its crystal structure overlaid with those of EPSPS of crop plants show that in addition to having the same function, they share the same structural fold and topology (Hammond et al. 2013). Homologous EPSPS proteins are ubiquitous in plant, yeast, and microbial food sources and have widely ranging degrees of amino acid sequence identity. All have a long history of safe use. No form of EPSPS, including CP4, has been reported to be toxic or allergenic.

The US EPA and other regulatory agencies support the use of glyphosate in GR crops and assure the public that it is safe when used according to label directions. Still, there is strong opposition that prevents their deployment in many crops. Businesses that own approved HR crop regulatory packages have a significant competitive advantage as the process of getting new approvals is too costly and too slow for most investors. For example, the commercialization of a single transgenic herbicide tolerance trait typically costs ~$136 million and takes over 13 years. Codeveloping a broad-spectrum herbicide in conjunction with a transgenic tolerance trait increases the risk and the cost, explaining why companies are shifting resources to less regulated methodologies such as gene editing (Crop Life America 2012, 2016). At the current rate that weeds evolve resistance, one new HR crop trait would not be enough to ensure sustainability.

Forecasts predicting that glyphosate would soon be the first pesticide to reach $10 billion annual sales have disappeared, but nobody predicts zero sales. Glyphosate is still the most broadly effective herbicide for most growers on most weeds (Abnewswire 2016). However, the spread of GR weeds is raising the cost of weed control, which creates more incentive for the industry to renew herbicide discovery efforts. Although the payoff for the simultaneous paired discovery of highly effective herbicides with a new MOA and associated trait would be very high, so are the risks associated with the high cost of discovery and development, long timelines, and the threat that non-target site resistance that could confer cross-resistance to a herbicide before it reaches the market. Growers and scientists agree that no weed management system used alone is sustainable for very long as weeds eventually evolve resistance to any single management tactic. This imperative is driving the industry to discover multiple new herbicide MoA and tolerance traits.

Today, scientists are exploring alternative ways to create glyphosate tolerance, such as wide-crossing from related resistant weed species, gene editing, gene shuffling, and new transgene options. One effort to displace CP4 EPSPS gene technology was to shuffle a gene encoding the acetyltransferase enzyme

(EC 2.3.1.13). The gene shuffling methodology resulted in very high crop tolerance to glyphosate (Castle et al. 2004; Green et al. 2009). Recently, an isolate of *Echinochloa colona* with low-level resistance to glyphosate showed elevated expression of an aldol-keto reductase capable of cleaving glyphosate to AMPA and glyoxylate (Pan et al. 2019). Though the resistance factor was modest, well-established enzyme optimization methods could identify amino acid substitutions that could greatly improve activity and be introduced into the native gene through CRISPR/Cas9-facilitated gene editing. The other known pathway for glyphosate degradation is through cleavage of the C-P bond, yielding phosphate and N-methylglycine (sarcosine), catalyzed by C-P lyase, described in Sect. 2. Because the pathway minimally requires seven gene products, it would be difficult to express in plants through transformation and impossible by gene editing. However, such a crop would have significantly reduced glyphosate residue with no nonnatural metabolites. A new metabolic trait that eliminates glyphosate residues in food crops such as wheat and sugarcane could be highly valuable, especially if it is non-transgenic.

New geographies for the introduction of GR crops should include developing countries, where they could satisfy a huge unmet need. In sub-Saharan Africa, for example, cassava, a staple crop relied upon by 500 million people, is normally hand-weeded by women and children. This primitive method often fails to optimize yield, is enormously time-consuming, and can result in spinal deformation (Gianessi 2013). GR and many other traits could be a great benefit to African farmers.

In the Americas, the "Roundup Ready Revolution" is over. Still, the use of glyphosate traits in combination with other traits could expand into some new regions and crops if public concerns about glyphosate and GR crops lessen. Growers that do not have GR crops yet can learn from the American experience and use glyphosate to expand the diversity of weed management practices to sustain the utility of the GR crop system. The new multiple HR crops enable more diverse and improved stewardship practices if growers follow advice from experts and label directions (Kaskey and Mulvany 2016; Heacox 2015).

7 Conclusion

No technology with the impact of glyphosate and GR crops is on the horizon (Westwood et al. 2018). Glyphosate-based crop systems will continue to be the mainstays of weed management in many areas, but they have lost value because they cannot keep up with the capacity of weeds to evolve resistance. Crops resistant only to glyphosate are not an acceptable options in most situations because of the evolution and spread of GR weeds. Relying on one weed management solution does not work anymore. Growers need multi-HR crops to combine glyphosate with other potent HR-enabled herbicides to control GR weeds. Traditional approaches by companies, sometimes called the pesticide and transgenic treadmills, cannot provide new solutions fast enough to match the speed that weeds evolve resistance (PAN 2016; Binimelis et al. 2009). Managing weeds using all currently available tactics in a systems approach is working for most growers, but nobody knows for how much longer. Hopefully, long enough to develop new weed management technologies.

In many market segments, it is challenging to buy seeds without a glyphosate trait. Many growers expect the glyphosate trait to be in the seed. The cost and time to introduce new HR crops is a high hurdle that slowed research for glyphosate trait combinations. Opposition to GR crops and the associated use of glyphosate and transgenic methods is still strong even after a quarter of a century of widespread use. The outlook for GR crops is at a nexus and depends on the following issues:

- Economic, e.g., input costs, farm income, and demands to increase production;
- Social, e.g., public opposition to glyphosate and GM crops;
- Environmental, e.g., regulations requiring minimum and no-tillage practices, drift control, and other label mandates;
- Biological, e.g., the continued evolution and spread of GR weeds;
- Technological, e.g., the effectiveness of new chemical and nonchemical weed management technologies to combat GR weeds;
- Sustainable, e.g., the continued utility of current technologies such as glufosinate, dicamba, 2,4-D, as well as HPPD- and PPO-inhibiting herbicides and their trait technologies;
- Regulatory, e.g., any removal or approval of GR crops or other technologies;
- Legal, e.g., resolution of the current and future environmental and human safety litigation; and
- Political, e.g., how public officials respond to activist pressure to restrict and even ban glyphosate and GM crops.

Conflict of Interest Statement The authors state that there are no conflicts of interest.

References

Abnewswire (2016) Global glyphosate market is expected to cross US$ 10.0 Billion by 2021. Abnewswire http://www.abnewswire.com/pressreleases/global-glyphosate-market-is-expected-to-cross-us-100-billion-by-2021-by-market-research-engine_52900.html

Allen J, Hinz J, Essner FJ, Van Wert S (2012) Introducing a new soybean event with glyphosate and HPPD tolerance. Abstr Weed Sci Soc Am 52:190

Baek Y, Bobadilla LK, Giacomini DA, Montgomery JS, Murphy BP, Tranel PJ (2021) Evolution of glyphosate-resistant weeds. Rev Environ Contam Toxicol. https://doi.org/10.1007/398_2020_55

Baerson SR, Rodriguez DJ, Tran M, Feng Y, Biest NA, Dill GM (2002) Glyphosate-resistant goosegrass: identification of a mutation in the target enzyme 5-enolpyruvylshikimate-3-phosphate synthase. Plant Physiol 129:1265–1275

Barry GF, Kishore GA (1995) Glyphosate tolerant plants. 5,463,175. US Patent

Barry G, Kishore G, Padgette S, Taylor M, Kolacz K, Weldon M, Re D, Eichholtz D, Fincher D, Hallas L (1992) Inhibitors of amino acid biosynthesis: strategies for imparting glyphosate tolerance to crop plants. In: Singh BK, Flores HE, Shannon JC (eds) Biosynthesis and molecular regulation of amino acids in plants. American Society of Plant Physiologists, Rockville, pp 139–145

Beckie HJ, Flower KC, Ashworth MB (2020) Farming without glyphosate. Plan Theory 9:96. https://doi.org/10.3390/plants9010096
Behrens MR, Mutlu N, Cjalrabprtu S, Dumitru R, Jiang WZ, LaVallee BJ, Herman PL, Clemente TE, Weeks DP (2007) Dicamba resistance: enlarging and preserving biotechnology-based weed management strategies. Science 316:1185–1188
Binimelis R, Pengue W, Monterroso I (2009) "Transgenic treadmill": responses to the emergence and spread of glyphosate-resistant johnsongrass in Argentina. Geoforum 40:623–633
Bøhn T, Millstone E (2019) The introduction of thousands of tonnes of glyphosate in the food chain – an evaluation of glyphosate tolerant soybeans. Foods 8:669. https://doi.org/10.3390/foods8120669
Boocock MR, Coggins JR (1983) Kinetics of 5-enolpyruvylshikimate-3-phosphate synthase inhibition by glyphosate. FEBS Lett 154:127–133
Bradley K (2018) July 15 dicamba injury update. Different year, same questions. University of Missouri Integrated Pest Management, Columbia. https://ipm.missouri.edu/ipcm/2018/7/July-15-Dicamba-injury-update-different-year-same-questions/
Brookes G, Taheripour F, Tyner W (2017) The contribution of glyphosate to agriculture and potential impact of restrictions on use at the global level. GM Crops Food 8:216–228
Castle LA, Siehl DL, Gorton DL, Patten PA, Chen YH, Certain S, Cho H-J, Duck N, Wong J, Liu D, Lassner MW (2004) Discovery and directed evolution of glyphosate tolerance gene. Science 304:1151–1154
Charles D (2001) Lords of the harvest: biotech, big money, and the future of food. Perseus Publishing, Cambridge, 348 pp
Chen J, Huang H, Zhang C, Wei S, Huang Z, Chen J, Wang X (2015) Mutations and amplification of EPSPS gene confer resistance to glyphosate in goosegrass (*Eleusine indica*). Planta 242:859–868
Crop Life America (2012) Getting a biotech crop to the market. Crop Life America, Washington. http://croplife.org/wp-content/uploads/2014/04/Fact-Sheet-Getting-a-Biotech-Crop-to-Market.pdf
Crop Life America (2016) A consultancy study for CropLife America. Crop Life America, Washington. http://croplifeamerica.org/wpcontent/uploads/2016/04/Phillips-McDougall-Final-Report_4.6.16.pdf
Dayan FE (2019) Current status and future prospects in herbicide discovery. Plan Theory 8(9):341. https://doi.org/10.3390/plants8090341
Délye C (2013) Unravelling the genetic bases of non-target-site-based resistance (NTSR) to herbicides: a major challenge for weed science in the forthcoming decade. Pest Manag Sci 69:176–187
Dong Y, Ng E, Lu J, Fenwick T, Tao Y, Bertain S, Sandoval M, Bermudez E, Hou Z, Patten P, Lassner M, Siehl D (2019) Desensitizing plant EPSP synthase to glyphosate: optimized global sequence context accommodates a glycine-to-alanine change in the active site. J Biol Chem 294:716–725
Duke SO (2011) Glyphosate degradation in glyphosate-resistant and -susceptible crops and weeds. J Agric Food Chem 59:5835–5841
Duke SO (2021) Glyphosate: uses other than in glyphosate-resistant crops, mode of action, degradation in plants, and effects on non-target plants and agricultural microbe. Rev Environ Contam Toxicol. https://doi.org/10.1007/398_2020_53
Duke SO, Powles SB (2008) Glyphosate: A once-in-a-century herbicide. Pest Manag Sci 64:319–325
Duke SO, Powles SB (2009) Glyphosate-resistant crops and weeds: now and in the future. AgBioforum 12:346–357
Eschenburg S, Healy ML, Priestman MA, Lushington GH, Schönbrunn E (2002) How the mutation glycine96 to alanine confers glyphosate insensitivity to 5-enolpyruvyl shikimate-3-phosphate synthase from *Escherichia coli*. Planta 216:129–135
Egan JF, Barlow KM, Mortensen DA (2014) A meta-analysis on the effects of 2,4-D and dicamba drift on soybean and cotton. Weed Sci 62:193–206

Elmore GA, Roeth FW, Nelson LA, Shapiro CA, Klein RN, Knezevic SZ et al (2001) Glyphosate-resistant soybean cultivar yields compared with sister lines. Agron J 93:408–412
EPA (2018) EPA announces changes to dicamba registration. https://www.epa.gov/newsreleases/epa-announces-changes-dicamba-registration
EPA (2020) EPA announces 2020 dicamba registration decision. https://www.epa.gov/newsreleases/epa-announces-2020-dicamba-registration-decision
Funke T, Han H, Healy-Fried ML, Fischer M, Schönbrunn E (2006) Molecular basis for the herbicide resistance of roundup ready crops. Proc Natl Acad Sci U S A 103:13010–13015
Funke T, Yang Y, Han H, Healy-Fried M, Olesen S, Becker A, Schönbrunn E (2009) Structural basis of glyphosate resistance resulting from the double mutation Thr-97 3 Ile and Pro-101 3 Ser in 5-enolpyruvylshikimate-3-phosphate synthase from *Escherichia coli*. J Biol Chem 284:9854–9860
Gianessi LP (2013) The increasing importance of herbicides in worldwide crop production. Pest Manag Sci 69:1099–1105
Green JM (2009) Evolution of glyphosate-resistant crop technology. Weed Sci 57:108–117
Green JM (2012) The benefits of herbicide-resistant crops. Pest Manag Sci 68:1323–1331
Green JM (2014) Current state of herbicides in herbicide-resistant crops. Pest Manag Sci 70:1351–1367
Green JM (2018) The rise and future of glyphosate and glyphosate-resistant crops. Pest Manag Sci 74:1035–1039
Green JM, Castle LA (2010) Transitioning from single to multiple herbicide-resistant crops. In: Nandula VK (ed) Glyphosate resistance in crops and weeds: history, development, and management. Wiley, Hoboken, pp 67–91
Green JM, Owen MDK (2011) Herbicide-resistant crops: utilities and limitations for herbicide-resistant weed management. J Agric Food Chem 59:5819–5829
Green JM, Hale T, Pagano MA, Andreassi JL II, Gutteridge SA (2009) Response of 98140 corn with gat4621 and hra transgenes to glyphosate and ALS-inhibiting herbicides. Weed Sci 57:142–148
Hager A (2019) Observations from the field: dicamba. The bulletin. University of Illinois Extension, Urbana. http://bulletin.ipm.illinois.edu/?page_id=1196&pf=4834
Hammond B, Kough J, Herouet-Guicheney H, Jez JM (2013) Toxicological evaluation of proteins introduced into food crops. Crit Rev Toxicol 43:25–42. https://doi.org/10.3109/10408444.2013.842956
Han H, Yu Q, Owen MJ, Cawthray GR, Powles SB (2016) Widespread occurrence of both metabolic and target-site herbicide resistance mechanisms in *Lolium rigidum* populations. Pest Manag Sci 72:255–263
Han H, Vila-Aiub MM, Jalaludin A, Yu Q, Powles SB (2017) A double EPSPS gene mutation endowing glyphosate resistance shows a remarkably high resistance cost. Plant Cell Environ 40:3031–3042
Heacox L (2015) Spray drift enters more complex era. CropLife. http://www.croplife.com/crop-inputs/adjuvants/spray-drift-enters-more-complex-era/
Healy-Fried ML, Funke T, Priestman MA, Han H, Schönbrunn E (2007) Structural basis of glyphosate tolerance resulting from mutations of Pro-101 in *Escherichia coli* 5-enolpyruvylshikimate-3-phosphate synthase. J Biol Chem 282:32949–32955
Heap I (2020) International survey of herbicide resistant weeds. http://www.weedscience.org/summary/Species.aspx. Accessed 14 May 2020
Hove-Jensen B, Zechel DL, Jochimsen B (2014) Utilization of glyphosate as phosphate source: biochemistry and genetics of bacterial carbon-phosphorus lyase. Microbiol Mol Biol Rev 78:176–197
Huffman JL, Riggins CW, Steckel LE, Tranel PJ (2016) The EPSPS Pro106Ser substitution solely accounts for glyphosate resistance in a goosegrass (*Eleusine indica*) population from Tennessee, United States. J Integr Agric 15:1304–1312
ISAAA (2020) GM approval database http://www.isaaa.org/gmapprovaldatabase/

Kabat B (2019) Who's afraid of roundup. Iss Sci Technol 36:64–73
Kaskey J, Mulvany L (2016) Monsanto seeds unleash unintended consequences across US farms. Bloomberg. http://www.bloomberg.com/news/articles/2016-09-01/a-soybean-killing-pesticide-spreads-across-america-s-farm-belt
Kaundun SS, Dale RP, Zelaya IA, Dinelli G, Marotti I, McIndoe E, Cairns A (2011) A novel P106L mutation in EPSPS and an unknown mechanism(s) act additively to confer resistance to glyphosate in a south African *Lolium rigidum* population. J Agric Food Chem 59:3227–3233
Kirkwood R, Hetherington R, Reynolds TL, Marshall G (2000) Absorption, localisation, translocation and activity of glyphosate in barnyard grass (*Echinochloa crus-galli* (L) Beauv): influence of herbicide and surfactant concentration. Pest Manag Sci 56:359–367
Kishore G, Padgette S, Fraley R (1992) History of herbicide-tolerant crops, methods of development and current state of the art – emphasis on glyphosate tolerance. Weed Technol 6:626–634
Komossa D, Gennity I, Sanderman H (1992) Plant metabolism of herbicides with C-P bonds: Glyphosate. Pestic Biochem Physiol 43:85–94
Li X, Nicholl D (2005) Development of PPO inhibitor-resistant cultures and crops. Pest Manag Sci 61:277–285
Li M, Tank H, Kennedy A, Zhang H, Downer B, Ouse D, Liu L (2013) Enlist duo herbicide: a novel 2,4-D plus glyphosate premix formulation with low potential for off-target movement. In: Bernards ML (ed) Pesticide formulations and delivery systems: 32nd volume, innovating legacy products for new uses. American Society for Testing Materials International, West Conshohocken, pp 125–161
Li J, Peng Q, Han H, Nyporko A, Kulynych T, Yu Q, Powles S (2018) Glyphosate resistance in *Tridax procumbens* via a novel EPSPS Thr-102-Ser substitution. J Agric Food Chem 66:7880–7888
Lu J, Dong Y, Ng EC, Siehl DL (2017) Novel form of the Michaelis-Menten equation that enables accurate estimation of (Kcat/Km)*Ki with just two rate measurements: utility in directed evolution. Protein Eng Des Sel 30:295–299
Meyer JJ (2006) Petition for the determination of nonregulated status for roundup RReady2Yield™ Soybean MON 89788. https://www.aphis.usda.gov/brs/aphisdocs/06_17801p.pdf
Mezzomo BP, Miranda-Vilela A-L, de Souza FI, Barbosa LCP, Portilho FA, Lacava ZGM, Grisolia CK (2013) Hematotoxicity of *Bacillus thuringiensis* as spore-crystal strains Cry1Aa, Cry1Ab, Cry1Ac or Cry2Aa in Swiss Albino Mice. J Hematol Thromb Dis 1:104
Miller B, Manley BA, Terpstra K, Vail G, Silverstone A, Allen J, Fischer J, Hinz J, Bloomberg J (2013) Development of next generation herbicide tolerant traits to enable enhanced weed management. Abstr Weed Sci Soc Am 53:279
Moehs CP, Austill WJ, Facciotti D, Holm A, Loeffler D, Lu Z, Mullenberg JC, Slade AJ, Steine MN, van Boxtel J, McGuire C (2020) Development of non-transgenic glyphosate tolerant wheat by TILLING. bioRxiv. https://doi.org/10.1101/2020.07.23.218883
Mulvany L, Decker S (2019) Post-chemical world takes shape as agribusiness goes green. Bloomberg. https://www.bloombergquint.com/business/agribusiness-goes-green-with-eco-friendly-fertilizer-pesticides
Nandula VK (2019) Herbicide resistance traits in maize and soybean: current status and future outlook. Plan Theory 8(9):337. https://doi.org/10.3390/plants8090337
Ngo TD, Krishnan M, Boutsalis P, Gill G, Preston C (2018) Target-site mutations conferring resistance to glyphosate in feathertop Rhodes grass (*Chloris virgata*) populations in Australia. Pest Manag Sci 74:1094–1100
Nicolia A, Ferradini N, Molla G, Biagetti E, Pollegioni L, Veronesi F, Rosellini D (2014) Expression of an evolved engineered variant of a bacterial glycine oxidase leads to glyphosate resistance in alfalfa. J Biotechnol 184:201–208
Padgette SR, Re DB, Gasser CS, Eichholtz DA, Frazier RB, Hironaka CM, Levine EB, Shah DM, Fraley RT, Kishore GM (1991) Site-directed mutagenesis of a conserved region of the 5-enolpyruvylshikimate-3-phosphate synthase active site. J Biol Chem 266:22364–22369

Padgette SR, Kolacz KH, Delanny X, Re DB, LaVallee BJ, Tinius CN, Rhodes WK, Otero YI, Barry GF, Eichholtz DA, Peschke VM, Nida DL, Taylor NB, Kishore GM (1995) Development, identification, and characterization of a glyphosate-tolerant soybean line. Crop Sci 35:1451–1461

PAN (2016) The pesticide treadmill. Pesticide Action Network. http://www.panna.org/gmos-pesticides-profit/pesticide-treadmill

Pan L, Yu Q, Han H, Mao L, Nyporko A, Fan L, Bai L, Powles SB (2019) Aldo-keto reductase metabolizes glyphosate and confers glyphosate resistance in *Echinochloa colona*. Plant Physiol 181:1–16

Pedotti M, Rosini E, Molla G, Moschetti T, Savino C, Vallone B, Pollegioni L (2009) Glyphosate resistance by engineering the flavoenzyme glycine oxidase. J Biol Chem 284:36415–36423

Perry ED, Ciliberto F, Hennessy DA, Moschini GC (2016) Genetically engineered crops and pesticide use in US maize and soybeans. Sci Adv 2(8):e1600850

Phillips MWA (2020) Agrochemical industry development, trends in R&D and the impact of regulation. Pest Manag Sci. https://doi.org/10.1002/ps.5728

Pollegioni L, Schonbrunn E, Siehl D (2011) Molecular basis of glyphosate resistance-different approaches through protein engineering. FEBS J 278:2753–2766

Powles SB (2008) Evolved glyphosate-resistant weeds around the world: lessons to be learnt. Pest Manag Sci 64:360–365

Que Q, Chilton M-DM, de Fontes CM, He C, Nuccio M, Zhu T, Wu Y, Chen JS, Shi L (2010) Trait stacking in transgenic crops: challenges and opportunities. GM Crops 1:220–229

Sammons RD, Gaines TA (2014) Glyphosate resistance: state of knowledge. Pest Manag Sci 70:1367–1377

Sammons RD, Giacomini E, Ostrander E, Silva A, Xiang B, Wang D (2016) Mechanisms of glyphosate resistance. Abstr Am Chem Soc 252:96

Schönbrunn E, Eschenburg S, Shuttleworth WA, Schloss JV, Amrhein N, Evans JN, Kabsch W (2001) Interaction of the herbicide glyphosate with its target enzyme 5-enolpyruvylshikimate 3-phosphate synthase in atomic detail. Proc Natl Acad Sci U S A 98:1376–1380

Sfiligoj E (2014) The resurgence of crop protection. CropLife. http://www.croplife.com/editorial/eric_sfiligoj/the-resurgence-of-crop-protection/

Shah DM, Rogers SG, Horsch RB, Fraley RT (1990) Glyphosate-resistant plants. US Patent 4,940,835

Siehl DL (1997) Inhibitors of EPSP synthase, glutamine synthetase and histidine synthesis. In: Roe RM et al (eds) Herbicide activity: toxicology, biochemistry and molecular biology. IOS Press, Amsterdam, pp 37–67

Siehl DL, Castle LA, Gorton R, Keenan RJ (2007) The molecular basis of glyphosate resistance by an optimized microbial acetyltransferase. J Biol Chem 282:11446–11455

Sost D, Amrhein N (1990) Substitution of Gly-96 to Ala in the 5-enolpyruvylshikimate-3-phosphate synthase of *Klebsiella pneumoniae* results in a greatly reduced affinity for the herbicide glyphosate. Arch Biochem Biophys 282:433–436

Spencer M, Mumm R, Gwynn J (2000) Glyphosate resistant maize lines. U. S. Patent 6040497

Stalker DM, Hiatt WR, Comai L (1985) A single amino acid substitution in the enzyme 5-enolpyruvylshikimate-3-phosphate synthase confers resistance to the herbicide glyphosate. J Biol Chem 260:4724–4728

Steinrucken HC, Amrhein N (1984) EPSPS of *Klebsiella pneumoniae* 2, inhibition by glyphosate. Eur J Biochem 143:351–357

Ungelesbee E (2019) A review of herbicide-tolerant soybean trait options for 2020. Farm Progress. https://www.dtnpf.com/agriculture/web/ag/crops/article/2019/10/17/review-herbicide-tolerant-soybean

Vencill WK, Nichols RL, Webster TM, Soteres JK, Mallory-Smith C, Burgos NR, Johnson WG, McClellan MR (2012) Herbicide resistance: toward an understanding of resistance development and the impact of herbicide-resistant crops. Weed Sci 60:2–30

Villareal-Chiu JF, Quinn JP, McGrath JW (2012) The genes and enzymes of phosphonate metabolism by bacteria, and their distribution in the marine environment. Front Microbiol 3:1–13

Westwood JH, Charudattan R, Duke SO, Fennimore SA, Marrone P, Slaughter DC, Swanton C, Zollinger R (2018) Weed Management in 2050: perspectives on the future of weed science. Weed Sci 66:275–285

Wright TR, Guomin S, Walsh TA, Lira JM, Cui C, Song P, Zhuang M, Arnold NL, Lin G, Russell SM, Cicchillo RM, Peterson MA, Simpson DM, Zhou N, Ponsamuel J, Zhang Z (2010) Robust crop resistance to broadleaf and grass herbicides provided by aryloxyalkanoate dioxygenase transgenes. PNAS 107:20240–20245

WSSA (2019) WSSA board issues statement concerning registration of glyphosate. Weed Science Society of America, Westminster. http://wssa.net/2019/08/wssa-board-issues-statement-concerning-registration-of-glyphosate/

Young BG (2015) Overlapping residual herbicides. FMC. http://www.fmccrop.com/Portals/_default/fmc_pdf/F100-37459_OverlapWhitePaper_MED.pdf?timestamp=1430496575694

Yu Q, Jalaludin A, Han H, Chen M, Sammons RD, Powles SB (2015) Evolution of a double amino acid substitution in the 5-enolpyruvylshikimate-3-phosphate synthase in *Eleusine indica* conferring high level glyphosate resistance. Plant Physiol 167:1440–1447

Evolution of Glyphosate-Resistant Weeds

Yousoon Baek, Lucas K. Bobadilla, Darci A. Giacomini, Jacob S. Montgomery, Brent P. Murphy, and Patrick J. Tranel

Contents

Abstract Widespread adoption of glyphosate-resistant crops and concomitant reliance on glyphosate for weed control set an unprecedented stage for the evolution of herbicide-resistant weeds. There are now 48 weed species that have evolved glyphosate resistance. Diverse glyphosate-resistance mechanisms have evolved, including single, double, and triple amino acid substitutions in the target-site gene, duplication of the gene encoding the target site, and others that are rare or nonexistent for evolved resistance to other herbicides. This review summarizes these resistance

Yousoon Baek, Lucas K. Bobadilla, Darci A. Giacomini, Jacob S. Montgomery, Brent P. Murphy, and Patrick J. Tranel contributed equally to this work.

Y. Baek · L. K. Bobadilla · D. A. Giacomini · J. S. Montgomery · B. P. Murphy · P. J. Tranel (✉)
Department of Crop Sciences, University of Illinois, Urbana, IL, USA
e-mail: ysbaek@illinois.edu; lucask3@illinois.edu; jsm2@illinois.edu; brentpm2@illinois.edu; tranel@illinois.edu

J. B. Knaak (ed.), *Reviews of Environmental Contamination and Toxicology Volume 255*,
Reviews of Environmental Contamination and Toxicology 255,
https://doi.org/10.1007/398_2020_55

mechanisms, discusses what is known about their evolution, and concludes with some of the impacts glyphosate-resistant weeds have had on weed management.

Keywords EPSPS · Evolution · Glyphosate · Herbicide resistance · Resistance mechanisms · Weed management

Abbreviations

AKR	Aldo-keto reductase
AMPA	Aminomethylphosphonic acid
C-P	Carbon-phosphorus
eccDNA	Extrachromosomal circular DNA
EPSPS	5-Enolpyruvylshikimate-3-phosphate synthase
FISH	Fluorescent in situ hybridization
GOX	Glyphosate oxidoreductase

1 Introduction

Glyphosate competes with phosphoenolpyruvate to bind the enzyme 5-enolpyruvylshikimate-3-phosphate synthase (EPSPS), preventing synthesis of the essential amino acids phenylalanine, tyrosine, and tryptophan (Steinrücken and Amrhein 1980; Schönbrunn et al. 2001). It was commercialized for use as a herbicide in 1974 for nonselective weed control (i.e., in non-crop areas or prior to planting) (Duke and Powles 2008). In this Volume of *Reviews of Environmental Contamination and Toxicology*, Duke (2021) provides a detailed review of the mode of action and use of glyphosate, and Green and Siehl (2021) discuss the development of glyphosate-resistant crops, which led to the unprecedented reliance on glyphosate for weed control.

Prior to the commercialization of these glyphosate-resistant crops, it was infamously argued in a 1997 paper (Bradshaw et al. 1997) that "the probability of evolution of glyphosate resistance seems low." To be fair, resistance to glyphosate does appear to arise spontaneously at a lower frequency than it does for other herbicides (Jander et al. 2003; Brotherton et al. 2007). Nevertheless, the probability that herbicide resistance occurs in weeds depends not only on the ease at which resistance to that herbicide evolves but also on the selection intensity imposed by that herbicide. In regards to glyphosate, perhaps not even Bradshaw et al. (1997) anticipated the unprecedented selection pressure that would be applied by this herbicide in the years following their publication.

After about a decade of very successful weed control, the beginning of the end of glyphosate as a stand-alone herbicide used in conjunction with glyphosate-resistant crops occurred around 2005, with the evolution of glyphosate-resistant populations of *Amaranthus palmeri* and *Amaranthus tuberculatus* (Culpepper et al. 2006; Legleiter and Bradley 2008). Although these were not the first two weeds to evolve glyphosate resistance (Heap 2020), they are driver weeds in USA cotton and soybean fields – the two crops for which glyphosate-resistant varieties were first rapidly adopted.

Heap and Duke (2018) provided a relatively recent and comprehensive description of the occurrence and distribution of 38 glyphosate-resistant weed species known at the time. Since then, 10 additional glyphosate-resistant weeds have been reported, bringing the total to 48, equally split between grass and broadleaf species (Heap 2020). Glyphosate-resistant weeds now have been documented in 30 countries (Table 1). Of these 30 countries, however, most (24) have reports of less than five species, with Australia, the USA, and Argentina being notable exceptions, with 19, 17, and 15 glyphosate-resistant species, respectively. The earliest glyphosate-resistant weed species reported, including *Lolium* spp., *Conyza* spp., and *Eleusine indica*, are also now the most widely distributed glyphosate-resistant species among different countries. Early and widespread occurrence of glyphosate-resistant populations of these species likely reflects some combination of these species' widespread occurrence, their propensity for gene flow [e.g., *Conyza spp.* seeds are wind dispersed over broad geographies (Weaver 2001)], and their innate abilities to evolve glyphosate resistance.

When considering the timeline of the appearance of glyphosate-resistant weeds, it is important to keep in mind that glyphosate selection might have occurred prior to the adoption of glyphosate-resistant crops (through traditional use of glyphosate), only after the adoption of such crops, or both, depending on the species. For example, glyphosate-resistant *Lolium rigidum* was reported in Australia in 1996, representing a clear case of glyphosate resistance occurring due to the traditional use of glyphosate. In contrast, *A. tuberculatus*, for example, occurs primarily in crop fields and germinates relatively late in the growing season (Costea et al. 2005). Consequently, there likely was a relatively limited selection for glyphosate-resistant biotypes of this species prior to 1996. *Conyza canadensis*, first reported glyphosate-resistant in 2000, likely was selected by glyphosate applied both traditionally and in glyphosate-resistant crops (VanGessel 2001), contributing to this species evolving resistance sooner than, e.g., *A. tuberculatus*.

It is interesting that in the USA and Brazil, where glyphosate-resistant crops were first widely adopted, most of the glyphosate-resistant weed species were reported within a decade after the adoption of those crops, with no new species having been reported from these countries since 2015. The recent lack of new glyphosate-resistant species in these countries cannot be explained entirely by local curtailing of glyphosate use: glyphosate often is used even in areas where glyphosate-resistant weeds exist to provide control of other weed species. For example, in the USA as recently as 2017, glyphosate was still used on three-fourths of the soybean hectares (https://www.nass.usda.gov). Perhaps evolutionary rescue of glyphosate selection is not possible, or highly improbable, in several weed species. Alternatively, after the evolution of an initial glyphosate-resistant weed species in a given field, glyphosate was more likely to be applied at higher use rates and in combination with one or more other herbicides, limiting the subsequent evolution of glyphosate-resistant species.

Table 1 Distribution of glyphosate-resistant weeds (listed by year first reported) and their known resistance mechanisms

No.	Weed	Year	Australia	USA	Argentina	Brazil	Canada	Spain	Columbia	Greece	Italy	Japan	Paraguay	Portugal	South Africa	China	Costa Rica	France	Israel	Malaysia	Mexico	New Zealand	Bolivia	Chile	Czech Republic	Hungary	Indonesia	Poland	South Korea	Switzerland	Turkey	Venezuela	Resistance mechanisms
1	*Lolium rigidum*	1996	X	X				X		X	X				X			X	X														RTU, STM
2	*Eleusine indica*	1997		X	X	X			X			X				X	X			X			X				X						STM, DTM, GD
3	*Conyza canadensis*	2000		X		X	X	X		X	X	X		X		X									X	X		X	X				RTU
4	*Lolium perenne ssp. multiflorum*	2001	X	X	X		X			X	X										X		X						X				RTU, STM, GD
5	*Conyza bonariensis*	2003	X	X	X	X		X	X	X				X	X				X														RTU
6	*Plantago lanceolata*	2003													X																		–
7	*Ambrosia artemisiifolia*	2004		X			X																										–
8	*Ambrosia trifida*	2004		X			X																										RTU, RR
9	*Parthenium hysterophorus*	2004		X					X																								RTU, STM
10	*Amaranthus palmeri*	2005		X	X	X																											STM, GD
11	*Amaranthus tuberculatus*	2005		X			X																										RTU, STM, GD
12	*Digitaria insularis*	2005			X	X							X																				RTU, DTM
13	*Hedyotis verticillata*	2005																		X													–
14	*Sorghum halepense*	2005		X	X																												RTU
15	*Echinochloa colona*	2007	X	X	X																											X	STM, MB
16	*Bassia scoparia*	2007		X			X																										GD
17	*Cynodon hirsutus*	2008			X																												–

18	*Lolium perenne*	2008			X								X							X											GD
19	*Urochloa panicoides*	2008	X		X																												–
20	*Conyza sumatrensis*	2009	X			X		X		X			X					X													X		RTU, STM
21	*Chloris truncata*	2010	X																														GD
22	*Leptochloa virgata*	2010																			X												RTU, STM
23	*Paspalum paniculatum*	2010															X																–
24	*Poa annua*	2010	X	X																													STM, GD
25	*Raphanus raphanistrum*	2010	X																														–
26	*Bromus diandrus*	2011	X																														GD
27	*Amaranthus spinosus*	2012		X																													GD
28	*Brassica rapa*	2012			X		X																										TG
29	*Amaranthus hybridus*	2013			X	X																											TTM
30	*Bidens pilosa*	2014																			X												RTU, STM, DTM
31	*Brachiaria eruciformis*	2014	X																														–
32	*Bromus rubens*	2014	X																														–
33	*Chloris elata*	2014				X																											RTU
34	*Sonchus oleraceus*	2014	X																														–
35	*Chloris virgata*	2015	X																														STM
36	*Helianthus annuus*	2015		X																													–
37	*Lactuca serriola*	2015	X																														–
38	*Salsola tragus*	2015		X																													–
39	*Hordeum murinum*	2016	X																														GD
40	*Tridax procumbens*	2016	X																														STM
41	*Bromus catharticus*	2017			X																												–
42	*Lactuca saligna*	2017	X																														–
43	*Avena fatua*	2018	X																														–
44	*Avena sterilis*	2016	X																														RTU
45	*Bidens subalternans*	2018										X																					DTM

(continued)

Table 1 (continued)

No.	Weed	Year	Australia	USA	Argentina	Brazil	Canada	Spain	Columbia	Greece	Italy	Japan	Paraguay	Portugal	South Africa	China	Costa Rica	France	Israel	Malaysia	Mexico	New Zealand	Bolivia	Chile	Czech Republic	Hungary	Indonesia	Poland	South Korea	Switzerland	Turkey	Venezuela	Resistance mechanisms
46	*Carduus acanthoides*	2019			X																												–
47	*Chloris radiata*	2019							X																								–
48	*Echinochloa crus-galli*	2019			X																												–
	Total		19	17	15	9	6	5	4	4	3	3	3	3	3	2	2	2	2	2	2	2	1	1	1	1	1	1	1	1	1	1	

Abbreviations: *RTU* reduced translocation and/or uptake, *STM* single target site mutation, *DTM* double target site mutation, *TTM* triple target site mutation, *GD* target-site gene duplication (or increased expression), *MB* metabolism, *RR* rapid response, *TG* Transgene escape. Species, year, and country data are from Heap (2020). References for resistance mechanisms are listed by No. **1** Lorraine-Colwill et al. (2002), Wakelin et al. (2004), González-Torralva et al. (2012a). **2** Baerson et al. (2002b), Zhang et al. (2015), Yu et al. (2015), Chen et al. (2015), Franci et al. (2020). **3** Feng et al. (2004), Koger and Reddy (2005), Ge et al. (2010), González-Torralva et al. (2012b, 2017). **4** Salas et al. (2012), González-Torralva et al. (2012a), Liu et al. (2016), Karn and Jasieniuk (2017), Brunharo and Hanson (2018). **5** Dinelli et al. (2008), Shaner (2009), Moretti et al. (2013), Kleinman and Rubin (2017). **7** Brewer and Oliver (2009). **8** Lespérance (2015), Nandula et al. (2015), Van Horn (2016), Moretti et al. (2018). **9** Bracamonte et al. (2016). **10** Gaines et al. (2010), Dominguez-Valenzuela et al. (2017). **11** Bell et al. (2013), Nandula et al. (2013), Lorentz et al. (2014). **12** de Carvalho et al. (2012). **14** Vila-Aiub et al. (2012). **15** Han et al. (2016), Goh et al. (2018), Pan et al. (2019), McElroy and Hall (2020). **16** Jugulam et al. (2014). **18** Yanniccari et al. (2017). **20** González-Torralva et al. (2014), Amaro-Blanco et al. (2018). **21** Ngo et al. (2018b). **22** Alcántara-de la Cruz et al. (2016b). **24** Brunharo et al. (2019). **26** Malone et al. (2016). **27** Nandula et al. (2014). **28** Pandolfo et al. (2018). **29** García et al. (2019), Perotti et al. (2019). **30** Alcántara-de la Cruz et al. (2016a). **33** Brunharo et al. (2016). **35** Ngo et al. (2018a). **39** Adu-Yeboah et al. (2019). **40** Li et al. (2018). **44** Fernández-Moreno et al. (2016). **45** Takano et al. (2020)

The most recent reports of new glyphosate-resistant weed species have come from other South American countries and Australia. In fact, of the 10 species added to the list since Heap and Duke's review (2018), half were in Australia and the other half were in the South American countries of Argentina, Columbia, or Paraguay. Glyphosate selection from both traditional use and in glyphosate-resistant crops is continuing to increase the number of glyphosate-resistant weeds.

Just as humans have to think "outside the box" when confronted with new challenges, weeds had to evolve "outside the box" when confronted with glyphosate selection. Hence, the combination of the difficulty in evolving glyphosate resistance and the intense glyphosate selection pressure resulted in diverse and unusual resistance mechanisms (Gaines et al. 2019). In this review, we discuss these diverse resistance mechanisms and what is known about their evolution. We conclude with a discussion of how the vast, real-world glyphosate evolutionary experiment has impacted weed management.

As is the case for any other herbicide, resistance mechanisms for glyphosate can be broadly grouped into target-site and nontarget-site mechanisms (Gaines et al. 2020). Historically, target-site resistance was described as a mutation in the gene encoding the protein that directly interacts with the herbicide, leading to a reduced affinity between the herbicide and its target site. Nontarget-site resistance includes all other mechanisms, primarily including herbicide detoxification (metabolism), reduced herbicide uptake, and reduced herbicide translocation. In general, nontarget-site mechanisms confer resistance by essentially reducing the concentration of herbicide that reaches the target site. A relatively new resistance mechanism, associated primarily with glyphosate resistance, is increased expression of the target site via gene duplication (discussed in Sect. 3.2). Duplication of the target-site gene has been categorized as another form of target-site resistance (Gaines et al. 2020). From a physiological perspective, however, resistance due to increased expression of the target site is more like nontarget-site resistance in that the net result in both cases is reduced concentration of herbicide per unit of target site. Additionally, from a genetic perspective, gene duplication in some cases results in resistance being inherited from multiple loci – as often is the case with nontarget-site resistance (Délye 2013) – whereas traditional target-site resistance involves a single locus. Nevertheless, in this review we will include *EPSPS* duplication as a form of target-site resistance.

2 Nontarget-Site Resistance

2.1 Uptake, Translocation, and Sequestration

The effectiveness of any herbicide is highly dependent on the active ingredient reaching the target site. The delivery of the herbicide to the target site is defined by the uptake and translocation of the herbicide in the plant, which, in turn, are dependent on factors such as plant cuticle physiology, herbicide formulation, environmental factors, and molecular properties of the herbicide (e.g., size and polarity) (Hess and Duke 1985).

The polar nature of glyphosate makes it poorly absorbed by leaves, but once absorbed, it can be rapidly translocated into plant meristems (Preston and Wakelin 2008). Glyphosate is mainly translocated via phloem following the source-to-sink pattern of photoassimilates. Translocation via xylem can also occur, but it rapidly goes back into the phloem and accumulates more in sink tissues (Bromilow et al. 1990).

Reduced translocation and absorption of glyphosate are known mechanisms of nontarget-site resistance documented in some weed species (Sprankle et al. 1975; Feng et al. 2004; Wakelin et al. 2004; Preston and Wakelin 2008; Shaner 2009; Vazquez-Garcia et al. 2020). Glyphosate uptake reduction occurs when chemical or morphological changes in the leaf cuticle or leaf shape reduce the amount of herbicide entering the plant. Most cases of reduced glyphosate uptake show a variation in leaf angle and cuticle properties and were observed in grass species (Michitte et al. 2007; Vila-Aiub et al. 2012; de Carvalho et al. 2012; Alcántara-de la Cruz et al. 2016b).

Reduced glyphosate translocation occurs when the herbicide molecules have limited or no movement to the plant meristem, a factor that can profoundly affect herbicide efficacy, and has evolved as a resistance mechanism. In some species, such as *Conyza* spp. and *Lolium* spp., the reduced translocation is attributed to a rapid-vacuolar sequestration mechanism (Ge et al. 2010, 2012). Such sequestration prevents translocation of the glyphosate molecules to meristematic tissue.

Vacuoles are degradative organelles, similar to lysosomes in animal cells, and are the largest organelles of plant cells, representing around 80% of the total cell space (Martinoia 1992). These large cell compartments serve as reservoirs for ions and metabolites and play fundamental roles in detoxification and maintaining cell homeostasis (Marty 1999). Studies have shown that active tonoplast transporters such as ABC transporters are possibly linked with the movement of glyphosate into the vacuoles, suggesting that ABC transporter genes regulate this resistance mechanism (Nol et al. 2012; Ge et al. 2014; Tani et al. 2015).

Environmental factors are also known to affect these key genes in the uptake and translocation of glyphosate. Studies suggest that glyphosate uptake may vary in different light regimes, showing greater uptake when conditions are optimum for high ATP levels (Kells and Rieck 1979; Devine et al. 1983; Ge et al. 2010). Temperature can also play a role in glyphosate uptake and translocation (Vila-Aiub et al. 2013; Palma-Bautista et al. 2019). Vacuole sequestration was shown to vary with temperature: with low temperature, glyphosate-resistant plants showed a reduction in the resistance level and herbicide retention in the vacuoles (Ge et al. 2011). Because vacuolar sequestration provides a relatively low level of resistance (Fig. 1), it potentially could be overcome by making applications when temperatures are low. Further studies of vacuolar sequestration are still required for a better understanding of this nontarget-site resistance mechanism at the molecular level.

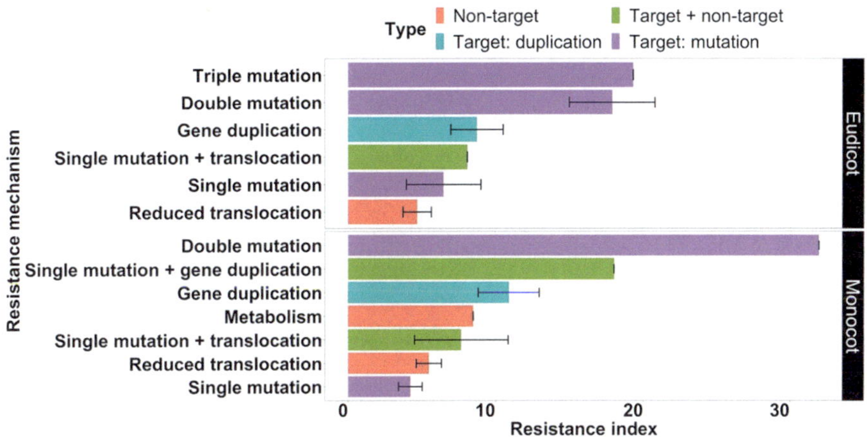

Fig. 1 Comparison of resistance index conferred by different glyphosate-resistance mechanisms. Mean resistance indices (±1 standard error) were calculated from resistance ratios obtained from the literature. Data aggregated from: (Baerson et al. 2002b; Wakelin et al. 2004; Culpepper et al. 2006; Yu et al. 2007, 2015; Perez-Jones et al. 2007; Dinelli et al. 2008; Lamego and Vidal 2008; Jasieniuk et al. 2008; Kaundun et al. 2011; Chandi et al. 2012; Salas et al. 2012; Vila-Aiub et al. 2012; de Carvalho et al. 2012; Gaines et al. 2012; González-Torralva et al. 2012a; Bell et al. 2013; Moretti et al. 2013; Nandula et al. 2013, 2014; Mohseni-Moghadam et al. 2013; Lorentz et al. 2014; Wiersma et al. 2015; Alcántara-de la Cruz et al. 2016b, a; Brunharo et al. 2016, 2019; Kleinman and Rubin 2017; Yanniccari et al. 2017; Amaro-Blanco et al. 2018; Morran et al. 2018; Ngo et al. 2018b, a; Pandolfo et al. 2018; Beres et al. 2018; Li et al. 2018; Brunharo and Hanson 2018; Takano et al. 2019; Perotti et al. 2019)

2.2 *Rapid Response (Phoenix Phenomenon)*

Because glyphosate's herbicidal activity involves plants starving for aromatic amino acids, it generally takes several days for plants to die after application. First documented in 2008, some biotypes of *Ambrosia trifida* have evolved a rapid-response glyphosate-resistance mechanism in which leaves treated with the herbicide quickly wither and fall from the plant (Brabham et al. 2011; Moretti et al. 2018; Van Horn et al. 2018). This rapid cell death limits the ability of the herbicide to move throughout the plant and, therefore, can be considered a "reduced translocation" mechanism. After shedding tissue containing glyphosate, the plant begins new growth, seemingly from the ashes of a dead plant, and hence the name "Phoenix" phenomenon. This rapid cell death also affects the efficacy of other herbicides included in the spray mixture, because translocation is generally inhibited (Harre et al. 2018). Though this mechanism is still not well understood, it can be reversed with the application of exogenous phenylalanine and tyrosine, indicating that it is somehow involved with a deregulation of the shikimate pathway (Moretti et al. 2018). Increased accumulation of reactive oxygen species following glyphosate application in leaf discs displaying the rapid-response phenotype when compared to sensitive leaf discs points to the possibility that an accumulation of reactive oxygen species plays a role in rapid cell death, though this remains to be elucidated. It is assumed that this resistance mechanism requires an actively metabolizing plant, given rapid-response plants do not display rapid cell death in the absence of light and sucrose (Moretti et al. 2018).

Recently, Queiroz et al. (2020) reported a similar resistance phenotype to the auxinic herbicide 2,4-D in *Conyza sumatrensis*, in which herbicide application results in hydrogen peroxide accumulation and rapid cell death. While this report means the rapid-response resistance mechanism no longer is unique to glyphosate, it is unknown how similar the two resistance mechanisms are at the molecular level.

2.3 Metabolism

Studies of metabolic-based herbicide-resistance mechanisms began occurring in earnest in the United Kingdom and Australia in the mid-1980s due to increasing cases of resistances particularly in *Alopecurus myosuroides* and *Lolium* spp. (Moss and Cussans 1985; Heap and Knight 1986). Documented cases have become more common in recent years with the innovation of biochemical and genetic tools that allow researchers to identify specific genes and metabolic pathways conveying such resistance. Given that glyphosate is a relatively slow-acting herbicide that causes depletion of aromatic amino acids, enhanced metabolism would be a highly effective mechanism of resistance; one in which plants would be able to detoxify the herbicide before the significant injury occurred.

Many known cases of metabolic herbicide resistance are due to mutated or overexpressed cytochrome P450, glucosyltransferase, or glutathione S-transferase enzymes (Yuan et al. 2007; Yu and Powles 2014). These enzymes belong to large protein families and have many roles in primary and secondary metabolism, with some having specificity to herbicide molecules. To date, there has been no report of a protein from either of these families that significantly interacts with glyphosate in plants. However, Van Etten et al. (2020) reported several genomic regions of *Ipomoea purpurea* that are associated with an increase in glyphosate tolerance and enriched for genes from these families. Further physiological validation to confirm the roles of these gene families in glyphosate metabolism may help elucidate previously reported variation of glyphosate tolerance within and among populations of this species (Baucom and Mauricio 2010; Kuester et al. 2015).

Two enzymes have been reported to metabolize glyphosate: glyphosate oxidoreductase (GOX), which cleaves a C-N bond within glyphosate, and carbon-phosphorus (C-P) lyase, which cleaves glyphosate's C-P bond (Liu et al. 1991; Van Eerd et al. 2003; also see Fig. 3 in Green and Siehl 2021). An unknown enzyme that acts similarly to GOX is suspected to be the primary catalyst for glyphosate detoxification in plants (Reddy et al. 2008). Along with the primary product of GOX-mediated detoxification of glyphosate, aminomethylphosphonic acid (AMPA), several other metabolites of glyphosate have been detected in higher plants, including glycine, glyoxylate, sarcosine, formaldehyde, and inorganic phosphate (Marshall et al. 1987; Duke 2011; Rojano-Delgado et al. 2012; Gomes et al. 2014). Formaldehyde and hydrogen peroxide are compounds associated with C-P lyase-mediated metabolism of glyphosate, and their phytotoxicity in plants may explain why C-P lyase has not evolved to be the primary catalyst of glyphosate degradation in plants (Mutters et al. 1993; Goyer et al. 2004; Reddy et al. 2008). Although most metabolites of GOX-mediated glyphosate degradation are common compounds in plants and are unlikely to cause damage, AMPA has some evidence of phytotoxicity in plants. For example, AMPA was shown to accumulate as a result of glyphosate application and to cause injury in glyphosate-resistant soybean (Hoagland 1980; Duke 2011). Gomes et al. (2014) hypothesized that AMPA's phytotoxic effects are the result of competitive inhibition of glycine decarboxylase, thereby inhibiting chlorophyll biosynthesis. However, a microbial GOX has been used as a transgene to successfully confer glyphosate resistance to tobacco and rape, indicating that these plant species – and likely others – possess the molecular machinery sufficient for further metabolizing any products of GOX-mediated glyphosate metabolism (Duke 2011; Pollegioni et al. 2011).

Previously, de Carvalho et al. (2012) showed increased metabolism of glyphosate in resistant varieties of *Digitaria insularis* when compared to sensitive varieties, but failed to tease this effect from other possible mechanisms in the population. Additionally, Rojano-Delgado et al. (2012) proposed that glyphosate metabolism worked in conjunction with limited uptake and translocation to convey glyphosate tolerance in *Mucuna pruriens* but failed to quantify these effects. Recently, an aldo-keto reductase (AKR) enzyme was found to metabolize glyphosate to AMPA and glyoxylate in an Australian population of *Echinochloa colona*, just as GOX does in bacteria (Pan et al. 2019). While no variation in coding sequence of this AKR delimited resistant and sensitive populations, increased expression was shown to be associated with resistance to glyphosate. To further verify this AKR as the causative agent of glyphosate resistance, rice was transformed with AKR cDNA from *E. colona*. Calli and seedlings overexpressing the transcript and displaying increased AKR activity were resistant to glyphosate (Pan et al. 2019). McElroy and Hall (2020) later revisited this population of *E. colona*, however, and discovered the presence of the Pro-106-Thr substitution encoded within *EPSPS*, a target site mutation previously shown to reduce EPSPS affinity for glyphosate in this species (Alarcón-Reverte et al. 2015; Han et al. 2016). This discovery obscures, but does not eliminate, the effect of increased expression of AKR on the evolution of glyphosate resistance in *E. colona*. In any case, the discovery of AKR's role in glyphosate metabolism emphasizes the need for future metabolism research efforts to treat all candidate genes as true candidates in lieu of searching solely for common herbicide metabolism genes such as cytochrome P450s or glutathione S-transferases. In short, metabolism of glyphosate seems to have the potential to be a viable mechanism of resistance, and it is surprising that more cases of metabolism-based resistance have not been documented.

3 Target-site Resistance

3.1 Insensitive Target Site

Within the context of glyphosate resistance, an insensitive target site occurs through modifications to the primary amino acid sequence of EPSPS (Heap and Duke 2018). When considering the total length of the enzyme (520 amino acids – GenBank accession AT2G45300), relatively few amino acids are associated with resistance. That there are few target-site mutations for glyphosate resistance is attributed to the similarity in how glyphosate and phosphoenolpyruvate bind the EPSPS enzyme (Schönbrunn et al. 2001). Such similarity means that structural changes that reduce EPSPS affinity for glyphosate likely will also reduce its affinity for the phosphoenolpyruvate substrate. Indeed, only three amino acid positions have been implicated in evolved herbicide resistance in weed species: Thr-102, Ala-103, and Pro-106 (Murphy and Tranel 2019). Amino acid substitutions at the Pro-106 position, to Ser, Leu, Thr, or Ala, alone are sufficient for resistance to glyphosate (Heap and Duke 2018; Morran et al. 2018; Brunharo and Hanson 2018). Substitutions at Thr-102 and Ala-103 generally have only been observed coexisting with Pro-106 substitutions. Previously, Thr-102 substitutions observed in combination with Pro-106 substitutions contained Ile as the substitute amino acid, and it was suggested that this Thr-102-Ile mutation would not occur on its own because of its negative effect on EPSPS enzyme activity (Sammons and Gaines 2014). Recently, however, a Thr-102-Ser substitution was identified to confer glyphosate resistance in the tetraploid *Tridax procumbens* (Li et al. 2018). Effects of substitution at Ala-103 are not well known, and this substitution has been observed only in a triple substitution referred to as TAP-IVS (Thr-102, Ala-103, and Pro-106 are substituted with Ile, Val, and Ser, respectively) in Argentinian *Amaranthus hybridus*, (García et al. 2019; Perotti et al. 2019). Green and Siehl (2021) in this same Volume provide further discussion of the effects of different amino acid substitutions on EPSPS kinetics, and a database of EPSPS amino acid changes conferring glyphosate resistance in weeds is maintained by Gaines and Heap (2020).

Experiences with resistance to other herbicide groups, particularly to inhibitors of photosystem II, acetolactate synthase, and acetyl-CoA-carboxylase, have indicated that single amino acid changes to herbicide target sites confer very robust levels of resistance relative to nontarget-site resistance mechanisms (Powles and Yu 2010). In the case of glyphosate, however, resistance derived from a single target-site substitution is often associated with weak resistance relative to other glyphosate-resistance mechanisms (Fig. 1). Consistently, single mutation events provide some of the lowest levels of resistance when compared to all other mechanisms in both grass and broadleaf weed species. In comparison, the documented double and triple substitutions to EPSPS confer resistance levels greater than those provided by nontarget-site mechanisms. This is consistent with attempts to develop glyphosate resistance traits in crops through site-directed mutagenesis. The pairing of Thr-102-Ile and Pro-106-Ser substitutions, which has evolved in weeds, also resulted in commercial resistant germplasm developed through site-directed mutagenesis (Dill 2005). Indeed, the introduction of single point mutations through ethyl methanesulfonate was widely unsuccessful in the creation of an acceptable resistance phenotype for commercial use, consistent with modest levels of resistance conferred by single amino acid substitutions in EPSPS.

3.2 EPSPS Gene Duplication

Beginning with its discovery in *A. palmeri* in 2010 (Gaines et al. 2010), EPSPS gene duplication has become a relatively common mechanism of glyphosate resistance. Thus far, three other broadleaf species (*A. tuberculatus*, *Amaranthus spinosus*, and *Bassia scoparia*) and six grass species (*Lolium perenne*, *Bromus diandrus*, *E. indica*, *Chloris truncata*, *Poa annua*, and *Hordeum glaucum*) have evolved glyphosate resistance via this mechanism (Salas et al. 2012; Nandula et al. 2014; Lorentz et al. 2014; Jugulam et al. 2014; Chen et al. 2015; Malone et al. 2016; Ngo et al. 2018a; Adu-Yeboah et al. 2019; Brunharo et al. 2019). These species appear to require differing levels of genomic copies for resistance. Three of these species, *A. tuberculatus*, *B. scoparia*, and *H. glaucum*, show resistance with relatively low numbers of *EPSPS* genomic copies (Fig. 2), often between 3 and 14, with a minimum of three copies needed to confer glyphosate resistance (Lorentz et al. 2014; Wiersma et al. 2015; Chatham et al. 2015; Godar et al. 2015; Adu-Yeboah et al. 2019). One study has reported plants with >15 *EPSPS* copies in *A. tuberculatus* (Dillon et al. 2017), but this appears to be the exception to the norm. EPSPS expression mostly correlates with *EPSPS* genomic copy number for *B. scoparia* and *A. tuberculatus* but does not correlate well with resistance level, with most resistant accessions showing similar levels of glyphosate resistance despite varying levels of *EPSPS* copies (Fig. 2). Few studies examine all potential glyphosate resistance mechanisms, so some of this disconnect between *EPSPS* copy number and resistance may be due to the presence of alternative mechanisms of resistance. For *H. glaucum*, there was no correlation between copy number and expression, but some evidence of correlation between copy number and glyphosate resistance (Adu-Yeboah et al. 2019).

In contrast, for all other species with this resistance mechanism, at least 10 *EPSPS* gene copies have been shown to be necessary for resistance. Around 10–36 copies have been documented in *B. diandrus* (Malone et al. 2016), 32–48 copies in *C. truncata* (Ngo et al. 2018a), and 33–37 in *A. spinosus* (Nandula et al. 2014). These mid-range levels of EPSPS copy numbers confer approximately the same level of resistance to glyphosate (compared to a sensitive control) as observed in *B. scoparia* and *A. tuberculatus*, with resistance between about 3- and 7-fold. Much higher *EPSPS* copy numbers have been observed in *A. palmeri* (35–160); (Gaines et al. 2010), *L. perenne* (11–151); (Salas et al. 2012), and *E. indica* (89); (Chen et al. 2015), with both *A. palmeri* and *L. perenne* demonstrating increasing levels of glyphosate resistance with increasing numbers of *EPSPS* gene copies (Fig. 2).

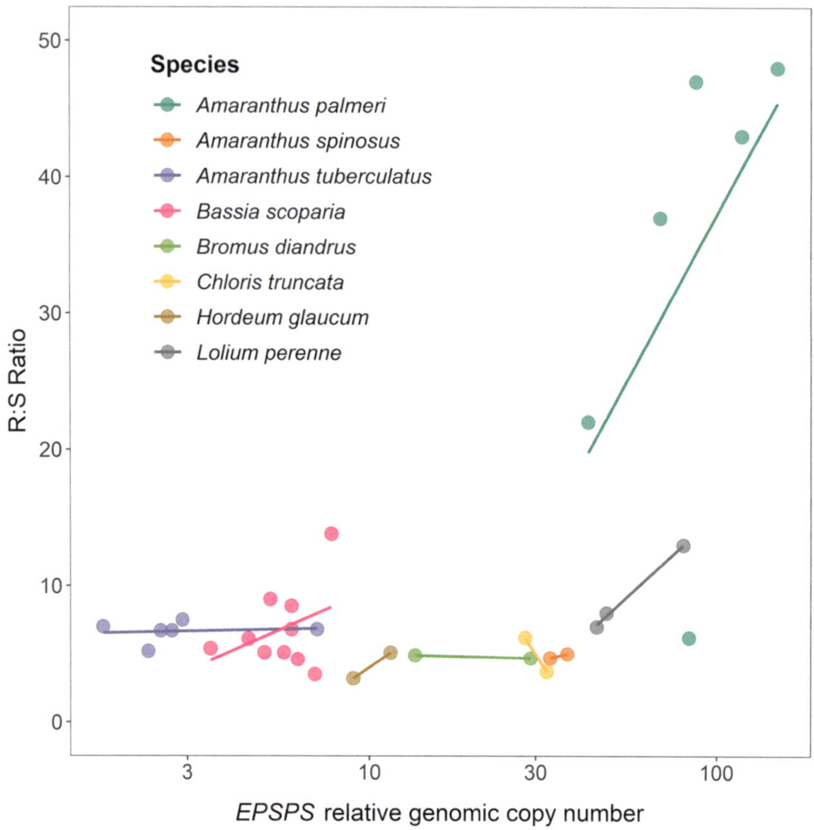

Fig. 2 Average *EPSPS* genomic copy number plotted against the resistant:susceptible (R:S) ratio for glyphosate-resistant populations. Each dot is a single population and each color indicates a different weed species, with linear regression lines plotted separately for each species. Data aggregated from (Gaines et al. 2010; Chandi et al. 2012; Salas et al. 2012; Nandula et al. 2014; Lorentz et al. 2014; Wiersma et al. 2015; Godar et al. 2015; Malone et al. 2016; Chahal et al. 2017; Ngo et al. 2018b; Singh et al. 2018; Adu-Yeboah et al. 2019)

Eleusine indica has shown a positive correlation between *EPSPS* gene copy number and expression, but whether this translates to higher levels of glyphosate resistance is not yet known (Chen et al. 2015). A population of *P. annua* was reported with 18-fold resistance to glyphosate but with only seven *EPSPS* copies (Brunharo et al. 2019). This was a novel case, however, in which, for the first time, it was reported that the duplicated *EPSPS* gene also encoded a glyphosate-resistant enzyme (Pro-106-Leu). The relative contribution of each (duplication and mutation) to glyphosate resistance is unknown, but they both likely contributed, because the magnitude of resistance was greater than that typically conferred by either mechanism alone (Fig. 1).

The correlation of *EPSPS* genomic copy number and the resistance phenotype has been investigated in multiple species, as indicated with some examples just discussed. At a population level, a positive correlation between genomic copy number and resistance has been reported within *B. scoparia*; however, this relationship does not appear to be linear (Godar et al. 2015; Gaines et al. 2016). In some cases, relationships may be population-specific, suggesting that each evolved event may follow a distinct relationship (Gaines et al. 2016). In fact, this is well supported by the meta-analysis shown in Fig. 2, because a diversity of relationships are observable among species. For instance, while a strong linear correlation is observed within *A. palmeri* data points, such a correlation is not consistent in other species. The relationship between genomic copy number and resistance should be established for each species, if not for each evolved instance of this mechanism. Breakdowns in the relationship between genomic copy number and resistance are not wholly unexpected. For example, an increase in genomic copy number is several steps removed from an increase in protein abundance. Consequently, demonstration of an elevated protein abundance is necessary to attribute increases in genomic copy with the resistance phenotype. And, as previously mentioned, the coexistence of one or more other resistance mechanisms within individual plants can be a confounding factor and typically can be ruled out only by further genetic analyses.

4 Distribution of Resistance Mechanisms Among Species

Of the 48 glyphosate-resistant weed species, there is strong evidence for the existence of a particular resistance mechanism in 29 of them (Table 1). Although only one glyphosate-resistance mechanism has been documented in 16 weed species, there are 9 species for which two mechanisms have been reported and 4 species in which three different mechanisms have been reported. Reduced uptake/translocation and single EPSPS amino acid substitutions are the most common mechanisms, with each having been reported in 14 different weed species. As mentioned above, gene duplication, although not known as a herbicide-resistance mechanism prior to glyphosate resistance, is now also a quite common glyphosate-resistance mechanism, being reported in 10 different weed species.

In general, there do not appear to be significant differences in the distributions of glyphosate-resistance mechanisms between grass and broadleaf weed species. In fact, the three most common categories of mechanisms shown in Table 1 (reduced uptake/translocation, single EPSPS substitution, and *EPSPS* duplication) are surprisingly evenly distributed, with *EPSPS* duplication showing the greatest deviation from 1:1 (4 broadleaf species:6 grass species). However, as noted above (Sect. 2.1), reduced glyphosate uptake tends to be more common in grass than in broadleaf species.

There are 19 reported glyphosate-resistant weeds for which resistance mechanisms have not yet been confirmed. It will be interesting to see what new glyphosate-resistant mechanisms might be found in these weeds. To be sure, there very well might be additional resistance mechanisms, which simply have not been identified yet, in the 29 species for which mechanisms have already been reported. And, of course, new glyphosate-resistant species certainly will be added to the list shown in Table 1. It should also be noted that the categories of resistance mechanisms listed in Table 1 underreport the variety of mechanisms at the molecular level. For example, as discussed in Sect. 3.1, a variety of single amino acid substitutions can confer glyphosate resistance, but they are all grouped together under the category of "single target site mutation" in Table 1. Additionally, quite a variety of molecular mechanisms associated with a variety of different genes could contribute to altered glyphosate uptake/translocation. Clearly, weeds have evolved diverse mechanisms to survive glyphosate, and more mechanisms likely await future discovery.

5 Evolutionary Origins of Resistance

As described in Sects. 2 and 3, glyphosate resistance can be mediated by a variety of mechanisms. These resistance mechanisms arise as a result of changes to one or more locations in the genome, resulting in structural or regulatory changes to gene products. Genetic changes that are beneficial (e.g., confer reduced sensitivity to glyphosate) are selected and increase in frequency in the selected populations. The source of the genetic differences that can be selected include standing genetic variation (i.e., they already exist in the population before the onset of selection), immigration from a different population or species, or new mutations. As is the case with resistance to other herbicides, the relative contribution of these sources for glyphosate-resistance evolution are largely unknown (Casale et al. 2019). Ultimately, a better understanding of the evolution of herbicide resistance could lead to novel strategies to mitigate it (Neve et al. 2009).

Naturally occurring plant tolerance cases to a given chemistry may provide insight into what mechanisms may evolve in the future. Several plant species have exhibited a natural tolerance to glyphosate, although the underlying mechanisms have been investigated in few cases. In both *Convolvulus arvensis* and lilyturf species, gene copy number has been attributed to at least part of the observed tolerance phenotype (Mao et al. 2016; Huang et al. 2019). However, these tolerance cases are frequently due to a combination of mechanisms. In lilyturf species, for example, EPSPS structural differences were also noted, relative to other plant EPSPS enzymes, due to multiple amino acid substitutions and deletions Both modeling and in vitro enzyme assays indicated that these structural differences resulted in reduced glyphosate sensitivity (Mao et al. 2016). In *C. arvensis*, a promoter-mediated overexpression, associated with glyphosate application, was also observed in addition to increased *EPSPS* copy number (Huang et al. 2019). Reduced glyphosate translocation was associated with increased tolerance in *Ipomoea lacunosa* (Ribeiro et al. 2015), whereas increased glyphosate metabolism is hypothesized to confer tolerance in *I. purpurea* (Van Etten et al. 2020). While there does not appear to be an overarching trend in tolerance mechanisms among species, similar mechanisms are observed across plant tolerance and resistance. The structure-based tolerance of lilyturf could be considered analogous to target-site resistance. Promoter-mediated overexpression, resulting in an increase in EPSPS protein abundance, also has been occasionally associated with evolved resistance (Baerson et al. 2002a). Gene copy number increase and reduced translocation both have been implicated in both tolerance and evolved resistance. The investigation of tolerance mechanisms to a given chemistry, even beyond the scope of glyphosate, can provide insight into what mechanisms might evolve in response to selection.

There have been a couple of cases in which a weed evolved glyphosate resistance via gene flow from a related species. In one case, the weed *Brassica rapa* acquired the transgene (*CP4 EPSPS*) conferring glyphosate resistance from cultivated rape (Warwick et al. 2003). This evolutionary path to glyphosate resistance in *B. rapa* subsequently has been shown to be a common event and has occurred in multiple countries (Simard et al. 2006; Pandolfo et al. 2018). Another case involves weed-to-weed gene flow, in which *A. spinosus* acquired *EPSPS* gene duplication that had evolved in *A. palmeri* (Nandula et al. 2014). These cases are the exception to the norm, however, and most weed species have evolved glyphosate resistance from either standing genetic variation or new mutations.

5.1 Nontarget-Site Mechanisms

In general, nontarget-site glyphosate resistance mechanisms are still poorly understood, and even less is known about their evolutionary origins. In regard to enhanced detoxification, because glyphosate is metabolized readily through multiple pathways in bacteria, horizontal gene transfer could certainly be a source of resistance, though no evidence exists for this having occurred. As discussed above, AKR likely plays a role in glyphosate resistance in *E. colona*, perhaps via enhanced expression, and remains the only plant protein proven to directly metabolize glyphosate (Duke 2019). The evolutionary origin of enhanced expression of AKR in *E. colona*, or of any other herbicide-metabolizing enzyme selected in weed populations, remains unknown. Now that AKR has been identified to metabolize glyphosate, evaluation of homologous genes in other weed species likely will follow and should reveal the potential of AKR to confer glyphosate resistance in other species.

Because inheritance studies have not yet been published regarding the rapid-response glyphosate-resistance mechanism, its genetic complexity is not known. Additionally, though similarities exist with a recently identified resistance mechanism to 2,4-D (Queiroz et al. 2020), it is unclear if these rapid response mechanisms have any evolutionary relatedness. The similarities with plant pathogen response (e.g., hypersensitivity and rapid cell death) suggest this mechanism evolved by somehow co-opting a pathway for plant defense against abiotic attack (Roden and Ingle 2009).

As discussed above, glyphosate resistance due to vacuolar sequestration might be mediated by an ABC transporter and has been most studied in *C. canadensis*. Just as enhanced herbicide metabolism can evolve through increased expression of a herbicide-metabolizing enzyme, sequestration could evolve through increased expression of an ABC transporter. A previous study found glyphosate resistance in *C. canadensis* to be mediated by a single gene (Zelaya et al. 2004), although the identity of that gene is unknown. Increased expression of both *EPSPS* and ABC transporters in a glyphosate-resistant *C. canadensis* biotype prompted Margaritopoulou et al. (2018) to investigate methylation of the *EPSPS* gene. Their finding of differential *EPSPS* methylation between resistant and sensitive biotypes suggests epigenetic changes could be playing an evolutionary role. The contribution of epigenetic changes to herbicide-resistance evolution in general, not just specifically to glyphosate, remains an unanswered question (Markus et al. 2018).

Nontarget-site herbicide resistance offers the field of weed science many novel research questions to be answered through a variety of omics-based approaches (Maroli et al. 2018; Patterson et al. 2019a). The recent establishment of an International Weed Genomics Consortium promises the development of reference genome assemblies for many of the world's most problematic weeds (Ravet et al. 2018). This effort will supplement other recent but less coordinated efforts to produce genomic resources for driver weed species, including *L. multiflorum* (Copetti et al. 2019), *A. tuberculatus* (Kreiner et al. 2019), *B. scoparia* (Patterson et al. 2019b), and *C. canadensis* (Laforest et al. 2020). The availability of these genomic resources enables genetic mapping of traits such as glyphosate resistance (Korte and Farlow 2013; Van Etten et al. 2020) and will complement previous transcriptomic studies designed to identify candidate genes that may be involved in herbicide resistance (Piasecki et al. 2019). The identification of genomic regions associated with the trait of interest, via a genetic mapping experiment, allows for the filtering of candidate genes identified via expression- or variant-based transcriptomic analyses and hedges against the possibility that the trait is ultimately controlled by some regulatory element located far from the genes that would be identified through expression-based transcriptomic approaches. These filtered candidates should be judged, based on physiological characteristics of the trait, and functionally validated via loss- or gain-of-function experiments (Sauka-Spengler and Barembaum 2008; Housden et al. 2017). Pan et al. (2019) provide a good model for functional validation of a glyphosate-resistance gene (AKR), but additional genetic study may have identified the second locus (*EPSPS*, see Sect. 2.3) contributing to glyphosate resistance. With the identification of the genes involved in nontarget-site glyphosate resistance, researchers will be able to better understand the evolutionary origins of such resistance and predict how likely it is that other species will evolve similar resistance mechanisms in the future.

5.2 *EPSPS* Gene Duplication

Because of the novelty and importance of *EPSPS* gene duplication as a resistance mechanism, its evolutionary origin is of great interest and has been addressed in several studies (Patterson et al. 2018). Except for the case of *A. spinosus*, wherein the *EPSPS* amplicon from *A. palmeri* introgressed into the *A. spinosus* population after a hybridization event (Nandula et al. 2014), *EPSPS* gene duplication evolved independently in each of these species. Accordingly, the mechanism of duplication and the length and content of the *EPSPS* amplicon varies across the different species. For two species with relatively low *EPSPS* copy numbers, *A. tuberculatus* and *B. scoparia*, cytogenomic analysis using fluorescent in situ hybridization (FISH) has shown the duplicated *EPSPS* genes are arranged as tandem repeats along one chromosome pair. In *B. scoparia*, these tandem repeats of *EPSPS* occurred at the distal end of one pair of homologous chromosomes, with approximately 40–70 kb between *EPSPS* genes and one copy inverted compared to the rest (Jugulam et al. 2014). The tandem arrangement of the *EPSPS* genes and their location in the telomeres suggests an unequal recombination-based mechanism of gene duplication since unequal crossing over occurs most frequently in telomeric regions of the chromosome and leads to tandem duplications. Similarly, in *A. tuberculatus*, the *EPSPS* repeats were found to occur at a single locus in one set of homologous chromosomes, but unlike in *B. scoparia*, these repeats were in the pericentromeric region of the chromosome, where recombination is less likely to occur (Dillon et al. 2017). Whether the mechanism of gene duplication in this species is also unequal recombination or some other form of chromosomal rearrangement or segmental duplication is unknown.

To further complicate the story, some *A. tuberculatus* individuals with higher *EPSPS* copy numbers (>15 copies) showed multiple *EPSPS* signals on an additional small chromosome (Dillon et al. 2017). Further cytogenomic work found this extra chromosome to be a ring chromosome that was derived from the pericentromeric region of the chromosome with multiple *EPSPS* gene duplications (Koo et al. 2018a). FISH assays of F_1 progeny showed variation in the size and *EPSPS* copy number of these ring chromosomes across different individuals and, surprisingly, additional *EPSPS* gene copies on other pairs of chromosomes, indicating reintegration of the ring chromosomes into the linear chromosomes through ectopic recombination (Koo et al. 2018a). The hypothesized model of ring chromosome formation includes breakage of the linear chromosome at two spots flanking the original *EPSPS* gene duplicates (perhaps via aneuploidy-triggered destabilization), followed by fusion of the broken chromosome ends into a shortened linear chromosome. The excised middle region containing one or more *EPSPS* genes then undergoes fusion of its proximal ends to form a ring chromosome, that may then form varying sizes of ring chromosomes via a breakage-fusion-bridge cycle model (Koo et al. 2018a). Work looking into the *EPSPS* gene duplication mechanism in *A. palmeri* has found similar results, with the additional *EPSPS* gene copies occurring on extrachromosomal DNA. In the initial report of gene duplication in this species, a FISH image showed *EPSPS* gene signals distributed across all 34 chromosomes of *A. palmeri* (Gaines et al. 2010), but a later study (Koo et al. 2018b) showed these gene signals were not actually on the linear chromosomes but were located on extrachromosomal circular DNA (eccDNA) tethered to the main chromosomes. Inheritance of these eccDNA molecules was highly variable and displayed unequal mitotic segregation, illustrating the need for glyphosate selection for retention of glyphosate-resistant plants with high numbers of *EPSPS* copies. Further work has highlighted that these eccDNA molecules are highly structured with 59 genes, 41 of which are expressed under glyphosate application, and a complex array of mobile genetic elements, repeat sequences, and clustered palindromes (Molin et al. 2017, 2020). The contribution of these additional genes/sequences to the overall resistance phenotype is unknown. Syntenic analysis using genomic assembly of closely related species (*Amaranthus hypochondriacus* and *A. tuberculutus*) suggested that the eccDNA was built from several regions across the genome, rather than derived from a single locus (Molin et al. 2020). Consequently, some of the genes (in addition to *EPSPS*) within the eccDNA may have been selected by glyphosate. An alternative hypothesis is that one or more genes in the eccDNA were selected in the evolutionary past by some other plant stress, and *EPSPS* happened to get captured within the amplicon, priming the species for the later evolution of glyphosate resistance.

In grass species with the *EPSPS* gene duplication mechanism, some recent publications have begun to shed light on the arrangement and origin of the *EPSPS* gene copies. In *L. perenne* ssp. *multiflorum*, FISH mapping of the *EPSPS* gene on somatic metaphase chromosomes revealed a similar pattern as that observed in *A. palmeri*, with *EPSPS* signals distributed across all chromosomes in plants with high *EPSPS* gene copy number (Putta 2017). As with *A. palmeri*, the signals appeared to be on the outer edges of the chromosomes, perhaps indicating a similar mechanism of gene duplication involving circular extrachromosomal DNA tethered to the main chromosomes, but conclusive evidence of this does not yet exist. Conversely, in *E. indica*, *EPSPS* gene copies in a resistant individual appeared to be restricted to two pairs of homologous chromosomes, as indicated by FISH work in this species (Chen et al. 2019). In *B. diandrus*, no FISH assays have yet been published, but inheritance work has shown F_2 offspring to have a range (3–30) of *EPSPS* gene copies, with all F_2 offspring showing an increase in the baseline copy number (Malone et al. 2016). If the *EPSPS* gene copies were inherited as a single locus, as would be expected in a tandem repeat model, 25% of the F_2s should have a single *EPSPS* copy, and the fact that this is not observed indicates these *EPSPS* gene copies likely occur on multiple chromosomes. For the other three grass species (*C. truncate*, *H. glaucum*, and *P. annua*), no cytogenetic or inheritance work has yet been completed and the mechanism of *EPSPS* gene duplication is unknown.

Gene duplication as a herbicide-resistance mechanism thus far has been reported in only one other case, resistance to acetyl-CoA-carboxylase inhibitors (Laforest et al. 2017). Why, then, has it repeatedly evolved for glyphosate resistance? As can be seen in Fig. 1, besides multiple amino acid substitutions in EPSPS, gene duplication confers the highest magnitude of resistance among the known resistance mechanisms evolved to date. Perhaps *EPSPS* duplication is the evolutionary "path of least resistance" for robust glyphosate resistance (Tranel 2017).

Recent population genetics analysis of glyphosate-resistance evolution in *A. tuberculatus* indicated that *EPSPS* duplication in this species – which appears to be due primarily to tandem duplications – independently occurred multiple times (Kreiner et al. 2019). In contrast, the *EPSPS*-containing eccDNA in *A. palmeri* was nearly identical among geographically dispersed populations, suggesting a single evolutionary origin (Molin et al. 2018). Conservation of the eccDNA among these populations suggests a relatively recent evolutionary event, arguing against the hypothesis mentioned above, that the amplicon was selected by some plant stress prior to glyphosate selection. Kreiner et al. (2019) presented evidence suggesting *EPSPS* duplication preexisted as standing genetic variation in *A. tuberculatus*, in contrast to the eccDNA in *A. palmeri* being a relatively recent event. Certainly, more work is needed, but comparison of these two species suggests that tandem duplication is a higher probability event than the eccDNA-based duplication. Why these two related species used different evolutionary paths to *EPSPS* duplication is unknown. One possibility is that tandem duplication may not have evolved as a glyphosate-resistance mechanism in *A. palmeri* because this species is inherently more sensitive than *A. tuberculatus* to glyphosate. Therefore, *A. palmeri* needed tens of copies of *EPSPS* for resistance, which was enabled only after evolution of the *EPSPS*-containing eccDNA. In fact, if the linear correlation between *EPSPS* copy number in *A. palmeri* and resistance magnitude shown in Fig. 2 is extrapolated, resistance would not be observed below 10 copies. As mentioned above, it is also possible that other genes within the eccDNA augment the glyphosate resistance conferred by *EPSPS* duplication.

5.3 Target-Site Mutations

The relative contributions of standing genetic variation versus new mutations for target-site resistance likely vary among herbicides. In the case of target-site resistance to glyphosate, repeated occurrence of double mutations and the occurrence of a triple mutation (discussed in Sect. 3.1) present additional evolutionary questions. These multiple-mutation alleles could preexist in a population as part of the standing genetic variation, or the multiple mutations could arise sequentially during the course of herbicide selection. In addition, the spontaneous occurrence of a double or triple-mutation allele (e.g., both or all three of the mutations occurring in a single generation) is formally possible, but the probability is so low that this route probably can be considered inconsequential (Ossowski et al. 2010). Sequential evolution could occur by a second mutation occurring in an allele that already has one mutation, or via recombination between two alleles each carrying one of the two mutations. Given the close proximity of the double and triple mutation sites in the gene, however, recombination between them will be exceedingly rare. Therefore, the two most likely evolutionary paths to the multiple-mutation alleles are either they existed prior to selection or a single-mutation allele increased in frequency as a result of herbicide selection, and then acquired one or more additional mutations.

If a multiple-mutation allele preexisted in the population, then one would expect it to have a limited fitness cost, because a large fitness cost would result in it having been purged from the population. From limited studies to date on fitness costs of multiple-mutation *EPSPS* alleles, however, at least some seem to have significant fitness costs (see Sect. 5.4). Additionally, if a multiple-mutation allele preexisted, one would expect to find this allele in essentially all resistant plants, i.e., occurrence of alleles containing only one of the mutations would be rare (since they would only come about via recombination or a mutation back to wild type). In an *E. indica* population with the Thr-102-Ile + Pro-106-Ser double mutation, both the double mutant and the single mutant Pro-106-Ser, but not the single mutant Thr-102-Ile, allele were found at high frequencies, leading the authors to conclude that the two mutations evolved sequentially (Yu et al. 2015).

In the cases of multiple-mutation *EPSPS* alleles in *Bidens subalternans* (double mutant) and *A. hybridus* (triple mutant), however, only the multiple-mutant alleles were observed (Perotti et al. 2019; Takano et al. 2020), which is consistent with the alleles preexisting in the population. Furthermore, because *B. subalternans* is tetraploid, it was suggested that fitness cost of the double-mutation allele could be masked by the second, wild type *EPSPS* gene (Takano et al. 2020), which could explain how such an allele persisted in the population prior to glyphosate selection. Because the multiple-mutation alleles confer higher resistance than the single-mutation alleles, there are caveats with the expectation that lack of finding the single-mutation alleles is evidence of the multiple-mutation alleles preexisting in the population. For example, with repeated selection of glyphosate, especially with high doses, the multiple-mutation alleles will be favored over the single-mutation alleles and, therefore, the single-mutation alleles will be purged over time. Thus, one must consider the glyphosate selection timeframe. In addition, if the multiple-mutation allele arose sequentially in one population, but then migrated to a second population, analysis of the second population would incorrectly lead to support of the hypothesis that the multi-mutation allele preexisted.

In summary, there is good evidence that multiple-mutation *EPSPS* alleles evolved from sequential events in at least some cases. More evidence is needed, however, to conclude that glyphosate resistance also has evolved via selection of multiple-mutation *EPSPS* alleles that preexisted as part of the standing genetic variation of a population.

5.4 Fitness Costs

In many organisms, the evolutionary adaptation to a new environment or to a new selection pressure is often accompanied by tradeoffs that can affect the general fitness of the organism, commonly referred to as fitness cost (Purrington 2000; Strauss et al. 2002; Vila-Aiub 2019). The presence of resistance alleles in a biotype can cause pleiotropic effects that will enhance some negative phenotypes, such as lower number and viability of seeds, less biomass, and less attraction to pollinators.

All of these effects can prevent the fixation of resistance alleles, making the adaptation process occur slower (Tian et al. 2003; Vila-Aiub 2019). On the other hand, studies have also shown that, in some cases, no fitness cost was observed due to the presence of herbicide-resistance alleles (Vila-Aiub 2019). Understanding fitness costs related to the presence of herbicide resistance traits is important to understand the evolution patterns that these traits will follow (Cousens and Fournier-Level 2018).

Studies to investigate fitness cost due to glyphosate resistance have shown different results according to the mechanism of resistance involved. In the case of target-site glyphosate resistance, there is generally a correlation between higher levels of resistance and greater fitness costs (Vila-Aiub et al. 2019). For example, substitution of two amino acids in EPSPS in *E. indica* was accompanied by a high fitness cost, whereas a single mutation in the same species—which provided lower resistance—conferred a negligible fitness cost (Yu et al. 2015; Han et al. 2017). Fitness studies of *EPSPS* gene duplication generally have identified little if any fitness costs, although costs may be higher in certain genetic backgrounds (Giacomini et al. 2014; Vila-Aiub et al. 2014; Martin et al. 2017; Osipitan and Dille 2019). That *EPSPS* duplication does not confer a large fitness penalty is particularly surprising in *A. palmeri*, given both the large number of copies in resistant plants and the size of the amplicon (Vila-Aiub et al. 2019). The *EPSPS* amplicon in *A. tuberculatus* also appears quite large (Kreiner et al. 2019) but, nevertheless, only modestly decreased in frequency in a multi-generational fitness study (Wu et al. 2017). Vila-Aiub (2019) provides a recent and more comprehensive review of fitness costs associated with glyphosate resistance. When considering fitness costs of herbicide-resistance mechanisms, it is important to keep in mind that those mechanisms that confer extremely high fitness penalties are unlikely to be selected. Consequently, our vantage point is skewed by studying only those mechanisms that have evolved in weed populations.

6 Impacts on Weed Management

Widespread adoption of glyphosate-resistant crops resulted in reliance on glyphosate for weed control in those crops and a dramatic drop in the use of alternative herbicides (Young 2006). A primary impact of glyphosate-resistant weeds has been a reversal of that trend. Initially, farmers typically responded to glyphosate-resistant weeds by increasing the glyphosate use rate (Weller et al. 2010). However, because glyphosate-resistant weeds often can withstand maximum labeled use rates, such an approach was largely futile. The second approach often was to use a tank mix, spraying a second herbicide with glyphosate. For example, in the case of glyphosate-resistant *A. tuberculatus* and *A. palmeri*, a herbicide that inhibits protoporphyrinogen oxidase often was added. This is reflected in the use of these herbicides in the USA declining precipitously, beginning in 1996, but then beginning to increase in 2013, coinciding with increasing occurrence of glyphosate-resistant *Amaranthus* populations (Dayan et al. 2018). Similar management responses, i.e., initially increasing the glyphosate rate, and then adding an appropriate tank-mix partner, were not restricted to USA farmers (Valverde 2010). Other chemical strategies such as returning to the use of soil-residual herbicides and rotating herbicides (e.g., not using glyphosate ever year) also were implemented in response to glyphosate-resistant weeds.

From the broader weed science industry perspective, a major impact of glyphosate-resistant crops was a decrease in herbicide discovery efforts (Duke 2012). Consequently, there are essentially no new herbicide options for farmers to turn to for combatting glyphosate-resistant weeds. This is particularly problematic for those weed populations that possess multiple resistance to other herbicides. Therefore, some farmers reluctantly responded to glyphosate-resistant weeds by implementing nonchemical strategies, including hand-weeding, tillage, and growing cover crops (Sosnoskie and Culpepper 2014; Duzy et al. 2016). Ironically, farmers are having to use diverse tactics to control glyphosate-resistant weeds, which are the same tactics that would have mitigated the evolution of these biotypes in the first place (Powles 2008).

Widespread adoption of glyphosate plus glyphosate-resistant crops also may have contributed to the range expansion of some of the weed species that evolved glyphosate resistance. *Conyza canadensis*, for example, was one of the first weeds to evolve glyphosate resistance, occurring originally in Delaware, USA (VanGessel 2001). Although at that time it was already a widespread weed in the USA (and elsewhere), long-distance wind dispersal of seeds with glyphosate resistance across a landscape heavily dominated by glyphosate-based weed management undoubtedly contributed to its invasiveness as a weed (Weaver 2001; Shah et al. 2014). Glyphosate resistance in both *A. tuberculatus* and *A. palmeri* also likely fostered their expansions. For example, glyphosate-resistant *A. tuberculatus* was identified in Canada, and at least one such population likely arrived via seed movement from the USA Midwest (Kreiner et al. 2019). Perhaps even more widespread dissemination of *A. palmeri* has occurred over the past few years, both within and beyond the USA, as a seed contaminant in, e.g., harvest equipment, livestock feed, and conservation-planting mixtures (Kistner and Hatfield 2018) and by migratory waterfowl (Farmer et al. 2017). To be sure, glyphosate resistance is not a prerequisite for the expansion of weed species, and maybe these weeds would have similarly expanded in a non-glyphosate scenario. However, it cannot be discounted that these weeds evolved glyphosate resistance in an era in which glyphosate was the sole means of chemical weed control in many fields, allowing populations of these species to explode in size. The increased population sizes increased the likelihood that seeds of these species would be disseminated.

Over the past few years, dicamba, coupled with dicamba-resistant crops, has been rapidly adopted in USA soybean and cotton production, largely to provide a solution for managing glyphosate-resistant weeds (Byker et al. 2013; Cahoon et al. 2015). It is unfortunate that glyphosate-resistant weeds have created such a demand for this technology, given the off-target concerns with dicamba, which are only exacerbated by wider adoption (Soltani et al. 2020).

The emergence of glyphosate resistance was the motivation behind an epidemiology approach to understand the spread of this resistance in *A. tuberculatus* (Evans et al. 2016). As expected, frequent use of glyphosate was identified as a key driver. However, this study also identified that, at least for glyphosate resistance in *A. tuberculatus*, the use of annual herbicide rotation was ineffective, whereas the use of herbicide mixtures was effective, at mitigating resistance evolution. A follow-up modeling study predicted that glyphosate-resistant *A. tuberculatus* evolution could have been even more effectively mitigated if practices such as herbicide mixing were coordinated at regional scales (Evans et al. 2018). Subsequently to these *A. tuberculatus* studies, a somewhat similar epidemiological approach was taken to proactively predict glyphosate-resistance evolution in *A. myosuroides* (Comont et al. 2019). Although glyphosate resistance has not yet been reported in this species, the study identified heritable variation for glyphosate sensitivity and that directional selection towards glyphosate resistance was occurring. Recently, there has been a call to increase the use of these types of epidemiological approaches to better predict, understand, and ultimately mitigate herbicide-resistance evolution in weeds (Comont and Neve 2020).

Currently, there is substantial interest in the development of novel, nonchemical, weed management technologies; much of this interest is largely (although not solely) attributable to glyphosate-resistant weeds. Examples of such new technologies include gene drives and robots (Neve 2018; McAllister et al. 2019). In retrospect, perhaps a positive outcome of the occurrence of glyphosate-resistant weeds will be spurred development of novel, nonchemical weed management strategies, which are particularly needed because the glyphosate-resistant crop era stifled the development of new herbicides.

Herbicide resistance is not a new phenomenon. In the 1990s, the widespread and rapid occurrence of resistance to inhibitors of acetolactate synthase taught us the importance of not relying on a single weed-control tactic (Tranel and Wright 2002). Apparently, that lesson was largely forgotten and then relearned through glyphosate-resistant weeds. Hopefully, this lesson will not be forgotten again.

7 Conclusion

Investigation of glyphosate-resistant weeds has revealed new mechanisms that weeds can evolve in response to intense herbicide selection. Although some of these mechanisms have been thus far associated exclusively or nearly exclusively with glyphosate resistance, now that they have been identified, it will be interesting to see if corresponding mechanisms for other herbicides are indeed rare, or simply have been overlooked. The source, i.e., new mutations vs. standing genetic variation, of adaptive glyphosate-resistance mechanisms remains largely unknown. Beyond herbicide resistance, the source of adaptive alleles is a fundamental and unresolved question in evolutionary biology. We suggest that glyphosate resistance, given its recent and rapid evolution, and the evolution of multiple adaptive mechanisms, provides an appropriate model system for this broad evolutionary question. There is a need to increase the use of genomics approaches to better understand resistance mechanisms to glyphosate, as well as to other herbicides. In particular, genetic mapping, which is just beginning to become a viable strategy with the availability of assembled weed genomes, offers great promise for elucidating previously intractable herbicide-resistance mechanisms. Such studies, together with epidemiological and population genetics approaches, should shed much light on glyphosate-resistance evolution. Lessons learned from studying the evolution of glyphosate-resistant weeds likely could be broadly translated to inform mitigation strategies for future herbicides. However, a major challenge posed by glyphosate-resistant weeds is that they evolved in an era coinciding with reduced research and development for alternative herbicides, ironically owing to the success of glyphosate/glyphosate-resistant crops. Consequently, there is a dearth of new herbicides to manage glyphosate-resistant weeds. Although only time will tell, glyphosate-resistant weeds should serve as a lasting example of the perils of relying on a single pest-management strategy.

Conflict of Interest The authors declare that they have no conflict of interest.

References

Adu-Yeboah P, Malone JM, Fleet B et al (2019) *EPSPS* gene amplification confers resistance to glyphosate resistant populations of *Hordeum glaucum* Stued (northern barley grass) in South Australia. Pest Manag Sci n/a. https://doi.org/10.1002/ps.5671

Alarcón-Reverte R, García A, Watson SB et al (2015) Concerted action of target-site mutations and high EPSPS activity in glyphosate-resistant junglerice (*Echinochloa colona*) from California. Pest Manag Sci 71:996–1007. https://doi.org/10.1002/ps.3878

Alcántara-de la Cruz R, Fernández-Moreno PT, Ozuna CV et al (2016a) Target and non-target site mechanisms developed by glyphosate-resistant hairy beggarticks (*Bidens pilosa* l.) populations from mexico. Front Plant Sci 7. https://doi.org/10.3389/fpls.2016.01492

Alcántara-de la Cruz R, Rojano-Delgado AM, Giménez MJ et al (2016b) First resistance mechanisms characterization in glyphosate-resistant *Leptochloa virgata*. Front Plant Sci 7. https://doi.org/10.3389/fpls.2016.01742

Amaro-Blanco I, Fernández-Moreno PT, Osuna-Ruiz MD et al (2018) Mechanisms of glyphosate resistance and response to alternative herbicide-based management in populations of the three *Conyza* species introduced in southern Spain. Pest Manag Sci 74:1925–1937. https://doi.org/10.1002/ps.4896

Baerson SR, Rodriguez DJ, Biest NA et al (2002a) Investigating the mechanism of glyphosate resistance in rigid ryegrass (*Lolium ridigum*). Weed Sci 50:721–730. https://doi.org/10.1614/0043-1745(2002)050[0721:ITMOGR]2.0.CO;2

Baerson SR, Rodriguez DJ, Tran M et al (2002b) Glyphosate-resistant goosegrass. Identification of a mutation in the target enzyme 5-enolpyruvylshikimate-3-phosphate synthase. Plant Physiol 129:1265–1275. https://doi.org/10.1104/pp.001560

Baucom RS, Mauricio R (2010) Defence against the herbicide RoundUp® predates its widespread use. Evol Ecol Res 12:131–141

Bell MS, Hager AG, Tranel PJ (2013) Multiple resistance to herbicides from four site-of-action groups in waterhemp (Amaranthus tuberculatus). Weed Sci 61:460–468

Beres ZT, Ernst EE, Ackley BA et al (2018) High levels of glyphosate resistance in *Conyza canadensis* from agricultural and non-agricultural sites in Ohio and Iowa. Sci Rep 8:1–8. https://doi.org/10.1038/s41598-018-28163-w

Brabham CB, Gerber CK, Johnson WG (2011) Fate of glyphosate-resistant giant ragweed (*Ambrosia trifida*) in the presence and absence of glyphosate. Weed Sci 59:506–511. https://doi.org/10.1614/WS-D-11-00050.1

Bracamonte E, Fernández-Moreno PT, Barro F, De Prado R (2016) Glyphosate-resistant *Parthenium hysterophorus* in the Caribbean islands: non target site resistance and target site resistance in relation to resistance levels. Front Plant Sci 7. https://doi.org/10.3389/fpls.2016.01845

Bradshaw LD, Padgette SR, Kimball SL, Wells BH (1997) Perspectives on glyphosate resistance. Weed Technol 11:189–198

Brewer CE, Oliver LR (2009) Confirmation and resistance mechanisms in glyphosate-resistant common ragweed (*Ambrosia artemisiifolia*) in Arkansas. Weed Sci 57:567–573

Bromilow RH, Chamberlain K, Evans AA (1990) Physicochemical aspects of phloem translocation of herbicides. Weed Sci 38:305–314

Brotherton JE, Jeschke MR, Tranel PJ, Widholm JM (2007) Identification of *Arabidopsis thaliana* variants with differential glyphosate responses. J Plant Physiol 164:1337–1345. https://doi.org/10.1016/j.jplph.2006.08.008

Brunharo CA, Patterson EL, Carrijo DR et al (2016) Confirmation and mechanism of glyphosate resistance in tall windmill grass (*Chloris elata*) from Brazil. Pest Manag Sci 72:1758–1764. https://doi.org/10.1002/ps.4205

Brunharo CACG, Hanson BD (2018) Multiple herbicide–resistant Italian ryegrass [*Lolium perenne* L. spp. *multiflorum* (Lam.) Husnot] in California perennial crops: Characterization, mechanism of resistance, and chemical management. Weed Sci 66:696–701. https://doi.org/10.1017/wsc.2018.50

Brunharo CADCG, Morran S, Martin K et al (2019) EPSPS duplication and mutation involved in glyphosate resistance in the allotetraploid weed species *Poa annua* L. Pest Manag Sci 75:1663–1670. https://doi.org/10.1002/ps.5284

Byker HP, Soltani N, Robinson DE et al (2013) Control of glyphosate-resistant horseweed (*Conyza canadensis*) with dicamba applied preplant and postemergence in dicamba-resistant soybean. Weed Technol 27:492–496. https://doi.org/10.1614/WT-D-13-00023.1

Cahoon CW, York AC, Jordan DL et al (2015) Palmer amaranth (*Amaranthus palmeri*) management in dicamba-resistant cotton. Weed Technol 29:758–770. https://doi.org/10.1614/WT-D-15-00041.1

Casale FA, Giacomini DA, Tranel PJ (2019) Empirical investigation of mutation rate for herbicide resistance. Weed Sci 67:361–368. https://doi.org/10.1017/wsc.2019.19

Chahal PS, Varanasi VK, Jugulam M, Jhala AJ (2017) Glyphosate-resistant Palmer amaranth (*Amaranthus palmeri*) in Nebraska: confirmation, *EPSPS* gene amplification, and response to post corn and soybean herbicides. Weed Technol 31:80–93. https://doi.org/10.1614/WT-D-16-00109.1

Chandi A, Milla-Lewis SR, Giacomini D et al (2012) Inheritance of evolved glyphosate resistance in a north carolina palmer amaranth (*Amaranthus palmeri*) biotype. Int J Agron. https://www.hindawi.com/journals/ija/2012/176108/. Accessed 5 Dec 2019

Chatham LA, Bradley KW, Kruger GR et al (2015) A multistate study of the association between glyphosate resistance and EPSPS gene amplification in waterhemp (*Amaranthus tuberculatus*). Weed Sci 63:569–577. https://doi.org/10.1614/WS-D-14-00149.1

Chen J, Huang H, Zhang C et al (2015) Mutations and amplification of *EPSPS* gene confer resistance to glyphosate in goosegrass (*Eleusine indica*). Planta 242:859–868. https://doi.org/10.1007/s00425-015-2324-2

Chen J, Cui H, Ma X et al (2019) Distribution differences in the *EPSPS* gene in chromosomes between glyphosate-resistant and -susceptible goosegrass (*Eleusine indica*). Weed Sci:1–28. https://doi.org/10.1017/wsc.2019.72

Comont D, Neve P (2020) Adopting epidemiological approaches for herbicide resistance monitoring and management. Weed Res n/a: https://doi.org/10.1111/wre.12420

Comont D, Hicks H, Crook L et al (2019) Evolutionary epidemiology predicts the emergence of glyphosate resistance in a major agricultural weed. New Phytol 223:1584–1594. https://doi.org/10.1111/nph.15800

Copetti D, Kuon J, Yates S et al (2019) A diploid phased genome assembly of the Italian ryegrass (Lolium multiflorum Lam.) cv. Rabiosa. 1 p. https://doi.org/10.3929/ETHZ-B-000353845

Costea M, Weaver SE, Tardif FJ (2005) The biology of invasive alien plants in Canada. 3. *Amaranthus tuberculatus* (Moq.) Sauer var. *rudis* (Sauer) Costea & Tardif. Can J Plant Sci 85:507–522. https://doi.org/10.4141/P04-101

Cousens RD, Fournier-Level A (2018) Herbicide resistance costs: what are we actually measuring and why? Pest Manag Sci 74:1539–1546

Culpepper AS, Grey TL, Vencill WK et al (2006) Glyphosate-resistant Palmer amaranth (*Amaranthus palmeri*) confirmed in Georgia. Weed Sci 54:620–626. https://doi.org/10.1614/WS-06-001R.1

Dayan FE, Barker A, Tranel PJ (2018) Origins and structure of chloroplastic and mitochondrial plant protoporphyrinogen oxidases: implications for the evolution of herbicide resistance. Pest Manag Sci 74:2226–2234. https://doi.org/10.1002/ps.4744

de Carvalho LB, Alves PL d CA, González-Torralva F et al (2012) Pool of resistance mechanisms to glyphosate in *Digitaria insularis*. J Agric Food Chem 60:615–622. https://doi.org/10.1021/jf204089d

Délye C (2013) Unravelling the genetic bases of non-target-site-based resistance (NTSR) to herbicides: a major challenge for weed science in the forthcoming decade. Pest Manag Sci 69:176–187. https://doi.org/10.1002/ps.3318

Devine MD, Bandeen JD, McKERSIE BD (1983) Temperature effects on glyphosate absorption, translocation, and distribution in quackgrass (*Agropyron repens*). Weed Sci 31:461–464

Dill GM (2005) Glyphosate-resistant crops: history, status and future. Pest Manag Sci 61:219–224. https://doi.org/10.1002/ps.1008

Dillon A, Varanasi VK, Danilova TV et al (2017) Physical mapping of amplified copies of the 5-enolpyruvylshikimate-3-phosphate synthase gene in glyphosate-resistant *Amaranthus tuberculatus*. Plant Physiol 173:1226–1234. https://doi.org/10.1104/pp.16.01427

Dinelli G, Marotti I, Bonetti A et al (2008) Physiological and molecular bases of glyphosate resistance in *Conyza bonariensis* biotypes from Spain. Weed Res 48:257–265. https://doi.org/10.1111/j.1365-3180.2008.00623.x

Dominguez-Valenzuela JA, Gherekhloo J, Fernández-Moreno PT et al (2017) First confirmation and characterization of target and non-target site resistance to glyphosate in Palmer amaranth (*Amaranthus palmeri*) from Mexico. Plant Physiol Biochem 115:212–218. https://doi.org/10.1016/j.plaphy.2017.03.022

Duke SO (2011) Glyphosate degradation in glyphosate-resistant and -susceptible crops and weeds. J Agric Food Chem 59:5835–5841. https://doi.org/10.1021/jf102704x

Duke SO (2012) Why have no new herbicide modes of action appeared in recent years? Pest Manag Sci 68:505–512. https://doi.org/10.1002/ps.2333

Duke SO (2019) Enhanced metabolic degradation: the last evolved glyphosate resistance mechanism of weeds? Plant Physiol 181:1401–1403. https://doi.org/10.1104/pp.19.01245

Duke SO (2021) Glyphosate: uses other than in glyphosate-resistant crops, mode of action, degradation in plants, and effects on non-target plants and agricultural microbes. Rev Environ Contam Toxicol (this volume). https://doi.org/10.1007/398_2020_53

Duke SO, Powles SB (2008) Glyphosate: a once-in-a-century herbicide. Pest Manag Sci 64:319–325. https://doi.org/10.1002/ps.1518

Duzy LM, Price AJ, Balkcom KS, Aulakh JS (2016) Assessing the economic impact of inversion tillage, cover crops, and herbicide regimes in Palmer amaranth *(Amaranthus palmeri)* infested cotton. Int J Agron 2016:1–9. https://doi.org/10.1155/2016/1524389

Evans JA, Tranel PJ, Hager AG et al (2016) Managing the evolution of herbicide resistance. Pest Manag Sci 72:74–80. https://doi.org/10.1002/ps.4009

Evans JA, Williams A, Hager AG et al (2018) Confronting herbicide resistance with cooperative management. Pest Manag Sci 74:2424–2431. https://doi.org/10.1002/ps.5105

Farmer JA, Webb EB, Pierce RA, Bradley KW (2017) Evaluating the potential for weed seed dispersal based on waterfowl consumption and seed viability. Pest Manag Sci 73:2592–2603. https://doi.org/10.1002/ps.4710

Feng PC, Tran M, Chiu T et al (2004) Investigations into glyphosate-resistant horseweed (*Conyza canadensis*): retention, uptake, translocation, and metabolism. Weed Sci 52:498–505

Fernández-Moreno PT, Alcantara-de la Cruz R, Cruz-Hipólito HE et al (2016) Non-target site tolerance mechanisms describe tolerance to glyphosate in *Avena sterilis*. Front Plant Sci 7. https://doi.org/10.3389/fpls.2016.01220

Franci J, Lam KW, Chuah TS, San Cha T (2020) Genetic diversity and in silico evidence of target-site mutation in the EPSPS gene in endowing glyphosate resistance in *Eleusine indica* (L.) from Malaysia. Pestic Biochem Physiol 104556

Gaines TA, Heap IM (2020) Mutations in Herbicide-Resistant Weeds to EPSPS Synthase Inhibitors. http://www.weedscience.com. Accessed 11 June 2020

Gaines TA, Zhang W, Wang D et al (2010) Gene amplification confers glyphosate resistance in *Amaranthus palmeri*. Proc Natl Acad Sci 107:1029–1034. https://doi.org/10.1073/pnas.0906649107

Gaines TA, Cripps A, Powles SB (2012) Evolved resistance to glyphosate in junglerice (*Echinochloa colona*) from the tropical Ord River region in Australia. Weed Technol 26:480–484. https://doi.org/10.1614/WT-D-12-00029.1

Gaines TA, Barker AL, Patterson EL et al (2016) *EPSPS* gene copy number and whole-plant glyphosate resistance level in *Kochia scoparia*. PLOS ONE 11:e0168295. https://doi.org/10.1371/journal.pone.0168295

Gaines TA, Patterson EL, Neve P (2019) Molecular mechanisms of adaptive evolution revealed by global selection for glyphosate resistance. New Phytol 223:1770–1775. https://doi.org/10.1111/nph.15858

Gaines TA, Duke SO, Morran S et al (2020) Mechanisms of evolved herbicide resistance. J Biol Chem 295(30):10307–10330. https://doi.org/10.1074/jbc.REV120.013572

García MJ, Palma-Bautista C, Rojano-Delgado AM et al (2019) The triple amino acid substitution TAP-IVS in the *EPSPS* gene confers high glyphosate resistance to the superweed *Amaranthus hybridus*. Int J Mol Sci 20:2396. https://doi.org/10.3390/ijms20102396

Ge X, d'Avignon DA, Ackerman JJ, Sammons RD (2010) Rapid vacuolar sequestration: the horseweed glyphosate resistance mechanism. Pest Manag Sci 66:345–348. https://doi.org/10.1002/ps.1911

Ge X, d'Avignon DA, Ackerman JJ et al (2011) Glyphosate-resistant horseweed made sensitive to glyphosate: low-temperature suppression of glyphosate vacuolar sequestration revealed by 31P NMR. Pest Manag Sci 67:1215–1221

Ge X, d'Avignon DA, Ackerman JJH et al (2012) Vacuolar glyphosate-sequestration correlates with glyphosate resistance in ryegrass (*Lolium* spp.) from Australia, South America, and Europe: a 31p nmr investigation. J Agric Food Chem 60:1243–1250. https://doi.org/10.1021/jf203472s

Ge X, d'Avignon DA, Ackerman JJ, Sammons RD (2014) In vivo 31P-nuclear magnetic resonance studies of glyphosate uptake, vacuolar sequestration, and tonoplast pump activity in glyphosate-resistant horseweed. Plant Physiol 166:1255–1268

Giacomini D, Westra P, Ward SM (2014) Impact of genetic background in fitness cost studies: an example from glyphosate-resistant Palmer amaranth. Weed Sci 62:29–37

Godar AS, Stahlman PW, Jugulam M, Dille JA (2015) Glyphosate-resistant kochia (*Kochia scoparia*) in Kansas: *EPSPS* gene copy number in relation to resistance levels. Weed Sci 63:587–595. https://doi.org/10.1614/WS-D-14-00145.1

Goh SS, Yu Q, Han H et al (2018) Non-target-site glyphosate resistance in *Echinochloa colona* from Western Australia. Crop Prot 112:257–263. https://doi.org/10.1016/j.cropro.2018.06.013
Gomes MP, Smedbol E, Chalifour A et al (2014) Alteration of plant physiology by glyphosate and its by-product aminomethylphosphonic acid: an overview. J Exp Bot 65:4691–4703. https://doi.org/10.1093/jxb/eru269
González-Torralva F, Gil-Humanes J, Barro F et al (2012a) Target site mutation and reduced translocation are present in a glyphosate-resistant *Lolium multiflorum* Lam. biotype from Spain. Plant Physiol Biochem 58:16–22. https://doi.org/10.1016/j.plaphy.2012.06.001
González-Torralva F, Rojano-Delgado AM, de Castro MDL et al (2012b) Two non-target mechanisms are involved in glyphosate-resistant horseweed (*Conyza canadensis* L. Cronq.) biotypes. J Plant Physiol 169:1673–1679
González-Torralva F, Gil-Humanes J, Barro F et al (2014) First evidence for a target site mutation in the EPSPS2 gene in glyphosate-resistant Sumatran fleabane from citrus orchards. Agron Sustain Dev 34:553–560. https://doi.org/10.1007/s13593-013-0163-8
González-Torralva F, Brown AP, Chivasa S (2017) Comparative proteomic analysis of horseweed (*Conyza canadensis*) biotypes identifies candidate proteins for glyphosate resistance. Sci Rep 7:42565. https://doi.org/10.1038/srep42565
Goyer A, Johnson TL, Olsen LJ et al (2004) Characterization and metabolic function of a peroxisomal sarcosine and pipecolate oxidase from Arabidopsis. J Biol Chem 279:16947–16953. https://doi.org/10.1074/jbc.M400071200
Green JM, Siehl DS (2021) History and outlook for glyphosate-resistant crops. Rev Environ Contam Toxicol (this volume). https://doi.org/10.1007/398_2020_54
Han H, Yu Q, Widderick MJ, Powles SB (2016) Target-site EPSPS Pro-106 mutations: sufficient to endow glyphosate resistance in polyploid *Echinochloa colona*? Pest Manag Sci 72:264–271. https://doi.org/10.1002/ps.4038
Han H, Vila-Aiub MM, Jalaludin A et al (2017) A double *EPSPS* gene mutation endowing glyphosate resistance shows a remarkably high resistance cost. Plant Cell Environ 40:3031–3042. https://doi.org/10.1111/pce.13067
Harre NT, Young JM, Young BG (2018) Glyphosate-induced antagonism in rapid response giant ragweed (*Ambrosia trifida*). Weed Technol 32:52–59. https://doi.org/10.1017/wet.2017.92
Heap I, Duke SO (2018) Overview of glyphosate-resistant weeds worldwide. Pest Manag Sci 74:1040–1049. https://doi.org/10.1002/ps.4760
Heap I, Knight R (1986) The occurrence of herbicide cross-resistance in a population of annual ryegrass, *Lolium rigidum*, resistant to diclofop-methyl. Aust J Agric Res 37:149–156. https://doi.org/10.1071/ar9860149
Heap IM (2020) The International Survey of Herbicide Resistant Weeds. http://www.weedscience.org. Accessed 13 Feb 2020
Hess FD, Duke SO (1985) Herbicide absorption and translocation and their relationship to plant tolerances and susceptibility. In: Weed physiology, vol 2, herbicide physiology. CRC Press, Boca Raton, pp 191–214
Hoagland RE (1980) Effects of glyphosate on metabolism of phenolic compounds: vi. effects of glyphosine and glyphosate metabolites on phenylalanine ammonia-lyase activity, growth, and protein, chlorophyll, and anthocyanin levels in soybean (*glycine max*) seedlings. Weed Sci 28:393–400. https://doi.org/10.1017/S0043174500055545
Housden BE, Muhar M, Gemberling M et al (2017) Loss-of-function genetic tools for animal models: cross-species and cross-platform differences. Nat Rev Genet 18:24–40. https://doi.org/10.1038/nrg.2016.118
Huang Z, Liu Y, Zhang C et al (2019) Molecular basis of natural tolerance to glyphosate in *Convolvulus arvensis*. Sci Rep 9:1–10. https://doi.org/10.1038/s41598-019-44583-8
Jander G, Baerson SR, Hudak JA et al (2003) Ethylmethanesulfonate saturation mutagenesis in *Arabidopsis* to determine frequency of herbicide resistance. Plant Physiol 131:139–146. https://doi.org/10.1104/pp.102.010397
Jasieniuk M, Ahmad R, Sherwood AM et al (2008) Glyphosate-resistant italian ryegrass (*Lolium multiflorum*) in California: Distribution, response to glyphosate, and molecular evidence for an altered target enzyme. Weed Sci 56:496–502. https://doi.org/10.1614/WS-08-020.1

Jugulam M, Niehues K, Godar AS et al (2014) Tandem amplification of a chromosomal segment harboring *5-enolpyruvylshikimate-3-phosphate synthase* locus confers glyphosate resistance in *Kochia scoparia*. Plant Physiol 166:1200–1207. https://doi.org/10.1104/pp.114.242826

Karn E, Jasieniuk M (2017) Nucleotide diversity at site 106 of *EPSPS* in *Lolium perenne* L. ssp. *multiflorum* from California indicates multiple evolutionary origins of herbicide resistance. Front Plant Sci 8:777

Kaundun SS, Dale RP, Zelaya IA et al (2011) A novel p106l mutation in epsps and an unknown mechanism(s) act additively to confer resistance to glyphosate in a South African *lolium rigidum* population. J Agric Food Chem 59:3227–3233. https://doi.org/10.1021/jf104934j

Kells JJ, Rieck CE (1979) Effects of illuminance and time on accumulation of glyphosate in johnsongrass (*Sorghum halepense*). Weed Sci 27:235–237

Kistner EJ, Hatfield JL (2018) Potential geographic distribution of Palmer amaranth under current and future climates. Agric Environ Lett 3:170044. https://doi.org/10.2134/ael2017.12.0044

Kleinman Z, Rubin B (2017) Non-target-site glyphosate resistance in *Conyza bonariensis* is based on modified subcellular distribution of the herbicide. Pest Manag Sci 73:246–253. https://doi.org/10.1002/ps.4293

Koger CH, Reddy KN (2005) Role of absorption and translocation in the mechanism of glyphosate resistance in horseweed (*Conyza canadensis*). Weed Sci 53:84–89

Koo D-H, Jugulam M, Putta K et al (2018a) Gene duplication and aneuploidy trigger rapid evolution of herbicide resistance in common waterhemp. Plant Physiol 176:1932–1938. https://doi.org/10.1104/pp.17.01668

Koo D-H, Molin WT, Saski CA et al (2018b) Extrachromosomal circular DNA-based amplification and transmission of herbicide resistance in crop weed *Amaranthus palmeri*. Proc Natl Acad Sci 115:3332–3337. https://doi.org/10.1073/pnas.1719354115

Korte A, Farlow A (2013) The advantages and limitations of trait analysis with GWAS: a review. Plant Methods 9:29. https://doi.org/10.1186/1746-4811-9-29

Kreiner JM, Giacomini DA, Bemm F et al (2019) Multiple modes of convergent adaptation in the spread of glyphosate-resistant *Amaranthus tuberculatus*. Proc Natl Acad Sci 116:21076–21084. https://doi.org/10.1073/pnas.1900870116

Kuester A, Chang S-M, Baucom RS (2015) The geographic mosaic of herbicide resistance evolution in the common morning glory, *Ipomoea purpurea*: Evidence for resistance hotspots and low genetic differentiation across the landscape. Evol Appl 8:821–833. https://doi.org/10.1111/eva.12290

Laforest M, Soufiane B, Simard M-J et al (2017) Acetyl-CoA carboxylase overexpression in herbicide-resistant large crabgrass (*Digitaria sanguinalis*). Pest Manag Sci 73:2227–2235. https://doi.org/10.1002/ps.4675

Laforest M, Martin SL, Bisaillon K et al (2020) A chromosome-scale draft sequence of the Canada fleabane genome. Pest Manag Sci 76:2158–2169. https://doi.org/10.1002/ps.5753

Lamego FP, Vidal RA (2008) Resistance to glyphosate in *Conyza bonariensis* and *Conyza canadensis* biotypes in Rio Grande do Sul, Brazil. Planta Daninha 26:467–471. https://doi.org/10.1590/S0100-83582008000200024

Legleiter TR, Bradley KW (2008) Glyphosate and multiple herbicide resistance in common waterhemp (*Amaranthus rudis*) populations from Missouri. Weed Sci 56:582–587. https://doi.org/10.1614/WS-07-204.1

Lespérance MA (2015) Programmed cell death and altered translocation cause glyphosate resistance in giant ragweed (*Ambrosia trifida* L.). PhD thesis

Li J, Peng Q, Han H et al (2018) Glyphosate resistance in *Tridax procumbens* via a novel EPSPS Thr-102-Ser substitution. J Agric Food Chem 66:7880–7888. https://doi.org/10.1021/acs.jafc.8b01651

Liu C-M, McLean PA, Sookdeo CC, Cannon FC (1991) Degradation of the herbicide glyphosate by members of the family rhizobiaceae. Appl Environ Microbiol 57:1799–1804

Liu M, Hulting AG, Mallory-Smith C (2016) Characterization of multiple herbicide-resistant Italian ryegrass (*Lolium perenne* ssp. *multiflorum*) populations from winter wheat fields in Oregon. Weed Sci 64:331–338

Lorentz L, Gaines TA, Nissen SJ et al (2014) Characterization of glyphosate resistance in *Amaranthus tuberculatus* populations. J Agric Food Chem 62:8134–8142. https://doi.org/10.1021/jf501040x

Lorraine-Colwill DF, Powles SB, Hawkes TR et al (2002) Investigations into the mechanism of glyphosate resistance in *Lolium rigidum*. Pestic Biochem Physiol 74:62–72

Malone JM, Morran S, Shirley N et al (2016) *EPSPS* gene amplification in glyphosate-resistant *Bromus diandrus*. Pest Manag Sci 72:81–88. https://doi.org/10.1002/ps.4019

Mao C, Xie H, Chen S et al (2016) Multiple mechanism confers natural tolerance of three lilyturf species to glyphosate. Planta 243:321–335. https://doi.org/10.1007/s00425-015-2408-z

Margaritopoulou T, Tani E, Chachalis D, Travlos I (2018) Involvement of epigenetic mechanisms in herbicide resistance: the case of *Conyza canadensis*. Agriculture 8:17. https://doi.org/10.3390/agriculture8010017

Markus C, Pecinka A, Karan R et al (2018) Epigenetic regulation – contribution to herbicide resistance in weeds? Pest Manag Sci 74:275–281. https://doi.org/10.1002/ps.4727

Maroli AS, Gaines TA, Foley ME et al (2018) Omics in weed science: a perspective from genomics, transcriptomics, and metabolomics approaches. Weed Sci 66:681–695. https://doi.org/10.1017/wsc.2018.33

Marshall G, Kirkwood RC, Martin DJ (1987) Studies on the mode of action of asulam, aminotriazole and glyphosate in Equisetum arvense L. (field horsetail). II: The metabolism of [14C]asulam, [14C]aminotriazole and [14C]glyphosate. Pestic Sci 18:65–77. https://doi.org/10.1002/ps.2780180107

Martin SL, Benedict L, Sauder CA et al (2017) Glyphosate resistance reduces kochia fitness: Comparison of segregating resistant and susceptible F2 populations. Plant Sci 261:69–79. https://doi.org/10.1016/j.plantsci.2017.04.010

Martinoia E (1992) Transport processes in vacuoles of higher plants. Bot Acta 105:232–245. https://doi.org/10.1111/j.1438-8677.1992.tb00293.x

Marty F (1999) Plant vacuoles. Plant Cell 11:587–599

McAllister W, Osipychev D, Davis A, Chowdhary G (2019) Agbots: Weeding a field with a team of autonomous robots. Comput Electron Agric 163:104827. https://doi.org/10.1016/j.compag.2019.05.036

McElroy JS, Hall ND (2020) *Echinochloa colona* with reported resistance to glyphosate conferred by aldo-keto reductase also contains a Pro-106-Thr EPSPS target-site mutation. Plant Physiol. https://doi.org/10.1104/pp.20.00064

Michitte P, De Prado R, Espinoza N et al (2007) Mechanisms of resistance to glyphosate in a ryegrass (*Lolium multiflorum*) biotype from Chile. Weed Sci 55:435–440

Mohseni-Moghadam M, Schroeder J, Ashigh J (2013) Mechanism of Resistance and Inheritance in Glyphosate Resistant Palmer amaranth (<i>Amaranthus palmeri<i>) Populations from New Mexico, USA. Weed Sci 61:517–525. https://doi.org/10.1614/WS-D-13-00028.1

Molin WT, Wright AA, Lawton-Rauh A, Saski CA (2017) The unique genomic landscape surrounding the *EPSPS* gene in glyphosate resistant *Amaranthus palmeri*: a repetitive path to resistance. BMC Genomics 18. https://doi.org/10.1186/s12864-016-3336-4

Molin WT, Wright AA, VanGessel MJ et al (2018) Survey of the genomic landscape surrounding the 5-enolpyruvylshikimate-3-phosphate synthase (*EPSPS*) gene in glyphosate-resistant *Amaranthus palmeri* from geographically distant populations in the USA. Pest Manag Sci 74:1109–1117. https://doi.org/10.1002/ps.4659

Molin WT, Yaguchi A, Blenner MA, Saski CA (2020) The eccDNA replicon: a heritable, extra-nuclear vehicle that enables gene amplification and glyphosate resistance in *Amaranthus palmeri*. Plant Cell. https://doi.org/10.1105/tpc.20.00099

Moretti ML, Hanson BD, Hembree KJ, Shrestha A (2013) Glyphosate resistance is more variable than paraquat resistance in a multiple-resistant hairy fleabane (*Conyza bonariensis*) population. Weed Sci 61:396–402. https://doi.org/10.1614/WS-D-12-00201.1

Moretti ML, Van Horn CR, Robertson R et al (2018) Glyphosate resistance in *Ambrosia trifida:* Part 2. Rapid response physiology and non-target-site resistance. Pest Manag Sci 74:1079–1088. https://doi.org/10.1002/ps.4569

Morran S, Moretti ML, Brunharo CA et al (2018) Multiple target site resistance to glyphosate in junglerice (*Echinochloa colona*) lines from California orchards. Pest Manag Sci 74:2747–2753. https://doi.org/10.1002/ps.5061

Moss SR, Cussans GW (1985) Variability in the susceptibility of *Alopecurus myosuroides* (black-grass) to chlortoluron and isoproturon. Asp Appl Biol 9:91–98

Murphy BP, Tranel PJ (2019) Target-site mutations conferring herbicide resistance. Plants 8:382. https://doi.org/10.3390/plants8100382

Mutters RG, Madore M, Bytnerowicz A (1993) Formaldehyde exposure affects growth and metabolism of common bean. Air Waste 43:113–116. https://doi.org/10.1080/1073161X.1993.10467112

Nandula VK, Ray JD, Ribeiro DN et al (2013) Glyphosate resistance in tall waterhemp (*Amaranthus tuberculatus*) from Mississippi is due to both altered target-site and nontarget-site mechanisms. Weed Sci 61:374–383. https://doi.org/10.1614/WS-D-12-00155.1

Nandula VK, Wright AA, Bond JA et al (2014) EPSPS amplification in glyphosate-resistant spiny amaranth (*Amaranthus spinosus*): a case of gene transfer via interspecific hybridization from glyphosate-resistant Palmer amaranth (*Amaranthus palmeri*). Pest Manag Sci 70:1902–1909. https://doi.org/10.1002/ps.3754

Nandula VK, Wright AA, Van Horn CR et al (2015) Glyphosate resistance in giant ragweed (*Ambrosia trifida* L.) from Mississippi is partly due to reduced translocation. Am J Plant Sci 6:2104

Neve P (2018) Gene drive systems: do they have a place in agricultural weed management? Pest Manag Sci 74:2671–2679. https://doi.org/10.1002/ps.5137

Neve P, Vila-Aiub M, Roux F (2009) Evolutionary-thinking in agricultural weed management. New Phytol 184:783–793. https://doi.org/10.1111/j.1469-8137.2009.03034.x

Ngo TD, Krishnan M, Boutsalis P et al (2018a) Target-site mutations conferring resistance to glyphosate in feathertop Rhodes grass (*Chloris virgata*) populations in Australia. Pest Manag Sci 74:1094–1100. https://doi.org/10.1002/ps.4512

Ngo TD, Malone JM, Boutsalis P et al (2018b) *EPSPS* gene amplification conferring resistance to glyphosate in windmill grass (*Chloris truncata*) in Australia. Pest Manag Sci 74:1101–1108. https://doi.org/10.1002/ps.4573

Nol N, Tsikou D, Eid M et al (2012) Shikimate leaf disc assay for early detection of glyphosate resistance in *Conyza canadensis* and relative transcript levels of EPSPS and ABC transporter genes. Weed Res 52:233–241

Osipitan OA, Dille JA (2019) No impact of increased *EPSPS* gene copy number on growth and fecundity of glyphosate-resistant kochia (*Bassia scoparia*). Weed Sci 67:22–28. https://doi.org/10.1017/wsc.2018.82

Ossowski S, Schneeberger K, Lucas-Lledo JI et al (2010) The rate and molecular spectrum of spontaneous mutations in *Arabidopsis thaliana*. Science 327:92–94. https://doi.org/10.1126/science.1180677

Palma-Bautista C, Alcántara-de la Cruz R, Rojano-Delgado AM et al (2019) Low temperatures enhance the absorption and translocation of 14C-glyphosate in glyphosate-resistant *Conyza sumatrensis*. J Plant Physiol 240:153009. https://doi.org/10.1016/j.jplph.2019.153009

Pan L, Yu Q, Han H et al (2019) Aldo-keto reductase metabolizes glyphosate and confers glyphosate resistance in *Echinochloa colona*. Plant Physiol 181:1519–1534. https://doi.org/10.1104/pp.19.00979

Pandolfo CE, Presotto A, Carbonell FT et al (2018) Transgene escape and persistence in an agroecosystem: the case of glyphosate-resistant *Brassica rapa* L. in central Argentina. Environ Sci Pollut Res 25:6251–6264. https://doi.org/10.1007/s11356-017-0726-3

Patterson EL, Pettinga DJ, Ravet K et al (2018) Glyphosate resistance and *EPSPS* gene duplication: convergent evolution in multiple plant species. J Hered 109:117–125. https://doi.org/10.1093/jhered/esx087

Patterson EL, Saski C, Küpper A et al (2019a) Omics potential in herbicide-resistant weed management. Plants 8:607. https://doi.org/10.3390/plants8120607

Patterson EL, Saski CA, Sloan DB et al (2019b) The draft genome of *Kochia scoparia* and the mechanism of glyphosate resistance via transposon-mediated *EPSPS* tandem gene duplication. Genome Biol Evol 11:2927–2940. https://doi.org/10.1093/gbe/evz198

Perez-Jones A, Park K-W, Polge N et al (2007) Investigating the mechanisms of glyphosate resistance in *Lolium multiflorum*. Planta 226:395–404. https://doi.org/10.1007/s00425-007-0490-6

Perotti VE, Larran AS, Palmieri VE et al (2019) A novel triple amino acid substitution in the *EPSPS* found in a high-level glyphosate-resistant *Amaranthus hybridus* population from Argentina. Pest Manag Sci 75:1242–1251. https://doi.org/10.1002/ps.5303

Piasecki C, Yang Y, Benemann DP et al (2019) Transcriptomic analysis identifies new non-target site glyphosate-resistance genes in *Conyza bonariensis*. Plants 8:157. https://doi.org/10.3390/plants8060157

Pollegioni L, Schonbrunn E, Siehl D (2011) Molecular basis of glyphosate resistance: Different approaches through protein engineering. FEBS J 278:2753–2766. https://doi.org/10.1111/j.1742-4658.2011.08214.x

Powles SB (2008) Evolved glyphosate-resistant weeds around the world: lessons to be learnt. Pest Manag Sci 64:360–365. https://doi.org/10.1002/ps.1525

Powles SB, Yu Q (2010) Evolution in action: plants resistant to herbicides. Annu Rev Plant Biol 61:317–347. https://doi.org/10.1146/annurev-arplant-042809-112119

Preston C, Wakelin AM (2008) Resistance to glyphosate from altered herbicide translocation patterns. Pest Manag Sci Former Pestic Sci 64:372–376

Purrington CB (2000) Costs of resistance. Curr Opin Plant Biol 3:305–308

Putta K (2017) Physical mapping of *EPSPS* gene copies in glyphosate resistant Italian ryegrass (*Lolium perenne* ssp. *multiflorum*). Thesis, Kansas State University

Queiroz ARS, de Delatorre CA, Lucio FR et al (2020) Rapid necrosis: a novel plant resistance mechanism to 2,4-D. Weed Sci 68:6–18. https://doi.org/10.1017/wsc.2019.65

Ravet K, Patterson EL, Krähmer H et al (2018) The power and potential of genomics in weed biology and management. Pest Manag Sci 74:2216–2225. https://doi.org/10.1002/ps.5048

Reddy KN, Rimando AM, Duke SO, Nandula VK (2008) Aminomethylphosphonic acid accumulation in plant species treated with glyphosate. J Agric Food Chem 56:2125–2130. https://doi.org/10.1021/jf072954f

Ribeiro DN, Nandula VK, Dayan FE et al (2015) Possible glyphosate tolerance mechanism in pitted morningglory (*Ipomoea lacunosa* L.). J Agric Food Chem 63:1689–1697. https://doi.org/10.1021/jf5055722

Roden LC, Ingle RA (2009) Lights, rhythms, infection: the role of light and the circadian clock in determining the outcome of plant–pathogen interactions. Plant Cell 21:2546–2552. https://doi.org/10.1105/tpc.109.069922

Rojano-Delgado AM, Cruz-Hipolito H, De Prado R et al (2012) Limited uptake, translocation and enhanced metabolic degradation contribute to glyphosate tolerance in *Mucuna pruriens* var. *utilis* plants. Phytochemistry 73:34–41. https://doi.org/10.1016/j.phytochem.2011.09.007

Salas RA, Dayan FE, Pan Z et al (2012) EPSPS gene amplification in glyphosate-resistant Italian ryegrass (*Lolium perenne* ssp. *multiflorum*) from Arkansas. Pest Manag Sci 68:1223–1230. https://doi.org/10.1002/ps.3342

Sammons RD, Gaines TA (2014) Glyphosate resistance: state of knowledge. Pest Manag Sci 70:1367–1377. https://doi.org/10.1002/ps.3743

Sauka-Spengler T, Barembaum M (2008) Chapter 12 Gain- and loss-of-function approaches in the chick embryo. In: Methods in cell biology. Academic Press, pp 237–256

Schönbrunn E, Eschenburg S, Shuttleworth WA et al (2001) Interaction of the herbicide glyphosate with its target enzyme 5-enolpyruvylshikimate 3-phosphate synthase in atomic detail. Proc Natl Acad Sci 98:1376–1380. https://doi.org/10.1073/pnas.98.4.1376

Shah MA, Callaway RM, Shah T et al (2014) *Conyza canadensis* suppresses plant diversity in its nonnative ranges but not at home: a transcontinental comparison. New Phytol 202:1286–1296. https://doi.org/10.1111/nph.12733

Shaner DL (2009) Role of translocation as a mechanism of resistance to glyphosate. Weed Sci 57:118–123

Simard M-J, Légère A, Warwick SI (2006) Transgenic *Brassica napus* fields and *Brassica rapa* weeds in Quebec: sympatry and weed-crop in situ hybridization. Can J Bot 84:1842–1851. https://doi.org/10.1139/b06-135

Singh S, Singh V, Lawton-Rauh A et al (2018) *EPSPS* gene amplification primarily confers glyphosate resistance among Arkansas Palmer amaranth (*Amaranthus palmeri*) populations. Weed Sci 66:293–300. https://doi.org/10.1017/wsc.2017.83

Soltani N, Oliveira MC, Alves GS et al (2020) Off-target movement assessment of dicamba in North America. Weed Technol 34:318–330. https://doi.org/10.1017/wet.2020.17

Sosnoskie LM, Culpepper AS (2014) Glyphosate-resistant Palmer amaranth (*Amaranthus Palmeri*) increases herbicide use, tillage, and hand-weeding in georgia cotton. Weed Sci 62:393–402. https://doi.org/10.1614/WS-D-13-00077.1

Sprankle P, Meggitt WF, Penner D (1975) Absorption, action, and translocation of glyphosate. Weed Sci 23:235–240

Steinrücken HC, Amrhein N (1980) The herbicide glyphosate is a potent inhibitor of 5-enolpyruvylshikimic acid-3-phosphate synthase. Biochem Biophys Res Commun 94:1207–1212. https://doi.org/10.1016/0006-291X(80)90547-1

Strauss SY, Rudgers JA, Lau JA, Irwin RE (2002) Direct and ecological costs of resistance to herbivory. Trends Ecol Evol 17:278–285

Takano HK, Mendes RR, Scoz LB et al (2019) Proline-106 *EPSPS* mutation imparting glyphosate resistance in goosegrass (*Eleusine indica*) Emerges in South America. Weed Sci 67:48–56. https://doi.org/10.1017/wsc.2018.71

Takano HK, Fernandes VN, Adegas FS et al (2020) A novel TIPT double mutation in EPSPS conferring glyphosate resistance in tetraploid *Bidens subalternans*. Pest Manag Sci 76:95–102. https://doi.org/10.1002/ps.5535

Tani E, Chachalis D, Travlos IS (2015) A glyphosate resistance mechanism in *Conyza canadensis* involves synchronization of *EPSPS* and ABC-transporter genes. Plant Mol Biol Report 33:1721–1730

Tian D, Traw MB, Chen JQ et al (2003) Fitness costs of R-gene-mediated resistance in *Arabidopsis thaliana*. Nature 423:74

Tranel PJ (2017) Herbicide-resistance mechanisms: gene amplification is not just for glyphosate. Pest Manag Sci 73:2225–2226. https://doi.org/10.1002/ps.4679

Tranel PJ, Wright TR (2002) Resistance of weeds to ALS-inhibiting herbicides: what have we learned? Weed Sci 50:700–712. https://doi.org/10.1614/0043-1745(2002)050[0700:RROWTA]2.0.CO;2

Valverde BE (2010) Glyphosate resistance in Latin America. In: Nandula VK (ed) Glyphosate resistance in crops and weeds. Wiley, Hoboken, pp 249–280

Van Eerd LL, Hoagland RE, Zablotowicz RM, Hall JC (2003) Pesticide metabolism in plants and microorganisms. Weed Sci 51:472–495. https://doi.org/10.1614/0043-1745(2003)051[0472:PMIPAM]2.0.CO;2

Van Etten M, Lee KM, Chang S-M, Baucom RS (2020) Parallel and nonparallel genomic responses contribute to herbicide resistance in *Ipomoea purpurea*, a common agricultural weed. PLOS Genet 16:e1008593. https://doi.org/10.1371/journal.pgen.1008593

Van Horn CR (2016) Investigating resistance and the rapid response to glyphosate in giant ragweed (*Ambrosia trifida* L.). Ph.D., Colorado State University

Van Horn CR, Moretti ML, Robertson RR et al (2018) Glyphosate resistance in *Ambrosia trifida*: Part 1. Novel rapid cell death response to glyphosate. Pest Manag Sci 74:1071–1078. https://doi.org/10.1002/ps.4567

VanGessel MJ (2001) Glyphosate-resistant horseweed from Delaware. Weed Sci 49:703–705

Vazquez-Garcia JG, Palma-Bautista C, Rojano-Delgado AM et al (2020) The first case of glyphosate resistance in Johnsongrass (*Sorghum halepense* (l.) pers.) in Europe. Plants 9:313. https://doi.org/10.3390/plants9030313

Vila-Aiub MM (2019) Fitness of herbicide-resistant weeds: current knowledge and implications for management. Plants 8:469

Vila-Aiub MM, Balbi MC, Distéfano AJ et al (2012) Glyphosate resistance in perennial *Sorghum halepense* (Johnsongrass), endowed by reduced glyphosate translocation and leaf uptake. Pest Manag Sci 68:430–436

Vila-Aiub MM, Gundel PE, Yu Q, Powles SB (2013) Glyphosate resistance in *Sorghum halepense* and *Lolium rigidum* is reduced at suboptimal growing temperatures. Pest Manag Sci 69:228–232

Vila-Aiub MM, Goh SS, Gaines TA et al (2014) No fitness cost of glyphosate resistance endowed by massive *EPSPS* gene amplification in *Amaranthus palmeri*. Planta 239:793–801. https://doi.org/10.1007/s00425-013-2022-x

Vila-Aiub MM, Yu Q, Powles SB (2019) Do plants pay a fitness cost to be resistant to glyphosate? New Phytol. https://doi.org/10.1111/nph.15733

Wakelin AM, Lorraine-Colwill DF, Preston C (2004) Glyphosate resistance in four different populations of *Lolium rigidum* is associated with reduced translocation of glyphosate to meristematic zones. Weed Res 44:453–459

Warwick SI, Simard M-J, Légère A et al (2003) Hybridization between transgenic *Brassica napus* L. and its wild relatives: *Brassica rapa* L., *Raphanus raphanistrum* L., *Sinapis arvensis* L., and *Erucastrum gallicum* (Willd.) O.E. Schulz. Theor Appl Genet 107:528–539. https://doi.org/10.1007/s00122-003-1278-0

Weaver SE (2001) The biology of Canadian weeds. 115. *Conyza canadensis*. Can J Plant Sci 81:867–875. https://doi.org/10.4141/P00-196

Weller SC, Owen MDK, Johnson WG (2010) Managing glyphosate-resistant weeds and population shifts in midwestern U.S. ropping systems. In: Nandula VK (ed) Glyphosate resistance in crops and weeds. Wiley, Hoboken, pp 213–232

Wiersma AT, Gaines TA, Preston C et al (2015) Gene amplification of 5-enol-pyruvylshikimate-3-phosphate synthase in glyphosate-resistant *Kochia scoparia*. Planta 241:463–474. https://doi.org/10.1007/s00425-014-2197-9

Wu C, Davis AS, Tranel PJ (2017) Limited fitness costs of herbicide-resistance traits in *Amaranthus tuberculatus* facilitate resistance evolution. Pest Manag Sci:9

Yanniccari M, Gómez-Lobato ME, Istilart C et al (2017) Mechanism of resistance to glyphosate in *Lolium perenne* from argentina. Front Ecol Evol 5. https://doi.org/10.3389/fevo.2017.00123

Young BG (2006) Changes in herbicide use patterns and production practices resulting from glyphosate-resistant crops. Weed Technol 20:301–307. https://doi.org/10.1614/WT-04-189.1

Yu Q, Powles S (2014) Metabolism-based herbicide resistance and cross-resistance in crop weeds: a threat to herbicide sustainability and global crop production. Plant Physiol 166:1106–1118. https://doi.org/10.1104/pp.114.242750

Yu Q, Cairns A, Powles S (2007) Glyphosate, paraquat and ACCase multiple herbicide resistance evolved in a *Lolium rigidum* biotype. Planta 225:499–513. https://doi.org/10.1007/s00425-006-0364-3

Yu Q, Jalaludin A, Han H et al (2015) Evolution of a double amino acid substitution in the 5-enolpyruvylshikimate-3-phosphate synthase in *Eleusine indica* conferring high-level glyphosate resistance. Plant Physiol 167:1440–1447. https://doi.org/10.1104/pp.15.00146

Yuan JS, Tranel PJ, Stewart CN (2007) Non-target-site herbicide resistance: a family business. Trends Plant Sci 12:6–13. https://doi.org/10.1016/j.tplants.2006.11.001

Zelaya IA, Owen MDK, VanGessel MJ (2004) Inheritance of evolved glyphosate resistance in *Conyza canadensis* (L.) Cronq. Theor Appl Genet 110:58–70

Zhang C, Li F, HE T et al (2015) Investigating the mechanisms of glyphosate resistance in goosegrass (*Eleusine indica*) population from South China. J Integr Agric 14:909–918

Ecotoxicology of Glyphosate, Its Formulants, and Environmental Degradation Products

Jose Luis Rodríguez-Gil, Ryan S. Prosser, Stephen O. Duke, and Keith. R. Solomon

Contents

Abstract The chemical and biological properties of glyphosate are key to understanding its fate in the environment and potential risks to non-target organisms. Glyphosate is polar and water soluble and therefore does not bioaccumulate, biomagnify, or accumulate to high levels in the environment. It sorbs strongly to

---

Electronic Supplementary Material The online version of this chapter (https://doi.org/10.1007/398_2020_56) contains supplementary material, which is available to authorized users.

J. L. Rodríguez-Gil (✉)
IISD – Experimental Lakes Area, Winnipeg, MB, Canada

Department of Environment and Geography, University of Manitoba, Winnipeg, MB, Canada
e-mail: jlrodriguez@iisd-ela.org

R. S. Prosser
School of Environmental Sciences, University of Guelph, Guelph, ON, Canada
e-mail: prosserr@uoguelph.ca

J. B. Knaak (ed.), *Reviews of Environmental Contamination and Toxicology Volume 255*,
Reviews of Environmental Contamination and Toxicology 255,
https://doi.org/10.1007/398_2020_56

particles in soil and sediments and this reduces bioavailability so that exposures to non-target organisms in the environment are acute and decrease with half-lives in the order of hours to a few days. The target site for glyphosate is not known to be expressed in animals, which reduces the probability of toxicity and small risks. Technical glyphosate (acid or salts) is of low to moderate toxicity; however, when mixed with some formulants such as polyoxyethylene amines (POEAs), toxicity to aquatic animals increases about 15-fold on average. However, glyphosate and the formulants have different fates in the environment and they do not necessarily co-occur. Therefore, toxicity tests on formulated products in scenarios where they would not be used are unrealistic and of limited use for assessment of risk. Concentrations of glyphosate in surface water are generally low with minimal risk to aquatic organisms, including plants. Toxicity and risks to non-target terrestrial organisms other than plants treated directly are low and risks to terrestrial invertebrates and microbial processes in soil are very small. Formulations containing POEAs are not labeled for use over water but, because POEA rapidly partitions into sediment, risks to aquatic organisms from accidental over-sprays are reduced in shallow water bodies. We conclude that use of formulations of glyphosate under good agricultural practices presents a *de minimis* risk of direct and indirect adverse effects in non-target organisms.

Keywords Adjuvants · AMPA · Ecotoxicology · Environmental fate · Glyphosate · POEA

Abbreviations

a.e.	Acid equivalents; use to normalize the reporting of concentrations of glyphosate where different salts or formulations have been used in different studies
AMF	Arbuscular mycorrhizal fungi
AMPA	Aminomethylphosphonic acid, a degradate of glyphosate
ANEOs	Alkylamine ethoxylates
ANVISA	Agência Nacional de Vigillância Sanitâria of Brazil
APVMA	Australian Pesticides and Veterinary Medicines Authority
ASE	Accelerated Solvent Extraction
C	Carbon

S. O. Duke
National Center for Natural Products Research, School of Pharmacy, University of Mississippi, Oxford, MS, USA
e-mail: sduke@olemiss.edu

K. R. Solomon
Centre for Toxicology, School of Environmental Sciences, University of Guelph, Guelph, ON, Canada
e-mail: ksolomon@uoguelph.ca

CAS	Chemical Abstracts Service
CLPP	Community-Level Physiological Profiling, a technique for characterizing function of mixed microbial communities
DGGE	Denaturing Gradient Gel Electrophoresis, a molecular technique for characterizing the composition, diversity, and function of microbial communities in the environment
DT50	Time for 50% of a substance in a matrix to dissipate and/or degrade
EFSA	European Food Safety Agency
EO	Ethoxylate groups as referring to ethoxylate-containing surfactants
Epilimnion	The upper layer of water in a temperature-stratified lake)
EPSPS	5-Enolpyruvylshikimate-3-phosphate synthase
FAO	Food and Agriculture Agency of the United Nations
FAR	Field application rate, applied to crops for pest management
FMOC Cl	Fluorenylmethyloxycarbonyl chloride, a reagent used to derivatize glyphosate during analysis
FSCJ	Food Safety Commission of Japan
GBH	Glyphosate-based herbicides (as used in weed management)
GLP	Good Laboratory Practice
GR	Glyphosate-resistant (crops)
JMPR	Joint FAO/WHO Meeting on Pesticide Residues
Kd	Soil-water partition coefficient
K_{OC}	Soil-water partition coefficient, normalized for content of organic carbon
K_{OM}	Soil-water partition coefficient, normalized for content of organic matter
LC50	Concentration that is lethal to 50% of the organisms exposed through the matrix
LD50	Dose that is lethal to 50% of the dosed organisms
LAI	Leaf area index, the one-sided area of green leaves per unit area of ground surface
LOD	Level of detection
LOQ	Level of quantification
LR50	Application rate lethal to 50% of the tested organisms
N	Nitrogen
NOEC	No observed effect concentration
NPDES	National Pollution Discharge Elimination System Permit
NT	No-tillage
PICT	Pollution-Induced Community Tolerance, a technique to characterize the induction of tolerance to pollutants in communities of organisms
PMRA	Pest Management Regulatory Agency of Canada
POEA	Polyoxyethylene amine, also Polyethoxylated tallow amine
POE-T	Polyoxyethylene (15) tallowamine, a formulant
QA/QC	Quality Assurance and Quality Control
SDS	Safety Data Sheet (AKA Material Safety Data Sheet)
USEPA	United States Environmental Protection Agency

1 Introduction

This review and risk assessment is focused on the environmental fate and effects of glyphosate and some of the formulants used in commercial products in non-target organisms. Potential effects of glyphosate in humans have been extensively reviewed and characterized by regulatory agencies and others and are not discussed in detail in this chapter. However, a brief summary of conclusions is provided for context. The most recent of these is the draft profile on glyphosate from the Agency for Toxic Substances and Disease Registry (ATSDR 2019). Since the classification of glyphosate by the International Agency for Research on Cancer as a "probably carcinogenic to humans (Group 2A)" (IARC 2015), the major focus of regulators has been on the carcinogenicity of glyphosate. In addition to regulatory reviews, several papers relevant to this have been published in the open literature and, with respect to carcinogenicity, most of these came to a different conclusion from IARC. This chronology is illustrated in Table 1. One of these regulatory reviews included assessment of risks to domesticated animals (EFSA 2018). This and studies on the exposures of humans to glyphosate were excluded from this chapter but are discussed in another review (Solomon 2020).

Table 1 Chronology of regulatory reviews and key publications on the carcinogenicity of glyphosate

Year	Agency/Institution	Classification/Conclusion	Reference
2015	IARC	Glyphosate is "probably carcinogenic to humans (Group 2A)"	(IARC 2015)
2017	EFSA	No hazard classification for carcinogenicity is warranted	(EFSA 2015, 2018)
2017	APVMA	Exposure does not pose a carcinogenic risk to humans	(APVMA 2017)
2017	ECHA	No hazard classification for carcinogenicity is warranted	ECHA
2016	FAO and WHO	Unlikely to pose a carcinogenic risk to humans from dietary exposure	(JMPR 2016)
2015	PMRA/Health Canada	Unlikely to pose a human cancer risk	(HC 2019; PMRA 2015b)
2016	Food Safety Commission of Japan	Glyphosate had no neurotoxicity, carcinogenicity, reproductive toxicity, teratogenicity, and genotoxicity.	FSCJ, 2016 #165
2016	New Zealand Environmental Protection Authority	Unlikely to be genotoxic or carcinogenic to humans	(NZ EPA 2016)
2017	USEPA	Not likely to be carcinogenic to humans	(USEPA 2017, 2019)

(continued)

Table 1 (continued)

Year	Agency/Institution	Classification/Conclusion	Reference
Publications in the literature			
2016	Intertek Panel	"The data do not support IARC's conclusion that glyphosate is a 'probable human carcinogen' and, consistent with previous regulatory assessments, further concluded that glyphosate is unlikely to pose a carcinogenic risk to humans"	(Williams et al. 2016, 2018) and related papers in the same issue of the journal.
2017	EFSA	In explaining the conclusions of the EFSA assessment "the EU assessment did not identify a carcinogenicity hazard"	Tarazona, 2017 #71
2020	Member of IARC Panel, Emory University, Atlanta, GA, USA	"The analyses conducted for this review clearly support the IARC's conclusion that there is sufficient evidence to say that glyphosate causes cancer in experimental animals"	(Portier 2020)
2020	Queen Mary University of London, UK	"Overall, there is no evidence in the animal studies to support the IARC conclusion that glyphosate is a probable human carcinogen"	(Berry 2020)
2020	Kenny Crump, Ruston, LA, USA	Of the epidemiology studies cited by IARC, "one could reasonably conclude that at least four of the case–control studies of glyphosate and NHL are contaminated by statistical bias, and consequently are not suitable for reaching conclusions about the potential ability of glyphosate to cause NHL"	(Crump 2020)
2020	Kenny Crump, Ruston, LA, USA	On the animal carcinogenicity studies evaluated by IARC "The present analysis provides new information on the potential carcinogenicity of glyphosate by being the first to provide results from statistical tests with correct false-positive rates. These tests found no strong or convincing evidence that glyphosate is an animal carcinogen"	(Crump et al. 2020)

The herbicide glyphosate, (CAS# 1071-83-6 [N-(phosphonomethyl)glycine] Fig. 1), has been on the market since the mid-1970s and is one of the most widely used pesticides in the world (Duke 2018, 2020). The widespread agricultural and

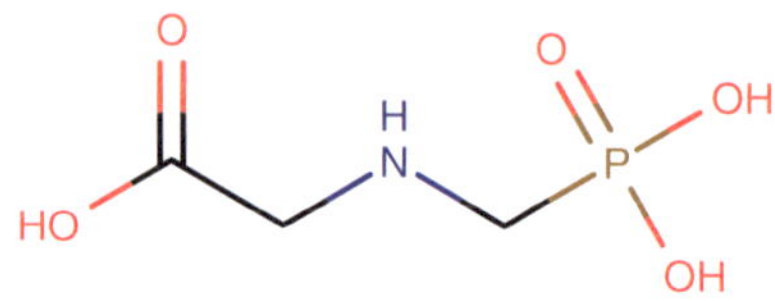

Fig. 1 Structure of glyphosate

non-agricultural uses of glyphosate have attracted considerable attention among the public and scientists (Solomon 2020).

In addition to the potential effects in humans, several reviews on the environmental fate and ecotoxicology of glyphosate and glyphosate-based herbicides (GBHs), or formulations, have been published in the literature. An early assessment of the ecotoxicology of Roundup® formulation of glyphosate was published in 2000 (Giesy et al. 2000) and, since that time, there have been several additional reviews and assessments. Solomon et al. (2007) reviewed the ecotoxicological risk associated with the use of glyphosate for the eradication of coca. Other reviews include (Annett et al. 2014; Blake and Pallett 2018; Borggaard and Gimsing 2008; Cedergreen and Streibig 2005; Matozzo et al. 2020; Pérez et al. 2011; Richmond 2018; Rott et al. 2018; Székács and Darvas 2018; Thiour-Mauprivez et al. 2019; Van Bruggen et al. 2018; Villamar-Ayala et al. 2019; Wagner et al. 2013). Rather than repeat what has been written in these reviews, the purpose of this chapter is focus on key properties of glyphosate and its formulations with respect to exposures, toxicity, and risks to non-target animals. Phytotoxicity to non-target plants is not included in this review because bioactivity is predicated on how and where glyphosate is used and applied, which can be managed by good agricultural practices (GAPs), such as the use of buffer zones. Furthermore, risks for non-target plant toxicity would also apply to other non-selective and some selective herbicides other than glyphosate.

Characterizing exposures is a critical component of risk assessment and concentrations in environmental samples are discussed in several sections of this review. There are a number of methods for analyzing glyphosate in environmental samples, but details of the analytical methods were not included in our review. However, Melo et al. (2018) have published a mini review of instrumental methods and recently published methods can be found in the following references (Byer et al. 2008; Carretta et al. 2019; Fritz-Wallace et al. 2020; Marek and Koskinen 2014; Okada et al. 2019; Pinto et al. 2018).

While not a systematic review per se, principles of this approach have been applied throughout this chapter and the focus is directed to the key properties of glyphosate and their influence on fate and toxicity. Whenever possible, the quality of the science presented in the studies cited, as well as their applicability and relevance, have been assessed following objective metrics such as those presented in (Moermond et al. 2017; Solomon and Stephenson 2017; Van Der Kraak et al. 2014). Direct statements regarding the quality and relevance of several studies are provided throughout this review.

2 Problem Formulation

2.1 Physical, Chemical, and Biological Properties

Glyphosate acid (Fig. 1) is fairly soluble in water (11.6 g/L, BCPC 2012) but the salts, such as the commonly used technical active ingredient isopropylamine (IPA) salt of glyphosate, are more soluble (1,050 g/L, BCPC 2012). Glyphosate acid has four pK_as, 10.1, 5.4, 2.2, and 2.0 (Stalikas and Konidari 2001) (Fig. 2) and, at typical pHs in agricultural and forestry soils (5–9), is usually dissociated into the ionic form (Fig. 2), which sorbs strongly via ionic forces to mineral particles (aluminum, iron oxides, and clay silicates, Borggaard and Gimsing 2008) in soils and/or sediments. In general, sorption to soil particles is directly dependent on the surface area of the minerals in the soil and the pH, with less glyphosate sorbed at low pH than high (Borggaard and Gimsing 2008). Once immobilized by sorption to soil particles, glyphosate is poorly absorbed by roots of plants and thus is not herbicidally active via soil exposure (Borggaard and Gimsing 2008). Therefore, commercial glyphosate-based herbicides (GBHs) must be applied directly to leaves of plants to be effective as an herbicide. Glyphosate is systemic and, after penetration into the leaf, it is translocated to other parts of the plant. This property allows GBHs to be used pre-plant for field preparation or for early post-plant weed control before the seedlings emerge in a large range of crops (Duke 2020). This sorption to soils and sediments results in rapid dissipation of glyphosate in natural waters such that exposures are usually short enough that much larger concentrations are required to elicit toxicity than in the absence of sediments (also see Sect. 2.2. below). For this reason, toxicity tests for aquatic organisms conducted in glass containers in the laboratory are poor predictors of responses in the field when sediments are a natural component of pools, ponds, and streams (Bernal et al. 2009). Glyphosate acid has a low vapor pressure (1.31×10^{-2} mPa, BCPC 2012) and a low Henry's constant ($< 2.1 \times 10^{-7}$ Pa m^3 mol^{-1}, BCPC 2012) and exposures in the environment via the atmosphere are very unlikely, except for drift of spray droplets or sorbed to particles of airborne dust. Measured environmental concentrations and toxicity values in this

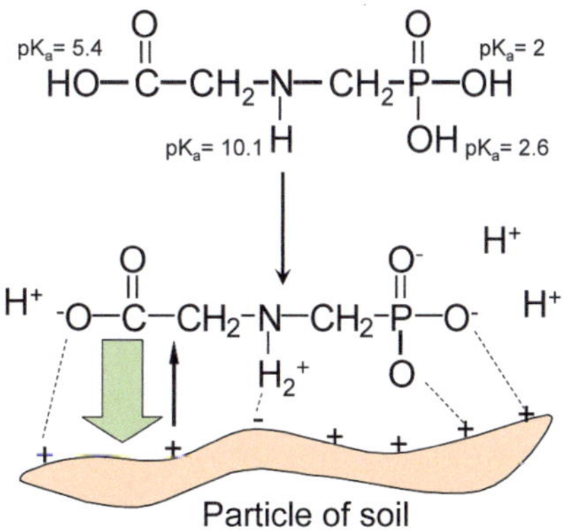

Fig. 2 Sites on the glyphosate molecule where sorption can potentially occur. Which sites are relevant is dependent on the pH of the soil (pKa values from Stalikas and Konidari 2001)

review are all expressed in μg/L to allow direct comparisons for characterizing hazards and risks.

Residues of glyphosate have been detected in rainwater and air (most probably sorbed to dust particles) but concentrations are very small (maximum reported value in air was 1.04×10^{-3} μg/m^3 and those in rainwater from an area of high use of glyphosate ranged from 1.2 to 67 μg/L, reviewed in Solomon 2016, 2020). Glyphosate in air and rainwater present a *de minimis* risk to humans (Solomon 2020) and would be unlikely to present a risk to organisms in the environment when compared to other potential sources that are discussed below (Sect. 3.1).

Because of strong sorption to soil particles, glyphosate does not leach significantly in most soils; however, in clay-soils, transport of glyphosate through macropores to tile drains and from there to surface water has been reported (Borggaard and Gimsing 2008). In a field study at the Helmholtz Centre for Environmental Research-UFZ (Falkenberg, Germany), duplicate stainless steel lysimeters (1 m square by 1.25 m deep) were used to measure the leaching of glyphosate in sandy loam soil typical of the Elbe valley planted with maize (Gros et al. 2020). The length of the study was one hydrological year and a glyphosate-based herbicide (GBH) containing $^{13}C_2$-^{15}N-glyphosate to allow tracing was applied to the surface of the soil following good agricultural practices at a nominal rate of 3.6 kg a.e./ha/year (the maximum rate allowed in Germany).[1] A potassium bromide tracer (40 kg KBr/ha) was applied to characterize flow of water in the lysimeters. Soils in the lysimeter were sampled with a soil-corer of unreported diameter to 5 cm depth before, directly after, 165, and 360 days after application. Five replicate samples were taken from each lysimeter. Deeper cores (0–30 and 30–60 cm) were taken only at the end of the study. Leachates were collected over the period of the study. The study year had greater than usual rainfall and the total volume of leachate was 203 and 215 L in the two lysimeters. Concentrations of glyphosate and AMPA in the soil and leachates were measured with HPLC-ESI-MS-MS after derivatization with *fluorenylmethyloxycarbonyl* chloride (FMOC Cl) with a level of detection (LOD) of 0.1 μg/L in the leachate. Isotopic ratios for ^{15}N/^{14}N and ^{13}C/^{12}C in soil, plants (roots, shoots, and cobs), and lyophilized leachate samples were measured with an elemental analyzer operated at the Institute for Nutritional Sciences, University of Gießen, Germany. During the study, the amount of glyphosate in the soil decreased to 3% of the initial amount. Leaching of KBr as a marker of movement of water in the soil column was detected 70 days into the study. Glyphosate and AMPA were not detected in the leachate indicating lack of leaching. The enrichment of ^{15}N and ^{13}C in the leachate over the period of the study indicated degradation in the soil column to more mobile degradation products other than AMPA. Enrichment of ^{15}N was observed in roots, shoots, and cobs of the maize but enrichment of ^{13}C was only

[1]Because technical glyphosate is used in an ion pair with one of several cations (e.g. potassium, isopropylamine, trimesium, etc.) concentrations of technical and formulated products should be expressed as glyphosate acid, or acid equivalents (a.e.). This normalization allows concentrations in the environment and GBHs to be easily compared.

observed in the roots. Statistical analysis of isotopes of carbon and nitrogen in plant matter was pseudoreplicated within the replicate test lysimeters. Overall, the evidence suggests that, in the absence of macropores and in agricultural soils, glyphosate does not leach to depths greater than 125 cm.

Glyphosate might be carried in runoff or surface flows of water. This is particularly relevant where glyphosate in applied for control of weeds in hard surfaces (e.g., paved areas) with fewer sites for sorption. Median concentrations of glyphosate measured in surface waters downstream from urban areas were 6 to 4 μg/L as compared to 1 μg/L upstream (Kolpin et al. 2006). Frequency of detection was ca. 80% downstream compared to 38% upstream. Similar observations have been reported from measurements of glyphosate in streams and wetlands in the State of Victoria in Australia (Okada et al. 2020). Analyses were conducted using a direct injection method on a LC-MS-MS. Isotopically labeled glyphosate (1,2-^{13}C,^{15}N) was used as an internal standard and the LOD was 0.25 μg/L. In this study, mean concentrations in rural streams were $\leq$ 0.3 μg/L with a frequency of detection of 4%: however, mean concentrations in urban streams and wetlands were 1.1 μg/L with a frequency of detection of 79 and 77%, respectively.

Glyphosate from treated areas may also be carried by surface runoff, and transported to surface water, sometimes in high concentrations (Coupe et al. 2012). Concentrations of glyphosate in rainfall-driven runoff of surface water were measured in three watersheds in the USA and one in France. Concentrations were measured after filtration through a 0.45 μm filter and therefore represent dissolved glyphosate. Analysis followed USGS protocols (USGS 2010) and the LOD was 0.02 μg/L. Between 2006 and 2008, median concentrations in Bogue Phalia, MS, USA ranged from 0.82 to 1.2 μg/L; between 2007 and 2008 in New Providence, IA, USA, medians ranged from 0.07 to 0.87 μg/L and, in Sugar Creek and Leary Weber Ditch, IN, USA in 2004 the medians were 0.32 to 1.1 μg/L. Values from an overland flow site near Sugar Creek and Leary Weber Ditch were much greater, with medians of 34 to 380 μg/L, measured in late May–early June, 2004 during a storm and shortly after application to crops. These high values were storm-driven, localized, and were not reflected in the nearby streams. The median concentration measured in Rouffach, France between 2003 and 2006 was 4.7 μg/L (LOD = 0.1 μg/L Coupe et al. 2012).

Since runoff water contains sediments, only measuring glyphosate in solution may under-represent total loads carried in runoff. In a study in the La Plata region of Argentina by Mac Loughlin et al. (2020), samples of runoff were collected and filtered through a 0.45 μm filter and glyphosate in the particulates and filtrate derivatized with FMOC-Cl then concentrations were measured by LC-MS with a recovery of 90% and a LOD of 0.03 μg/L in water and 0.06 μg/kg in particulates. The median concentration (in solution) in water was 3.1 (range, 0.2–17) μg glyphosate a.e./L while the median concentration sorbed to particulates was 3,735 (range, 245–35,620) μg a.e./kg. Clearly a significant proportion of the mass of glyphosate in surface runoff is sorbed to particulates and is likely less biologically available.

2.2 *Fate of Glyphosate in the Environment*

2.2.1 Fate of Glyphosate in Plants and Animals

When used in management of vegetation, GBHs are normally applied as a spray that is directed to the leaves of the target plants. Not all the spray is deposited on the leaves and some reaches the surface of the soil, either directly or as a result of wash-off if rainfall occurs shortly after spraying. Some residues of glyphosate may enter soil through exudation from roots of treated plants or from litter from sprayed plants. A study by Laitenen et al. (2007) reported that as much as 12% of the glyphosate applied to *Chenopodium quinoa* Willd plants was exuded by roots into the soil 6 weeks after spraying. Glyphosate might also leach from litter of sprayed plants; Mamy et al. (2016) reported that glyphosate taken up by glyphosate-resistant (GR) and non-GR canola was lost very slowly from plant residues in the field, increasing the overall persistence of glyphosate in sprayed fields. This effect was more pronounced in GR canola than non-GR canola because these plants were not killed and took up more glyphosate. Thus, a significant fraction of the glyphosate entering the plant can enter the soil if it is not metabolized to aminomethylphosphonic acid (AMPA, CAS# 1066-51-9) by the plant (Fig. 3). The degradation of glyphosate via sarcosine (Fig. 3) has been reported in soil microorganisms but is not a major route of dissipation (Kishore and Jacob 1987).

ATSDR (2019) has summarized the toxicokinetics of glyphosate in mammals. Some glyphosate is absorbed from the gastrointestinal (GI) tract but most (two thirds) is excreted in the feces. Respiratory absorption is assumed to be high but absorption through skin is low (< 1%, Bo Nielsen et al. 2009). Once in the blood, is distributed throughout the body but does not accumulate in any organs or in lipids. It is excreted in the urine with a relatively short half-life ($t\frac{1}{2}$) of 8–15 h. In the rat, glyphosate acid dosed by gavage in corn oil was poorly absorbed in the intestine (≈23%, Anadón et al. 2009) but, once absorbed, it was rapidly excreted with a $t\frac{1}{2}$ of 14.4 h with no evidence of bioaccumulation or biomagnification; however, some glyphosate (6.5%) was metabolized to AMPA.

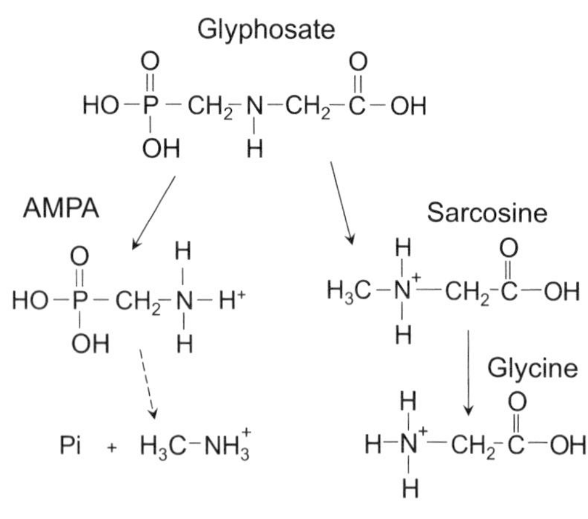

Fig. 3 In the environment, glyphosate degrades via two pathways. The sarcosine pathway leads to glycine, which is used my microorganisms. The terminal residues of the pathway via aminomethylphosphonic acid (AMPA) are inorganic phosphorus and nitrogenous compounds that are used as nutrients by microbiota

Based on a study in human volunteers dosed by mouth with glyphosate mixed with food, the half-life ($t½$) by urinary excretion was 8–10 h (Zoller et al. 2020). Only 1% of the dose of glyphosate was excreted in the urine suggesting that uptake from the gut in humans is less than in other animals and/or that the food matrix slowed or inhibited uptake. AMPA was detected in the urine and represented 0.2% of the dose of glyphosate suggesting only minimal metabolism of glyphosate to AMPA in humans. Whether this metabolism in rats and humans is mediated by organs in the body or microbiota in the gut is uncertain.

2.2.2 Fate of Glyphosate in Soil

When applied for weed management in agriculture and forestry, the amount of glyphosate reaching the surface of the soil depends on the timing of the application in relation to the development of the weeds (or the crop for genetically engineered glyphosate-resistant [GR] crops). The amount of pesticide reaching the soil is the proportion of spray that is not intercepted by the plants. Spray intercepted by and retained on the plant is related to the shape of the plant (its architecture), the characteristics of the surface of the leaf, the type of nozzle, and the contact angle of the spray. Spray not intercepted by the plants is inversely proportional to the proportion of soil covered by plants, and the leaf area index (LAI). In pre-plant and early post-plant applications of GBHs, the amount of glyphosate reaching the surface of the soil would most depend on the weed pressure and the LAI, so deposition would be highly variable in space and time and not easily predictable. In GR crops, or for the use of glyphosate as a ripening agent (Xu et al. 2019), the deposition of glyphosate on the soil under the plant canopy will be more dependent on the LAI, which, in turn, will be influenced by the structure of the plants and the spacing between the plants. For early application in the development of GR crops, more glyphosate will likely reach the surface of the soil than later applications, but for ripening or other preharvest applications deposition on soil will be small. Papers reporting the deposition of glyphosate on the canopy in comparison to soil were not found in a search of the literature.

The strong sorption of glyphosate to particles of soil and sediments (Fig. 2) rapidly reduces biological availability for herbicidal effects, non-target toxicity, and for biodegradation, which primarily takes place in the pore water (Borggaard and Gimsing 2008). Because of sorption in most soils, much of the glyphosate applied to crops is confined to the upper layers of the soil. Okada et al. (2016) reported that 46–58% of the glyphosate was found in the upper 5 cm of soil and 0–4.3% was found to have leached beyond 10 cm in the three soils from Argentina. The soils were from three experimental stations in: (1) Córdoba Province (coarse-silty, mixed, thermic Entic Haplustoll of the Oncativo series) (2) Entre Ríos Province (a fine, mixed, thermic Acuic Argiudoll from the Tezanos Pinto series), and (3) Buenos Aires Province (a fine, thermic, illitic, Typic Argiduoll from the Pergamino series) Two soils, each from a different cropping regimen were collected at each site. Leaching was conducted in soil columns treated with a GBH and leached with 4.16 mL leachate/h for 7 days. Analysis of the leachate was by derivatization with 9-fluorenylmethylchloroformate (FMOC-Cl). After partitioning with dichloromethane, the aqueous phase was analyzed by LC-MS-MS. The LOD and LOQ in leachate were 0.1 and 0.5 μg/L, respectively. Adsorption was increased with increasing clay content and cation exchange capacity and decreased with increasing pH and phosphorus content of soil.

Other than the type of soil, some agricultural practices can affect leaching of glyphosate in soils. For example, phosphate fertilizers compete for glyphosate-sorption sites and can therefore displace glyphosate and mobilize it to some extent in soil, especially in sandy soil with few sorption sites (Bott et al. 2011). In laboratory studies, five soils (Arenosol, Acrisol, Ferralsol, Luvisol, and Regosol) were treated with a GBH (Roundup UltraMax®) at rates ranging from 720 to 1,440 g glyphosate a.e./ha. After an incubation period, phosphate was added to the soils at rates of 0, 20, 40, 80, or 240 mg P kg/soil as $Ca(H_2PO_4)_2$. Six soybean seeds were then planted in the soils and plants sampled at 10 and 25 days after sowing and growth parameters and shikimate were measured in the roots. Effects and symptoms of glyphosate were greatest in Arenosol and then declined in order of Acrisol ≈ Ferralsol > Luvisol > Regosol. Symptoms increased with increasing amounts of phosphate added to the soil, confirming the effects of addition of phosphate.

Rate of degradation of glyphosate in field soils is variable, depending on many factors, such as the level of aeration, temperature, and pH. For example, Nguyen et al. (2018) reported that, over 32 d, mineralization of ^{14}C-glyphosate (labeled on the phosphonomethyl group) to $^{14}CO_2$ in 21 agricultural soils from Germany and Slovenia with different properties ranged from 7 to 70%. Cumulative mineralization correlated with soil exchangeable acidity (H^+ and Al^{3+}), exchangeable Ca^{3+} ions, and aluminum lactate-extractable K. In an analysis of the literature, Blake and Pallett (2018) reported the average *t*½ of glyphosate in soil as 30 d, (ranging from 5.7 to 40.9 d). Others have reported the *t*½ of glyphosate in soil to range from 1.0 to 68 days (EFSA 2015) but half-lives can be relatively long with DT50s up to 173 d (BCPC 2012). These half-lives may appear long but the strong sorption to soil particles results in very short *t*½ of bioavailability for uptake by soil organisms.

In plants, glyphosate is slowly degraded with AMPA (Fig. 3), being the major metabolite (BCPC 2012). AMPA is also the major metabolite in soil and water (Bento et al. 2016; Wang et al. 2016). A comprehensive study of 317 agricultural soils from Europe found glyphosate (LOD = 0.02 mg/kg) in 21% and AMPA (LOD = 3 mg/kg) in 42% of soils (Silva et al. 2018). Analysis was by LC-MS-MS with isotopically labeled internal standards, glyphosate (1,2-^{13}C, ^{15}N) and AMPA (^{13}C, ^{15}N) to normalize for matrix effects. Extraction from soil was with 0.6 M KOH and derivatized with FMOC-Cl. Other details of the method are provided in Bento et al. (2016). Concentrations only rarely exceeded 500 μg/kg of soil, and concentrations of AMPA were almost always higher than those of glyphosate.

Although not experimentally studied in the long term, it appears that glyphosate does not accumulate in soils to high levels. Results of a model predicted that, if used repeatedly, glyphosate would build up to a plateau in soil in about 10 years with a peak value of 3.1×10^3 μg/kg for annual crops tilled to 20 cm and 6.2 mg/kg for perennial crops tilled to 5 cm (EFSA 2015). Repeated use of pesticides can lead to selection for soil microbiota that degrade the chemicals, leading to shorter half-lives in that soil (e.g., Yale et al. 2017). This phenomenon is termed accelerated degradation, but it has not yet been documented for glyphosate in the field. Accelerated degradation might be expected in GR crops where glyphosate has been used for decades; however, the studies have not been conducted to establish whether this has occurred in the field or not. A laboratory study showed that repeated application of glyphosate over a short time can shift soil microbial community composition, but there was no evidence of accelerated degradation (Lancaster et al. 2010). Two studies have shown that glyphosate can increase the rate of degradation of other herbicides in soil (atrazine in a Brazilian Oxisol soil and fluometuron in Weswood silty clay loam, respectively) (Bonfleur et al. 2010; Lancaster et al. 2008). It is possible that accelerated degradation does not occur because of minimal bioavailability of glyphosate in pore water and that it is the phosphorus from the degradation of glyphosate that stimulates growth of and metabolism by microbiota in general. This is discussed further in Sect. 3.2.2.

2.2.3 Fate of Glyphosate in Water

Most GBHs are not registered for over-water uses; however, some can be used for this purpose (Solomon and Thompson 2003). In the USA, several formulations of glyphosate are registered for use over water for the control of emergent or floating plants, for example, the common reed (*Phragmites australis subsp. americanus*) and water hyacinth (*Eichhornia crassipes*) an invasive species endemic in South America. These formulations include Roundup Custom®, Rodeo®, Eraser AQ®, Aquamaster®, AquaNeat®, Refuge®, and AquaPro®. Some of these (e.g., Rodeo and Roundup Custom) only contain active ingredients and suitable adjuvants are added before application. Others (e.g., AquaPro) are formulated with adjuvants that present low risk to aquatic organisms. In the USA and Canada (and probably other countries), permits are required when proposing to apply GBHs overwater. This permitting process minimizes risks of misuse.

Under good application practice, spray drift of GBHs is not likely an important source of biologically significant residues in water; however, if followed by rain, spraying of weeds in hard surfaces might result in runoff (Kolpin et al. 2006). If heavy rainfall follows application of the GBH in agricultural fields, transport of glyphosate to surface waters is also possible (see discussion in Sect. 2.2.2). When applying GBHs by air in forestry, inadvertent deposition may occur in small forest pools and wetlands. However, analysis of unfiltered samples from replicated forest pond microcosms treated with a forestry-specific GBH (VisionMAX®) showed that glyphosate dissipated rapidly from initial mean aqueous concentrations of 3,100 and 600 μg/L to levels < 100 μg/L within four days (Edge et al. 2012). Analysis was by LC/MS-MS using a published method (Hao et al. 2011). Aqueous samples are directly injected with no sample concentration or derivatization steps. The internal standard was $^{13}C,^{15}N$-glyphosate and the instrument detection limits for glyphosate and AMPA were 1 and 2 μg a.e./L, respectively. The half-lives of glyphosate in another forestry cosm study were characterized after two applications of VisionMAX at nominal initial concentrations of 210 and 2.88×10^3 μg glyphosate a.e./L in each of 2 years, 2009 and 2010 (Edge et al. 2014). Analysis was as described above (Hao et al. 2011). Half-lives for the four applications at the lower concentration ranged from 1 to 1.9 days. For the higher concentration, *t*½ s for the four applications ranged from 0.75 to 1.31 days. Given that the cosms were shallow (15–30 cm) forest ponds, the very rapid dissipation from water was probably driven mostly by sorption to sediments.

Others have shown that glyphosate is degraded rapidly by microorganisms found in water and aquatic biofilms. In a study in Lake Greifensee (a stratified lake in Switzerland) concentrations of glyphosate and AMPA were observed to begin increasing in the epilimnion (the upper layer of water in the lake) in the early spring. Concentrations of glyphosate increased from 0.015 μg/L in March to 0.145 μg/L in July as result of inflow from rivers and streams from agricultural watersheds. Concentrations of AMPA also increased from 0.07 to 0.13 μg/L in the same time period. Analysis of both compounds was by derivatization with FMOC-Cl and then analysis with LC-MS-MS. Internal standards ($^{13}C_2{}^{15}N$-glyphosate and $^{13}C_2{}^{15}ND_2$-AMPA) were used to normalize for matrix effects. Recovery from water was 97–103% and the LOQ was 0.005 μg/L for both analytes (Poiger et al. 2017). In July and August, concentrations of glyphosate in the epilimnion decreased rapidly to < LOQ. This coincided with increased temperatures in the epilimnion, greater biomass of phytoplankton, and low concentrations of free phosphorus. Half-life for dissipation ranged from 2–4 days. Metagenomic analysis identified phytoplankton known to degrade glyphosate and AMPA. The authors suggested that these organisms were utilizing the phosphorus from the breakdown of glyphosate for growth. High rates of degradation were not observed in a laboratory study of the dissipation of glyphosate (as the IPA salt) and a GBH (Roundup Classic®; Monsanto Europe) in natural waters collected from the Danube river and Lake Balaton in Hungary, little dissipation was reported over a period of seven days (Klátyik et al. 2017). Initial treatment concentrations were in the range of 80–100 μg/L and analysis was by derivatization with FMOC-Cl and analysis with HPLC-Fluorescence Detection (FLD) with a LOD of 5 μg/L. Samples below the LOD were analyzed with LC-MS-MS with a LOD of 0.001 μg/L. Degradation in the presence of biofilms was also measured (see discussed below).

Microbiota in biofilms of surfaces in surface waters have been observed to degrade glyphosate, in the study on degradation of glyphosate in natural waters (Klátyik et al. 2017), biofilms from the River Danube and Lake Balaton. Biofilms were allowed to colonize and grow on glass plates in the field sites for 42 days and then transferred to 15-L aquaria in the laboratory. The water in the aquaria was totally replaced every 7 days, at which time glyphosate was added to the aquaria as IPA salt and GHB (Roundup Classic®) and degradation monitored using the analytical methods described above. There was no replication of treatments in the aquaria with biofilms although the authors stated that the five biofilm plates in each aquarium were replicates, they were, in fact, pseudoreplicated, a weakness in the experimental design. Rapid dissipation of the IPA salt was reported within 30 min of the initial treatment of the aquaria with biofilms and water from the Danube, which the authors attributed to adsorption to the biofilm matrix. There was little dissipation in the first 7 days, but after subsequent treatments, dissipation became more rapid and biomass of the biofilms increased. For the GBH treatments, the initial (30 min) dissipation was less pronounced, biomass decreased after 14 days and increased again after 28 days. Rate of dissipation increased up to the end of the study (35 days). In water and biofilms from Lake Balaton, initial and final dissipation rates of IPA and GBH were smaller and biomass in the treated and control biofilms decreased, probably as a result of the weakness in the design and the lack of replication of the treatments.

In another study on biofilms from France, different results were reported (Carles et al. 2019). Microbiota collected in biofilms in the Artière River in France were placed in thrice-replicated microcosms in the laboratory. Water was replaced every 3 days to maintain nutrient status. Degradation of glyphosate was measured at two exposures, 6.5 and 67 μg/glyphosate a.e./L. Analysis of glyphosate and AMPA was by HPLC FLD, (internal method M_ET143) and the LOQ for glyphosate and AMPA was 0.1 μg a.e./L. For biofilms from the upstream site (less nutrients in the water), dissipation of glyphosate was more rapid at 6.5 μg/glyphosate a.e./L ($t½ = 2.3 \pm 0.7$ d) than the high (67 μg/a.e./L, $t½ = 19 \pm 2.4$ d). When phosphorus (1,000 μg P/L) was added to the cosms, $t½$ increased to 12.8 ± 0.9 d and 206 ± 81 d, respectively. Concentrations of AMPA increased in the cosms as the glyphosate was degraded. The authors concluded that, where phosphorus was in low concentration in the water, microbiota in the biofilm were using glyphosate as a source of phosphorus for growth. In a follow-up study, the effect of exposure to light on degradation of glyphosate in biofilms was characterized (Artigas et al. 2020). The objective of the study was to test that hypothesis that degradation of glyphosate was influenced by dissolved organic matter released by autotrophs in the biofilm. The study was done in vitro with cultures of microbiota enriched by prior exposure to glyphosate at 100 μg a.e./L. One concentration of glyphosate was used (stated by the authors as 0.62 mM) equivalent to 105×10^3 μg/L, which is unrealistic. Analysis was described in detail by the authors. Samples were centrifuged, derivatized with FMOC-Cl and quantified via HPLC FLD. LODs and LOQs were not reported but were presumably the same as those in Carles et al. (2019), discussed above. There was no significant difference in the rate of degradation of glyphosate between light and dark-exposed biofilm. Half-lives were 9.41 and 9.26 d, respectively and not significantly different ($p > 0.05$). As production of DOM was not measured, the original question addressed in the study is moot. However, when the antibiotic chloramphenicol was added as a positive control (500 μg/L), no degradation of glyphosate was observed, indicating that bacteria and/or cyanobacteria were responsible for most of the degradation of glyphosate.

The strong sorption of glyphosate to soil greatly impedes movement to either surface or ground water. Once in surface water, glyphosate sorbs to sediment where it is degraded by microbes similar to its degradation in soil (Wang et al. 2016). Biodegradation by aquatic organisms might also take place. In laboratory studies with treatments of glyphosate acid in 2-L aquaria, the presence of the invasive mussel (*Limnoperna fortunei* Dunker) was reported to significantly reduce ($p < 0.05$) the $t½$ of glyphosate in water (from 34 to 28 days) in the presence of large mussels (Di Fiori et al. 2012). Methods of analysis for glyphosate were incompletely described other than a reference to Pessagno et al. 2008, which stated that the analysis was conducted by ion chromatography on a DIONEX DX-100 chromatograph. LOD and other quality control data were not provided. The authors attributed this to sorption to the mussel shells and to metabolism by microbes associated with the bivalve; however, glyphosate also could have chelated with the calcium in the shells. The exposure concentration in these experiments ranged from 10×10^3 to 40×10^3 µg/L which is unrealistic but only two mussels of the 120 exposed died, suggesting a lack of sensitivity to glyphosate. In a study conducted in brackish-water microcosms, Janßen et al. (2019) reported that 99% of the glyphosate (with an unrealistic initial concentration of 14×10^3 µg/L) was eliminated after 20 weeks. Analysis of glyphosate was by derivatization with FMOC-Cl and quantification on a LC–MS-MS. Isotopically labeled internal standards (1-2-^{13}C2, ^{15}N glyphosate) and (^{13}C ^{15}N AMPA) were used to normalize for matrix effects but the LOD was not reported. The measured $t½$ was less than a week, with degradation to AMPA being the primary pathway of dissipation.

Residues of AMPA in soils where GBHs have been used are most likely a result of degradation of the parent material; however, this is not necessarily the case in surface waters. As has been pointed out, AMPA is a degradate of glyphosate as well as of aminopolyphosphonates used in treatment of water, for descaling, and other industrial activities (Grandcoin et al. 2017), as well as some detergents (Botta et al. 2009). For this reason, residues of AMPA in surface waters should not be stoichiometrically combined with those of glyphosate when estimating environmental loads of glyphosate.

2.3 Mechanism of Action and Selectivity of Glyphosate

Glyphosate is a systemic herbicide in plants The mechanism of action of glyphosate in target plants is mediated by strong inhibition of 5-enolpyruvyl shikimate-3-P synthase (EPSPS), an enzyme in the pathway for synthesis of aromatic amino acids (Schönbrunn et al. 2001). EPSPS is expressed in plants and some microorganisms but is not reported to be expressed in animals. The EPSPS in plants is more sensitive to inhibition by glyphosate than the form of EPSPS expressed in fungi and some microbes. There is no evidence that glyphosate affects any enzyme other than EPSPS in plants, as it takes concentrations 50-fold higher to have a toxic effect on plants with a glyphosate-resistant EPSPS than isogenic plants with only susceptible EPSPS (Nandula et al. 2007). Generally, glyphosate has little to no bactericidal or fungicidal activity and growth of these organisms is only inhibited at concentrations well above environmentally realistic levels (Franz et al. 1997). However, some studies have shown that glyphosate is active against fungal rust pathogens, such as *Puccinia triticina* and *Puccinia striiformis* at recommended FARs (Dill et al. 2010). Animals lack the pathway for synthesis of aromatic amino acids, and, for this reason, glyphosate is of low toxicity to animals and other organisms that do not express this pathway. The selectivity between animals and plants lowers the risk of adverse effects to aquatic and terrestrial fauna.

2.4 The Need for Formulants and Surfactants

Pesticide active ingredients often have intensive properties[2] (solubility, partitioning, etc.) that reduce absorption in the target organism. For example, cuticular waxes on the leaves of plants are hydrophobic and this effectively reduces the rate of penetration of the very polar glyphosate into target plants. For this reason, most GBHs contain co-formulants that change the intensive properties of the commercial product to enhance efficacy and penetration into the target organism. By necessity, formulants have different intensive properties from technical active ingredients and they generally have different fates in the environment. However, some of these formulants also have different biological properties from glyphosate and are toxic, particularly to non-target aquatic organisms. As an example, one of the more popular formulations of glyphosate (Roundup) contains the formulant, polyoxyethylene amine (POEA; CAS No. 61791-26-2, see Sect. 4 for the structure and more information on fate and toxicity of this material).

Different formulations of GBHs might contain different formulants, which results in differences in toxicity, in particular, to non-target aquatic organisms. As has been pointed out (Mesnage et al. 2019), this results in confusion when assessing risks because formulations of the same active ingredient in the same concentration differ in toxicity to non-target organisms. This phenomenon is not unique to glyphosate and has been reported for other pesticides and their formulants (Nagy et al. 2020). This is most relevant if GBHs are registered for use over water or their use might result in contamination of surface waters by direct overspray, such as in forestry or in the control of illicit crops (Edge et al. 2012; Solomon et al. 2007, 2009). Agricultural GBHs, which are the most widely used commercial products such as Roundup, are not registered for overwater use, so the toxicity of the formulated product is not relevant to assessment of risks to aquatic organisms from these products. Other GBHs are available for control of emergent aquatic plants. These GBHs might contain formulants of lesser toxicity or none at all, in which case, adjuvants are added at the time of use (Solomon and Thompson 2003). Although many aquatic

[2]Intensive properties are physical and chemical properties of a substance that are independent of concentration, such as density, partitioning, solubility, and reactivity with other substances. Extensive properties are those that are dependent on concentration such as mass and toxicity.

bioassays are conducted with formulated products like Roundup, they lack environmental relevance owing to the prohibition on aquatic uses of this agricultural formulation. Although drift from agricultural sprays might contaminate water, concentrations tested in bioassays should reflect concentrations of likely environmental residues, not application rates over terrestrial crops. We excluded about 45 papers published between 2007 and 2020 from this review for this very reason.

2.5 Direct and Indirect Effects on Non-target Organisms

GBHs can have indirect effects on non-target organisms that are mediated through the effects on target plants that are beneficial, part of the food chain, or provide habitat for other organisms. Indirect effects on vegetation are not considered adverse within the cropped area of the field but could be in the field margins or where off-field drift occurs (Prosser et al. 2016). However, indirect effects are not unique to use of a particular herbicide because mechanical control of weeds will result in similar losses of food or habitat for non-target organisms. For this reason, we have not included extensive discussion of indirect effects in this review, unless they are the direct result of the chemical and biological properties of glyphosate.

3 Ecotoxicology of Glyphosate and Risks in Non-target Organisms

3.1 Toxicity and Risks to Aquatic Organisms

3.1.1 Toxicity

The toxicity of glyphosate (technical active) to aquatic organisms is well documented in the literature and has been included in a curated environmental toxicity database (Connors et al. 2019). The database was accessed (https://envirotoxdatabase.org) and the data for glyphosate were downloaded. The data were sorted to select out the 24-to 96-h LC and EC50 values for apical endpoints, such as mortality, and population responses, such as growth. Because glyphosate in surface waters sorbs rapidly to sediments, concentrations decrease quickly (Edge et al. 2012), and acute toxicity data (exposures between 48 and 96 h) are the most appropriate to assessing risks. The derivation of a Species-Sensitivity distribution (SSD) followed the methodologies described in Rodríguez-Gil et al. (2018) and are explained in detail in the Supplemental Information (SI) for this chapter. Briefly, toxicity data were fitted, as interval data, via maximum likelihood estimation to six possible distributions (log-normal, log-logistic, Weibull, Pareto, gamma, or exponential). The one providing the best fit (based on comparisons of the Akaike Information Criteria and visual assessment of the fit) was chosen for following

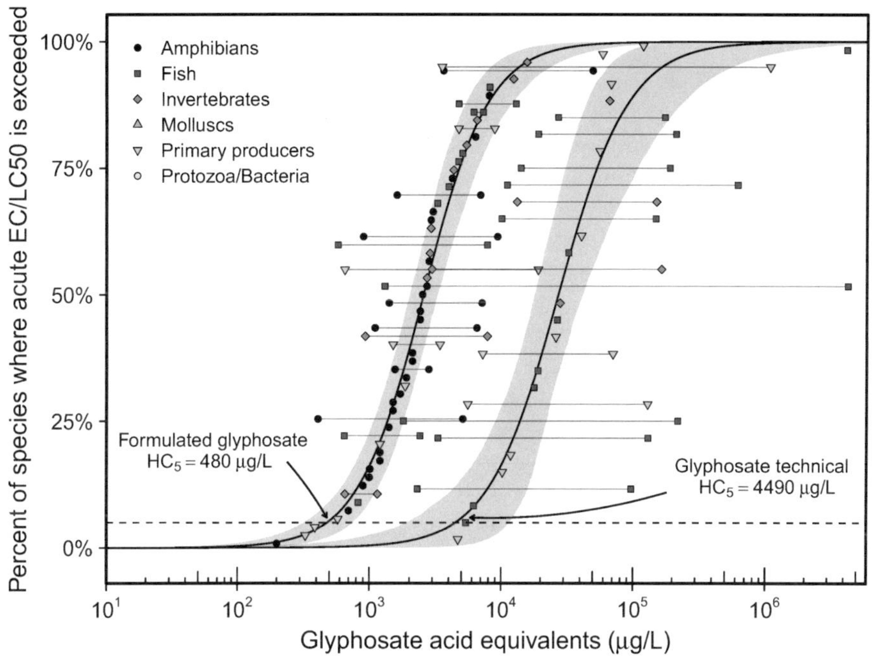

Fig. 4 Species sensitivity distributions (SSDs) for the acute toxicity of glyphosate active ingredient (right curve) and glyphosate formulated with POEA (left curve) to aquatic organisms. All data normalized to acid equivalents (a.e.) to allow direct comparison. Data for glyphosate technical from https://envirotoxdatabase.org/ (2019) and data for formulated glyphosate from (Currie et al. 2015). Species with several available toxicity datapoints are represented as a range between the minimum and maximum of these values. Shaded area represents the 95% confidence interval of the distribution. The horizontal dashed line represents 5% of the species, the commonly used threshold for risk assessment. Distribution parameters and centile intercepts are provided in SI Tables 2 and 3

analyses. Where there were multiple data entries for a single species, the data was fitted as a range representing the lowest and highest values. Fitting of the data to the distribution did not require the calculation of rank positions; however, for plotting and visualization purposes, whenever more than one value was available for one species, the geometric mean of the available values were calculated and used to compute the rank. For comparison, data for glyphosate formulated with POEA (from Currie et al. 2015) was also processed in this manner and plotted together. As a metric for toxicity that is often used in risk assessment, the fifth centile of the distributions (HC5) is shown in Fig. 4. All data for toxicity values and exposures were normalized to acid equivalents (a.e.) to allow direct comparison. The estimated HC5 for the formulated product (480 μg/L) was approximately 10 times lower than that of the active ingredient (4×10^3 μg/L).

3.1.2 Exposures in Surface Waters

Over the past decades, numerous studies have been published reporting glyphosate concentrations in surface waters, many of them quite extensive. Like in previous sections of this review, here we will focus our attention on reporting some of the most recent data. Additionally, we point the reader toward other recent reviews focusing on exposure alone such as Solomon (2016, 2020).

A survey of concentrations of glyphosate in 52 lakes from the Pampean region of Argentina was carried out in 2015 (Castro Berman et al. 2018). This is a region of intense agricultural activity and widespread use of glyphosate on soybean, wheat, maize, and sunflowers. Analysis was conducted using a HPLC-MS after derivatization of glyphosate with FMOC-Cl. Isotopically labeled glyphosate was used as an internal standard. Analytical procedures were well described in the paper. Based on one sample from each lake, concentrations of glyphosate in water ranged from the LOD (0.3 μg a.e./L) to 4.5 μg a.e./L and the median was the LOD. Frequency of detection was low (7 of 52 lakes were > LOD). The maximum concentration in suspended particles was reported as 0.13 μg a.e./L with only 3 of 52 samples above the LOD (0.02 μg a.e./L). Concentrations in sediment were detected more frequently (11 of 52 samples above the LOD: 2 μg a.e./kg) and a maximum of 20 μg a.e./kg. In a study in Córdoba, Argentina, samples of water from streams, ponds, and unconfined aquifers were analyzed for glyphosate and AMPA (Lutri et al. 2020). Samples were taken in November 2017, a month of high rainfall (118 mm). Analysis of glyphosate and AMPA was conducted using an Ultra high-pressure LC-MS-MS after addition of isotopically labeled glyphosate (1,2-^{13}C, ^{15}N) and derivatization with FMOC Cl. The LOD was 5 μg/kg for soil and 0.1 μg/L for water and particulates. Of the three streams, two measurements were > LOD (0.1 μg/L) and one was 0.2 μg glyphosate a.e./L, none had concentrations of AMPA > LOD. The two ponds had concentrations of 0.5 and 0.7 μg glyphosate a.e./L and one had AMPA at 0.7 μg/L. The one ditch had a concentration of 167 μg glyphosate a.e./L and 49 μg/L AMPA. Of the 19 unconfined aquifers, three had concentrations of glyphosate ranging from 1.2 to 2 μg a.e./L and these were associated with agricultural land. In another study carried out in the Quequén Grande River watershed in Argentina, glyphosate was measured in rainwater from Jan 2013–Feb 2014 (Lupi et al. 2019). Concentrations were greatest in Jan 2013 with 1 μg/L in one rainfall event and 2.2 μg/L in another. The authors suggested that glyphosate in rainfall was scavenged from spray drift into rainfall; however, concentrations of non-volatile AMPA (7 and 1.2 μg/L) were greater than glyphosate, suggesting that residues were carried by dust particles from disturbed soil. The measured concentrations of glyphosate were less than those summarized in (Solomon 2020). A survey of concentrations of glyphosate (and several other pesticides) was conducted in the St. Lawrence River and tributaries in Canada between the 9th and 16th of July 2017 (Montiel-León et al. 2019). Samples were filtered through a 0.3 μm filter and analyzed by ultrahigh-performance liquid chromatography-electrospray ionization tandem mass spectrometry after derivatization with FMOC Cl. Recovery for glyphosate ranged from 84–123%. Of the 64 samples, 84% contained detectable residues of glyphosate (method LOD = 0.002 μg/L). The median value was 0.027 μg/L and the greatest concentration was 3 μg/L, in a sample from the mouth of the Nicolet River. These values were all less than the 95th centile of values reported from surface waters in the USA (see below).

Using data from the U.S. Geological Survey (USGS) National Water Quality Network for Rivers and Streams (NWQN) data set from 2015 to 2017, Medalie et al. (2020) characterized concentrations of glyphosate in streams in the USA in relation to the use of land near (15 km radius) to the sampling site. Concentrations were expressed as mean maximum 21-day moving averages. Concentrations in water near undeveloped sites (11) were significantly smaller than those from mixed, developed, and agricultural sites (9, 23, and 27, respectively), which were not significantly different from each other. Mean max-21 concentrations of glyphosate were all less than 1 μg/L but the ranges were different: from LOD to 1.3 μg a.e./L in the undeveloped sites and LOD–2.2, LOD–5.6, and LOD to 6.1 μg/L from the mixed, developed, and agricultural sites, respectively. These observations indicate that inputs of glyphosate were driven mostly from nearby use. The concentrations reported in the above studies fell within the range of values reported in surface water waters in the USA (see below). Risks to aquatic organisms can be expressed deterministically by comparison of one exposure value to one toxicity value (generally the lowest available) via a quotient (i.e., risk quotient [RQ] or also as a hazard quotient [HQ]). Alternatively, risk can be expressed probabilistically, in which case the likelihood one or more exposures exceeding one or more toxicity values is used as a measure of risk. There are large datasets for toxicity of glyphosate to aquatic organisms and concentrations in surface water, as such, risks for this compound are best characterized probabilistically. However, one must be careful in the selection of data to be used in this process. Although there are many publications that report analyses of glyphosate in surface waters, many are not suitable for probabilistic risk assessment because they lack appropriate quality control and raw data are not provided.

One of the most complete and better curated datasets on concentrations of glyphosate in surface waters is that compiled by the US National Water Quality Monitoring Council (NWQMC 2019). This dataset is large, has good quality control, provides raw data, and is accessible to the public. Data from the NWQMC for the NWIS parameter codes 62722 and 99960 was compiled covering water samples collected in the US between 2001 and 2019. These data were used to generate an Environmental Exposure Distribution (EED) (Fig. 5). Processing and plotting of this EED followed similar approaches to those previously described for the derivation of the SSDs (Sect. 3.1.2) and are reported in detail in the Supplemental Information. Other regions have similar data sets and that for France is accessible to the public (NAIADES 2020). In order to provide a wider geographical coverage, we also considered this data set for our analysis. The data for glyphosate in raw water were downloaded from the Naïdes database and the data (raw water and filtrate of raw water) were then processed for probabilistic characterization and an EED (Fig. 5) was generated in the same way as data from the NWQMC. This data set was larger than that of the NWQMC (n = 107,487 after processing).

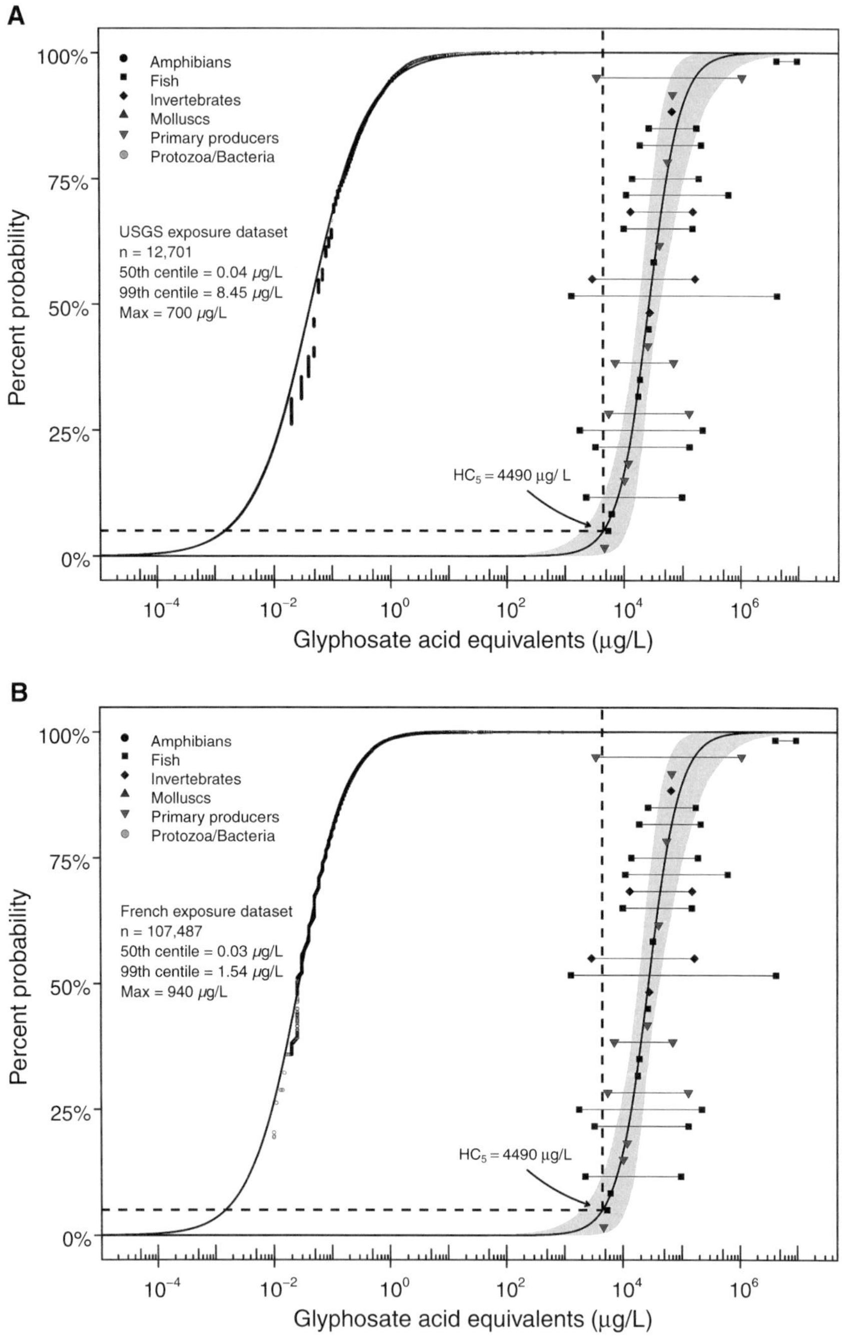

Fig. 5 Environmental exposure distributions (EEDs) (left curve) of concentrations of glyphosate (a.e.) in surface waters of the USA (**a**) and France (**b**). Species sensitivity distribution (SSD) of toxicity data for technical glyphosate from Fig. 4, above. Dashed line indicates the 5% of species as well as the intercepts with the SSD and EED curves. Shaded area indicates the 95% confidence interval of the distribution. This shaded area is also represented in the EED, but the confidence interval is so small that it is not visible on the graphic. Distribution parameters and centile intercepts are provided in SI Tables 2 and 3

The fifth centile from the distribution of toxicity values of glyphosate technical (glyphosate technical SSD, Fig. 4) was used as a point of departure to characterize risk. Exceedances from this point of departure were calculated from the derived exposure distributions (EEDs). For both datasets, the level of exceedance was zero. None of the available toxicity values were below the maximum measured concentration for either the USGS (700 μg/L) or French (940 μg/L) datasets. Concentrations of glyphosate measured in surface waters of France were generally lower than those from the US, despite the maximum measured concentration for this dataset being higher (940 μg/L). The data from both jurisdictions indicates *de minimis* acute risk to aquatic organisms from the exposure to glyphosate technical.

3.2 Risks from Exposure of Terrestrial Invertebrates to Glyphosate

Risk is a function of exposure and toxicity. The former is best characterized via samples collected in the field while toxicity is best characterized in studies conducted under conditions where realistic exposure levels are included in the range of concentrations tested. Topical exposures to beneficial insects, such as pollinators (e.g., honeybees), could occur if a GBH is used to spray weeds or GR crops that are attractive to these insects (i.e., during flowering) and oral exposures could occur via contamination of pollen or nectar, either directly or by translocation in the phloem: However, good agricultural practice would be to avoid spraying any pesticides on crops that are attractive to pollinators. On the other hand, other beneficial arthropods such as predators and parasites of arthropod pests of crops might have a greater probability of being sprayed as they would be in the crop before, during, and after flowering. There are no specific data on the effect of glyphosate on these types of interactions, but they would need to be considered when practicing IPM. Several guideline tests on the effects of glyphosate and GBHs on non-target organisms and environmental processes were included in the evaluations conducted by EFSA (2015). Details of the studies were not presented in their report, but they are assumed to be studies required of the registrant and to be conducted according to established protocols and under good laboratory practice (GLP) and quality assurance and quality control (QA/QC). These observations are included in the discussions below.

3.2.1 Toxicity of Glyphosate to Bees

For honeybees (*Apis mellifera L*), exposures to glyphosate might occur because of spray deposition on the integument, contact with treated surfaces, contamination of food sources such as nectar and pollen, or via contaminated honey and pollen in hive products. It is important to consider the route of exposure in honeybees foraging for food for the hive. These worker bees do not consume pollen or nectar directly but bring these back to the hive to produce honey and pollen-products such as beebread.

These products are mostly used to feed the developing larvae, the foragers, and the queen (via royal jelly processed by worker bees from hive products). In its review of glyphosate and potential risks to honeybees, EFSA (2015) concluded that the glyphosate and GBHs did not present a hazard to honeybees for intended uses and use rates. Other studies on honeybees were recently reviewed with a focus on behavior and cognitive effects (Farina et al. 2019). However, these authors did not critically assess behavioral responses in relation to realistic exposures or apical endpoints for the colony.

Exposures Compared to residues of glyphosate in surface waters, there is a paucity of data on residues in nectar, pollen, and honey. In a greenhouse study, honeybees were "forced" to forage on flowering plants (*Phacelia tanacetifolia* Benth.) recently sprayed with glyphosate (Thompson et al. 2014). GBH (Monsanto, St. Louis, MO, USA, a soluble concentrate, batch no GLP-0810-19515-A, containing 30.68% glyphosate a.e. as the IPA salt) was applied at the highest single recommended rate of 2.88 kg glyphosate a.e./ha for the UK and the hives were then sampled for residues in freshly collected nectar and in larvae at various times after application. The analytical methods were described in detail in the paper. Quantification was by HPLC-tandem mass spectrometry (MS-MS). An internal ^{13}C standard was used for the analyses and recovery ranged from 92–102%. LOD was 0.3 mg/kg and the LOQ was 1 mg/kg. Mean measured residues of glyphosate in nectar collected by the bees were 1,000 μg a.e./kg (± SE 150) seven days after treatment and in larvae were 11,900 (± SE 3800) and 5,300 μg a.e./kg (± SE 2100) 4 and 7 days after treatment, respectively. Residues in nectar collected from foragers and in pollen from pollen traps fitted to the hives declined from day-1 to day-7 after treatment.

In a survey of residues of glyphosate in commercial samples of honey from several countries, residues were above the limit of quantitation (LOQ = 15 μg/kg[3]) in 59 of the 69 samples analyzed (Rubio et al. 2014). Analysis was by ELISA using a commercial kit. The residues in the 59 positive samples of honey ranged from 17 to 163 μg a.e./kg with a mean of 66 μg a.e./kg. These values were less than those reported for nectar by Thompson et al. (2014) but the exposure conditions in the Thompson et al. study were worst case and dilution with uncontaminated nectar could not occur because the bees were confined to contaminated nectar only. In addition, the age of the honey and any degradation in residues over time was not considered. A similar survey of residues of glyphosate by ELISA in commercially produced honey from the state of Hawaii revealed detectable residues in 16 of 59 samples (Berg et al. 2018). The highest concentration measured was 228 μg a.e./kg and the mean for the samples > LOD was 27 μg a.e./kg (± SE 9.3). As might be expected, geographical analysis of the data from the island of Kauai indicated that

[3]Concentration of chemicals in syrup and honey are usually recorded as weight per unit weight because it difficult to measure volume in liquids as viscose as honey. The authors provide the LOD in ng/ml, so it is assumed that the units of ppb are weight/volume. The value here was derived by assuming that the concentrations reported in the honey are based on weight/volume and the density of the honey was 1.5 g/ml. This allows direct comparison with other published data.

greater concentrations of glyphosate in honey were associated with more intense agricultural activity.

In a field study in Belgium, exposure of honeybees to glyphosate and AMPA was measured in bee-relevant matrices (El Agrebi et al. 2020). Beehives from non-commercial bee-keepers were sampled for beebread (179 hives), wax from the brood chamber (100 hives) and a mixture of wax and honey from the super (10 hives selected from those with high residues of glyphosate in the wax). Analysis was well described by the authors. Glyphosate was extracted with water and quantification was by high-performance liquid chromatography-electrospray ionization tandem mass spectrometry (HPLC-ESI-MS-MS). Isotopically labeled internal standard (^{13}C-glyphosate) was added prior to extraction. Recovery ranged from 72%–113% with a relative standard deviation (RSD) of 0.1–4.5%. The LOQ was 10 μg a.e./kg for glyphosate and AMPA in all three matrices. The maximum measured residue in beebread was 700 μg a.e./kg (median of 26), the maximum residue in wax was 320 μg a.e./kg (median of 46 μg/kg. Residues in honey were < LOQ. On the basis of calculated hazard quotients based on an LD50 in honeybees ≥100 μg/bee, the authors concluded that these concentrations did not present a risk to honeybees.

Effects Because honeybees are social insects with a single reproductive female and high levels of redundancy in the hive, effects of chemicals on honeybees are best characterized by higher-tier tests at the level of the colony (Solomon and Stephenson 2017; USEPA 2014). In addition to measuring exposures to glyphosate, Thompson et al. (2014) also measured responses of honeybee broods exposed to glyphosate (IPA salt) via feeding with 50% sucrose solution amended with glyphosate. There were three replicate colonies for the control, a positive control (fenoxycarb) and the three concentrations of glyphosate IPA used were equivalent to 301, 150, and 75 mg a.e./L of sucrose solution (201, 100, and 50 mg a.e./kg) for 15 d. There were no differences in consumption of sucrose solution between the control and the three concentrations of glyphosate and concentrations were confirmed by analysis. Duration of feeding was 17 days and the brood were observed for 16 d after the start of exposure. There were no significant effects of glyphosate observed on survival and development of broods, no effects on weights of pupae and no significant effects on survival of adults. These results indicate that, at the level of the colony, exposure to glyphosate at concentrations in honey or nectar ≤200 mg a.e./kg has no measurable colony-level effects in honeybees.

In a 10-day guideline chronic feeding study on honeybee workers, bees were fed with solutions of glyphosate (IPA salt) ranging from 256 to 10,000 mg IPA salt/kg of sugar syrup (Monsanto USA 2017).[4] There was no significant increase in mortality at any of the tested concentrations. The study was conducted under GLP with

[4]It should be noted that summaries of GLP studies that have been conducted for glyphosate active ingredient and formulated products for submission by the registrant to regulatory agencies are publicly available and full study reports can be requested at the Bayer Transparency in Crop Science website https://www.cropscience-transparency.bayer.com/en/safety-results.

QA/QC and all raw data were provided. The daily dose per bee at the maximum exposure was 180 μg IPA salt/bee/day. This is equivalent to a no observed effect daily dose of glyphosate of 134 μg/a.e./bee/day. These results are consistent with the lack of effects reported in the whole-hive study (Thompson et al. 2014), discussed above. Various experiments on the effects of GBH (Roundup PROMAX®, containing 660 g potassium salt of glyphosate/L, equivalent to 540 g/L glyphosate a.e.) on honeybees were reported in a recent publication (Motta et al. 2020). Workers were exposed directly to spray (0.05 to 3% GBH in water. The authors state that "The final concentrations used in the experiments were achieved by considering the initial concentration of glyphosate acid in the formulation." However, it is not clear if this was a dilution of the formulation or if this concentration was glyphosate a.e. Analytical confirmation of exposure concentrations were not reported for any of the controlled-exposure experiments, a weakness in this study. Survival was significantly decreased at concentrations ≥0.5% GBH, (equivalent to 5,000 or 2,700 mg a.e./L, depending on how the dilutions were expressed). In another experiment, colonies were exposed orally to the GBH in 0.5 M sugar syrup at a concentration of 0.1% (equivalent to 1 or 0.54×10^6 μg/L or 0.933×10^6 or 0.504×10^6 μg a.e./kg depending on the expression of the dilutions). Some colonies were challenged with the pathogenic bacterium *Serratia marcescens*. There were no effects of GBH alone, but survival decreased when honeybees were challenged with *S. marcescens*. In this experiment, the measured range of concentrations of glyphosate in honey from exposed hives was reported as 700 to 1,500 μg/mL (Fig. 3 F in Motta et al. 2020). This is equivalent to 0.5×10^6–1.07×10^6 μg/kg assuming a density of honey of 1.4 kg/L. If these conversions are correct, the exposures via consumption syrup or honey were not realistic as compared to those reported from the literature (above).

In addition to studies on honeybees, the studies on the acute toxicity of glyphosate to the bumble bee (*Bombus terrestris* L.) and to the solitary bee *Osmia bicornis* L. have been conducted. In a GLP study, oral and contact exposures of the bumble bee to glyphosate IPA salt were non-toxic at the maximum dose tested (1,000 μg IPA salt /bee). The 48-h no observed effect dose for glyphosate by the oral and the contact route was reported as ≥461 μg a.e./bee (Monsanto Europe 2017a) (see footnote 4). For *O. bicornis,* a similar study indicated that the no observed effect dose for glyphosate by the contact route was ≥461 μg a.e./bee (Monsanto Europe 2017b) (see footnote 4). Testing of individual bees was appropriate in the case of these two species as they do not have the colony structure of honeybees and are thus more likely to be directly exposed to nectar than honeybees. Several well-conducted and well-described studies (described above) on the potential effects of glyphosate on honeybees have shown no significant effects at environmentally realistic concentrations of glyphosate. Collectively, these studies indicate that honeybees and other pollinators are not sensitive to glyphosate residues in bee-relevant food items.

Several studies on sublethal effects of glyphosate on honeybees have been published in the literature. These studies are summarized in Table 2. One study (Faghani 2018) was excluded from the analysis because insufficient data were

Table 2 Studies reporting sublethal effects of glyphosate in honeybees (*Apis mellifera*)

Exposure doses and/or concentrations of glyphosate	Results	Comments	Reference
Bees were exposed 0, 2.5, and 5 mg glyphosate a.e./L (source and purity not specified) dissolved in sugar. Tests were gustatory responsiveness, acquisition of proboscis extension response (PER), and persistence of PER. Nonelemental olfactory learning (NEOL) and foraging and dancing behavior were tested at 2.5 mg glyphosate/L only	Gustatory responsiveness was significantly reduced (from a mean of 3 to 2) after 15-day exposure. Per was significantly reduced with two trials but not three and there were no differences in persistence of PER. Acquisition of NEOL was significantly reduced. There were no differences in six measures of foraging and dancing	The concentrations tested in these studies were unrealistic and the results cannot be used for assessment of risks. Apical responses at the level of the colony were not measured	(Herbert et al. 2014)
Concentrations of 2.5, 5, and 10 mg a.e./L (0.125, 0.25 and 0.5 μg a.e./bee). Bees were fed on glyphosate dissolved in sugar syrup and then released from a novel site and followed by harmonic radar technology. Exposures were not verified by analysis	Forager bees fed syrup containing 10 mg a.e./L took significantly more time ($p < 0.05$) than control and smaller concentrations	The experimental design forced the bees to consume the treatment. When foraging, bees would normally not consume collected nectar during the return flight to the hive. Concentrations were unrealistic	(Balbuena et al. 2015)
Bee larvae were fed 0.8, 4, and 20 mg a.e./L in artificial diet. Exposures were not verified by analysis	Survival of larvae was significantly decreased ($p < 0.05$) at 4 and 20 mg/L. Body mass was significantly reduced only at 4 mg/L Species diversity and richness of gut microbiota were significantly reduced at exposures of 20 mg/L only	The concentrations at which potentially adverse effects were observed were 50-fold greater than maximum values reported from honey produced under normal agricultural conditions and are unsuitable for risk assessment	(Dai et al. 2018)
Bees in hives were fed Roundup diluted in sugar syrup at a concentration of 2.16 mg/kg for 28 days. Only one concentration was used, and it appears that only one replicate was tested. It is not clear if concentrations were reported as a.e. or the IPA salt of glyphosate. Exposure concentrations were not verified by analysis	The ultrastructure of the hypopharyngeal glands (which produce royal jelly) was reported to be altered by exposure to Roundup but a quantitative analysis was not conducted. Production of royal jelly was decreased in the Roundup treated hive but not significantly	It appears that only one replicate hive was used per treatment. Statistical analyses were probably compromised by pseudoreplication as the hive is the experimental unit	(Faita et al. 2018)

(continued)

Table 2 (continued)

Exposure doses and/or concentrations of glyphosate	Results	Comments	Reference
Groups of 60–80 worker bees of varying age were tested in small cages. One exposure concentration 2.5 mg a.e./L in sugar syrup (equivalent to about 2 mg a.e./kg. It appears that there were four replicate cages per treatment. Exposure concentrations were not verified by analysis	The presence of glyphosate in the sugar syrup did not change mortality compared to the controls Consumption of syrup was decreased in bees exposed to glyphosate for bees 4, 9, and 14 days old Score for responsiveness of bees to sucrose was decreased in bees exposed to glyphosate. Differential olfactory learning of honeybees was slightly reduced in bees exposed to glyphosate	Exposures were greater than would be expected from contamination of honey. The relevance of these changes in behavior was not investigated at the level of the hive. And the data are not suitable for risk assessment	(Goñalons and Farina 2018)
Exposure concentration was not clearly stated but the authors said that "3 μL of formulated glyphosate (Roundup Original DI®) containing 370 g/L glyphosate a.e. were dissolved in 10 μL of water to which bee larvae were exposed." This would be equivalent to 1110 μg glyphosate a.e./10 μl or 1110 μg/bee. This also contained an unspecified amount of surfactant (s) and adjuvant(s). Only one concentration was tested. Exposure concentrations were not verified by analysis	None of the larvae exposed to this concentration of formulated glyphosate survived for longer than about 7 h	Given the dose of glyphosate used and the fact that it was mixed with surfactant(s) and adjuvant (s), it is not surprising that the larvae died. This was an unrealistic exposure and not suitable for risk assessment	(Seide et al. 2018)
Larvae were exposed via food in vitro to glyphosate (not stated if this was pure glyphosate a.e., IPA salt or formulated product) at three ranges of concentration. Initial exposure concentrations were 1.25, 2.5, and 5 mg/L (equivalent to 12.5,	Exposures were to larvae from three colonies in year-1 and three in year-2. Survival of larvae was variable, and no consistent concentrations response was observed. Brood fed with 1.25 and 5 mg glyphosate/L had more larvae with delayed	The study was poorly and incompletely described, and the responses varied between colonies. Given the experimental design and the variability between colonies the results are uninterpretable	(Vázquez et al. 2018)

(continued)

Table 2 (continued)

Exposure doses and/or concentrations of glyphosate	Results	Comments	Reference
25, and 50 ng/larva). Every 24 h, more larval medium was added so that the exposure dose increased over time (see Table S1 in the paper). Accumulated doses after 144 h were 200, 400 and 800 ng/larva. Exposure concentrations were not verified by analysis	moult and weighed less than the controls. These effects were not observed at 2.5 mg/L, which the authors reported as non-monotonic responses		
Worker honeybees were exposed in a Potter Spray-Tower to 6–7 concentrations of 42 pesticides products (including the GBH, Roundup PowerMAX®) to determine lethal concentration and dose. Exposure concentrations were not verified by analysis	The GBH was the least toxic of the pesticides tested with an unbelievably high LD50 of 3.5×10^{31} μg AI/bee (reported in Table 2 in the paper). How this dose was estimated Is not clear	The LD50 reported for GBH is more than several tons per bee, which is obviously an error. Data were unusable for risk assessment	(Zhu et al. 2015)
Larvae of honeybees were fed with an artificial diet containing 0–2.5 mg glyphosate a.e./L from post-hatch to 120 h (concentration not verified but the authors state that the dose per larva was 275 ng/day. Exposures were in vitro Gene-expression profiles were determined from analysis of pooled samples of control and exposed larvae	There was no significant decrease in survival compared to the controls but the proportion of successful mounting of the larvae was significantly less than the controls Expression of several genes was modified by exposure; these were genes related to immunity, plant–herbivore interaction, epigenetic mechanisms of disrupted microbiota, and detoxification	The concentration tested in these studies was unrealistic and the results cannot be used for assessment of risks. Responses at the level of the colony were not measured	(Vázquez et al. 2020)

provided on the products and exposures used in the study. In another (Boily et al. 2013), the effect of exposure to glyphosate-treated corn on activity of cholinesterase (not the target site for glyphosate) in honeybees was investigated. Exposures to glyphosate were not quantified, so the study was not included. Several of these studies listed in Table 2 were poorly designed and/or poorly reported and could not be used for risk assessment. For others, the concentrations selected for testing were much greater than those determined in honey from commercial beehives and the relevance of the results could not be determined.

Other studies (all those in Table 2) that were poorly described and/or designed indicated adverse effects, mainly on sublethal endpoints. None of the studies listed in Table 2 measured effects on apical endpoints such survival, growth, development, and reproduction and several were used unrealistically large concentrations of glyphosate or GBHs.

Few studies on the potential effects of glyphosate on honeybees addressed the effects of residues in food items for honeybees at the level of the colony and none characterized all the colony-level responses that would be relevant to the sustainability of the colony. These are listed below (adapted from Stephenson and Solomon 2017) and include:

- Mortality.
  - Adults, workers, drones, pupae, larvae, queen.
- Colony strength.
  - Hive weights, number of workers, total number of adult honeybees, overwintering performance, rates of food consumption rates, etc.
- Colony development.
  - Queen development, brood development, numbers of eggs, larvae, capped cells, pupae, honey, nectar, pollen stores, etc.
- Colony health.
  - Infestation with *Varroa* mite, viruses, or disease.
- Foraging.
  - Intensity and activity.
- Flight dynamics.
  - Intensity and activity.
- Behavior.
  - Trembling, agitation, immobilization, incoordination, hyper- or hypo-responsiveness, etc.
- Productivity of the hive.
  - Hive weights, honey production, etc.

3.2.2 Toxicity and Risks to Other Terrestrial Invertebrates

In its review of glyphosate, EFSA (2015) reported laboratory and field toxicity data for GBH in the standard terrestrial arthropods; *Aphidius rhopalosiphi* de Stefani-Perez (a predatory wasp), *Typhlodromus pyri* Scheuten (mites), and *Aleochara bilineata* Gyllenhal (a predatory beetle). Toxicity values were all below the trigger values of concern. These guideline tests assess short-term responses, such as mortality, as well as longer term responses, such as reproduction.

Spiders In a review of the effects of various pesticides in spiders (Pekár 2012), no direct lethal effects were reported on spiders sprayed at field application rates (FAR) with GBHs (source and content of active ingredient not specified). In their review of the effects of pesticides in field margins Prosser et al. (2016) noted that there have been many studies on the effects of pesticides in general on spiders in field margins. From four older studies with GBH, they noted a lack of direct toxicity to spiders and concluded that any effects observed were indirect and related to changes in the structure of vegetation in the margins of the field (Prosser et al. 2016). There have been relatively few recent papers on the potential effects of glyphosate on spiders. However, indirect effects, mediated through changes in vegetation were reported for web-spinners but not for cursorial (hunting) species of spiders. For other spiders, sublethal effects were noted for some species in relation to avoidance of surfaces treated with GBH at FAR and efficiency of capture of prey (Pekár 2012). The GBH (Buccaneer Plus®, Monsanto, St. Louis, MO, USA, containing 480 g/L glyphosate IPA and POEA surfactant applied in laboratory studies at the FAR was reported to interfere with predatory behavior of two species of wolf spiders, *Tigrosa helluo* Walckenaer and *Pardosa milvina* Hentz (Rittman et al. 2013). Whether this is a short-term reduction or not and the role of formulants in GBH was not determined. In studies on two other wolf spiders (*P. lugubris* Walckenaer and *P. alacris* C. L. Koch) it was reported that spraying with Roundup Klasik Pro® (Monsanto Europe S.A, containing 360 g/L of glyphosate [IPA presumed]) did not affect predatory behavior (Niedobová et al. 2019). However, mixtures of this GBH and two tank-mixed adjuvants Wetcit® and Agrovital®, (mixed or alone) did affect predatory behavior but only in the first two hours after application (Niedobová et al. 2019). Spiders were sprayed directly in the laboratory at the FAR.

From these studies, it appears that some short-term effects could result from exposure of spiders to GBH in the crop at the FAR. Whether this would be the case in field margins exposed to rates of application < FAR and with interception of drift by plants only at the edge of the margin (Prosser et al. 2016) is uncertain at this time.

Amphibians There have been few studies on the toxicity of GBH to terrestrial stages of amphibians. In this scenario the use of a formulated product is appropriate because terrestrial stages of amphibians could be exposed via direct overspray. In laboratory studies on juvenile and adult terrestrial stages of frogs from Colombia, Bernal et al. (2009), reported LC50 and LC1 (equivalent to a no observed acute effect concentration [NOEC] derived from extrapolation of the concentration–response relationship). The nominal LR50[5] values from direct spraying of the frogs ranged from application rates of 4.5–22.8 kg a.e./ha for *Rhinella typhonius*

[5]Application rate lethal to 50% of the tested organisms.

L., *R. granulosa* Spix, *R. marina* L., *Engystomops pustulosus* Cope, *Scinax ruber* Laurenti, *Centrolene prosoblepon* Boettger, *Pristimantis taeniatus* Boulenger (adults), and *Dendrobates truncatus* Cope (adults). The GBH used was Gly-41 (C.A.C. Ltda, Bogotá, Colombia), which contained POEA surfactant and 1% of an agricultural adjuvant, Cosmo-Flux F411 (CosmoAgro, Palmira, Colombia), containing light petroleum oil. The most sensitive NOER[6] was 0.32 kg a.e./ha and the least sensitive was > 7.38 kg a.e./ha for the poison-dart frog, *D. truncatus* (LR50 ranged from 4.5 to 22.8 kg a.e./ha). The authors concluded that risks to adult frogs in the field would likely be lower because of less exposure. Sites >10 m outside the aerial spray swath and with no vegetative cover would receive deposition rates less than the LR1. Under actual conditions of use in aerial spraying in Colombia, interception by trees and other vegetation would further reduce deposition and risks to adult frogs. A study on the toxicity of directly applied GBH was conducted on adults and neonates of *Eleutherodactylus johnstonei* Barbour (Meza-Joya et al. 2013). *E. johnstonei* is invasive species in several South American countries and is a terrestrial-breeding frog, the eggs of which are laid in a foam nest where development and metamorphosis takes place. Juvenile (neonate) frogs emerge from the foam nest and there is no aquatic stage. Toxicity tests were conducted using the same formulation of GBH and added adjuvants as used by Bernal et al. (2009). Tests were conducted in glass containers and the spray mixture was applied as a mist. The nominal 96-h LR50s for adult males and females were 4.9 and 5.4 kg a.e./ha (in the range reported for other frogs by Bernal et al. 2009) while that for the neonates was 1.2 kg a.e./ha. The exposure scenario (glass exposure vessels) in these studies was unrealistic and a worst-case that probably resulted in greater uptake by the frogs. In the field, the substrate would be soil or organic matter, which would sorb the GBH and reduce bioavailability. In the tests conducted by Bernal et al. (2009), soil and leaf-litter were used a substrate for testing and, had the authors used this exposure scenario, LR50s would likely have been higher. Nevertheless, the results reported by Meza-Joya et al. (2013) do not suggest significant risks to these frogs. In two other studies that reported testing of adult frogs (Lajmanovich et al. 2015; Mann and Bidwell 1999) exposures were via water, which is unrealistic for characterizing toxicity to terrestrial stages of amphibians and cannot be used for risk assessment.

Few chronic toxicity studies have been carried out on the effects of glyphosate or GBHs in the laboratory. Given the lack of persistence of glyphosate and GBHs in realistic field-pond exposure scenarios (see Sect. 2.2.3), responses in laboratory studies are not observed in the field. For example, 48-d exposures of developing tadpoles of *Lithobates sylvaticus* LeConte (wood frog) to a forestry GBH (VisionMAX®, Monsanto Winnipeg, MB, Canada, containing 356 g glyphosate a. e./L as the IPA salt and POEA surfactant) at measured concentration ranging from 14 to 2,730 μg a.e./L in glass aquaria in the laboratory revealed several responses related to development, growth, and, at the greatest concentration, survival

[6]Estimated maximum application causing no observable effects in the tested organisms.

(Navarro-Martin et al. 2014). Similar studies on the same species in the laboratory exposed to the GBHs (VisionMAX and Roundup WeatherMax®; an agricultural formulation (Monsanto, Winnipeg, MB, Canada, containing 540 g glyphosate a.e./L as the potassium salt), and technical glyphosate IPA) (Lanctôt et al. 2014) resulted in several responses such as changes in weight and length, time to metamorphosis, changes in the expression of mRNAs related to development, and mortality. There were no observed effects on sex ratios or development of the gonads. Target concentrations were 210 and 2,890 μg a.e./L for Roundup WeatherMax, 2,890 μg a.e./L for Vision and 2,890 μg a.e./L for glyphosate IPA. There were two treatments 14 days apart and concentrations were maintained to simulate dissipation after each treatment. The section below describes the lack of effects observed in similar studies when they were conducted under realistic field conditions.

Relatively chronic or few partial life-cycle studies on the effects of GBH on amphibians have been conducted under realistic field conditions but these studies did not show responses that were observed in the laboratory (discussed above). The results of a forestry microcosm study on development and maturation of larval frogs in Canada showed that treatment of field enclosures with a forestry-use GBH (VisionMAX) had no adverse effects of tadpoles of the green frog (*Lithobates clamitans*, Latreille) (Edge et al. 2012). Natural forest ponds (cosms) were divided by an impermeable barrier and one half was treated with VisionMax and the other half served as a control. In August 2009, two concentrations were applied to the cosms, 550 μg a.e./L, which was based on the 99th centile concentration of glyphosate measured in shallow forest ponds during operational applications in forests. The other concentration was 2,880 μg a.e./L, the maximum concentration expected from a direct overspray of water 15 cm deep with the highest recommended rate of application. Treatments were replicated five times. The development of the tadpoles was characterized with follow-up in 2010 to capture the entire development of the larva cohort. Treatment with either concentration did not reduce survival or growth of the larvae of *L. clamitans* and no adverse effects were observed in the benthic invertebrate community. A similar study on the wood frog where the cosms were treated twice with Roundup WeatherMax was conducted over 2 years (2009 and 2010) (Lanctot et al. 2013). The target concentrations were 210 and 2,890 μg a.e./L. The time between treatments in 2009 was 24 days and in 2010, 28 days. Based on the results, the authors concluded that there was little evidence that exposure to this GBH affected abundance, growth, and development of tadpoles However, only one replicate cosm was used for each treatment, so the measurements were pseudoreplicated and lacked statistical power. A second study on the wood frog, *L. sylvaticus* began in 2009 (Edge et al. 2014), 24 similar split cosms were set up and one side of six each treated with GBH (Roundup WeatherMax) at two target concentrations (210 and 2,880 μg a.e./L). The treatments were applied twice (about 4 weeks apart in May and June of 2009) and this repeated in 2010 in April and May. Twelve of the split cosms were used to study the effects of addition of nutrients and GBH but these results are not discussed here. Mortality in free swimming tadpoles of *L. sylvaticus* was not observed in both years and the treated larvae were larger (10%, but not significantly so) than the controls. In the second year, green frogs, which had colonized the cosms, were observed to be significantly more numerous ($p = 0.05$). All of these results indicated that, in controlled field treatments, GBHs did not cause adverse effects on apical endpoints (survival, development, and growth) in green- and wood-frogs. In a follow-up analysis of the results of these studies on GBH (Edge et al. 2020), it was proposed that the main drivers of effects on communities of macrophytes, benthic invertebrates, and amphibians were driven by the size of wetland and ephemerality, not GBHs or nutrients. Fate of glyphosate in these field cosms was characterized and is discussed in Sect. 2.2.3.

The results of these studies indicate that terrestrial stages of amphibians are not at risk from direct exposures to GBHs, even if these contain POEA. Larval aquatic stages of amphibians are included in assessment of aquatic animals and, if they are in shallow pools, such as in forests sprayed from the air, they experience smaller exposures because of adsorption to sediments and risk is reduced (Bernal et al. 2009).

Soil Organisms In the OECD guidelines for laboratory testing of soil organisms (e.g., Tests 207, 220, 222, 226, and 232 OECD 2020) the test substance is well mixed with the soil matrix. This procedure is predicated on the assumption that the test material moves vertically and is evenly distributed through the soil column. This assumption is appropriate for pesticides that are mobile in soil but is unrealistic for glyphosate. When applied in the field, glyphosate will remain close to the surface where concentrations will be greater than those deeper in the soil. Therefore, complete mixing with the soil column is unrealistic and studies such as (Pochron et al. 2020), where glyphosate and GBHs were mixed with the entire column of soil were not included in the review. In its evaluation of glyphosate and GBHs, EFSA (2015) noted toxicity values for the IPA salt of glyphosate after 14-d exposures of the soil mite, *Hypoaspis aculeifer* Canestrini (14-day NOEC = 473 mg a.e./kg dry soil), and the springtail, *Folsomia candida Willem* (587 mg a.e./kg dry soil).

A few other studies have investigated the effects of application of glyphosate formulations to soil organisms. In a study on the use of a GBH (Roundup® Biactive™, Bayer, Australia, containing 360 g/L glyphosate IPA and propoxylated quaternary ammonium surfactant) for restoration of tropical rainforests in Australia, effects on soil- and litter-dwelling macro-arthropods were investigated (Nakamura et al. 2008). The GBH was applied as a spray at a nominal rate of 7.2 kg a.e./ha to 3 m × 3 m quadrants in five different study sites, which is an unrealistic exposure. Controls were treated with water only. Sites were sampled for litter-arthropods shortly before application, 2–4 and 86–103 days after application. The authors reported only minimal effects on composition of communities of arthropods as compared to control at either time after application. The one concentration tested was double the normal application rate of 3.24 kg a.e./ha typically used in forestry, but the lack of adverse effects suggested *de minimis* risk. A study in France investigated the effects of a GBH on the land snail *Helix aspersa* Muller (Druart et al. 2011). Newly hatched snails were exposed for 168 days to soil and food treated with GBH (Bypass® Dow Agrosciences, France, containing 360 g/L glyphosate

IPA) at the recommended FAR for vineyards [2.16 kg a.e./ha] and 10-times that rate (21.6 kg a.e./ha). Concentrations of glyphosate measured in soil were 2.6 ± 0.13 (SD) and 42.6 ± 2.6 mg a.e./kg dry soil. No effects on fresh body mass, shell dry mass, mortality, and percentage of adults were observed after 168 days at either exposure.

In its evaluation of glyphosate, EFSA (2015) noted low toxicity for 14-d exposures to glyphosate a.e. in the earthworm *Eisenia fetida* Savigny (LC50 = 5,600 mg a.e./kg d.w. soil) and also for a GBH (LC50 > 388 mg a.e./kg dry soil. The NOEC for a 56-d reproduction study for the IPA salt in the same species was > 473 mg a.e./kg dry soil.

In a study on the effects of treatment of soil with GBH on the earthworm, *E. fetida,* a decrease in reproduction was reported (Santadino et al. 2014). Earthworms (6 adults) were exposed in 28 × 14 cm pots of soil containing an unspecified weight of soil and finely chopped plant matter was placed on top of the soil as a source of food for the worms. After 6-days of acclimation, pots were treated with GBH (Roundup® SL, Monsanto, Argentina, containing 480 g/L glyphosate potassium salt) at a nominal rate of 6 L/ha formulation (FAR) and at a double rate of 12 L/ha. Formulation was applied in 50 mL of distilled water. There were eight replicates for each treatment. Worms were fed every 10 d and soil moisture was maintained at 80%. At 12, 21, 28, and 40 d after acclimation, two pots were examined, and the number of adult earthworms, their weights, number of cocoons, and presence of and number of young earthworms were recorded. There were only two replicates at each sampling time, which is a weakness in the design and exposure concentrations were not verified, also a weakness. Raw data were not provided, and analysis of the data was by use of a population model and ANOVA. The authors reported that the number of earthworm eggs increased with increasing concentration of glyphosate. However, the matrix population model indicated that glyphosate reduced the fertility of the eggs. Interpretation of the results for risk assessment was not possible because of lack of raw data and weakness in the design of the study. A recent greenhouse study on the effects of glyphosate on the earthworm (*Lumbricus terrestris* L.) showed no effects when applied to soil columns at a nominal rate equivalent to 1.08 kg a.e./ha (Nuutinen et al. 2020). The GBH used in the study was Rodeo XL®, (Monsanto EU) which contains 360 g/L glyphosate potassium salt and is free of surfactants; however, a commercial surfactant (isodecyl alcohol ethoxylate) was applied with the Rodeo at a rate equivalent to 0.5 L/ha. Exposures were in PVC cylinders filled with 10.8 kg of sieved field soil and there were 12 replicates at one rate of application and 12 controls, each containing 2 earthworms. Exposure was for 2 months. At the end of the exposure, no mortalities were observed, there were no significant differences in change in mass of the worms and the production of cocoons in the treated cylinders (31) was not significantly different from the controls (28). Mean exposure concentrations in the top 25 cm of treated soil at the end of the study were 0.52 mg/kg, similar to values measured in the agricultural soils in the EU where glyphosate was used.

Two studies from the same laboratory reported on the effects of glyphosate on the earthworm, *L. terrestris* and interactions with arbuscular mycorrhizal fungi (AMF) (Zaller et al. 2014) and on the effects of a GBH on the feeding strategies of the earthworms *L. terrestris* and *Aporrectodea caliginosa* Savigny (Gaupp-Berghausen et al. 2015). In the study of the interaction between GBH and AMF (Zaller et al. 2014), treatment experimental units (plastic pots) were filled with 12 L of steam-sterilized field soil (3 h at 100 °C). AMF treatments received an inoculum of *Glomus mosseae* (*Funneliformis mosseae* (T.H. Nicolson & Gerd.) in March 2011. There were three replicates for a total of 24 units. Only one rate of inoculum was used. One month later, the units were planted with seedlings of white clover (*Trifolium repens* L.). Units were in a greenhouse with daytime temperatures of 20 °C and night temperatures of 15 °C and 14 h light:10 h dark. Nine months later, four *L. terrestris* (16.6 ± 2.1 g per unit) were added. Five days after adding the earthworms, the glyphosate units were sprayed with a home-garden GBH (Roundup® Speed, Scotts Celaflor, Mainz, Germany, which contains 7.2 g glyphosate a.e./L, 9.55 g/L nonanoic acid[7] [pelargonic acid, which is also used as a contact herbicide] as well as other unspecified formulants). Label directions were followed. After 14 d, the experimental units were disassembled and effects on worms and AMF evaluated. Treatment with the GBH had no significant effect ($p > 0.05$) on activity during the experiment or mass ($p > 0.05$) of *L. terrestris* at the end of the experiment. Treatment with herbicide had a negative effect on colonization of roots by AMF at all three depths measured ($p < 0.05$).

The methods used in the second study (Gaupp-Berghausen et al. 2015) were similar but experimental units were seeded with three species of plants: orchard grass *Dactylis glomerata* Lam., white clover *T. repens*, and the common dandelion *Taraxacum officinale* F.H. Wigg. At 21 d post-planting, earthworms were added to the experimental units (12 each); five adult *L. terrestris* (vertically burrowing (anecic) worms) or ten adult/sub-adult *A. caliginosa* (horizontally burrowing (endogeic) worms) or a control with no earthworms. At 56-d post-planting, treated experimental units were sprayed first with 7.2 mL of a GBH "Roundup® Alphée" (Scotts Celaflor, Mainz, Germany containing 7.2 g glyphosate a.e. and unspecified formulants) on two consecutive days (in total 14.4 mL), and then, two days later, with 10 mL of another GBH, 'Roundup Speed, which, as mentioned above, contains 7.2 g glyphosate a.e./L, 9.55 g/L nonanoic acid as well as other unspecified formulants. Why there were three applications at such short intervals was not explained. Activity of worms and various chemical parameters were observed. Destructive sampling of the experimental units occurred 32 d after the last application of GBH when cocoons of the earthworms were counted, and hatching enumerated after another 105 d of incubation in uncontaminated soil. Number of earthworm casts decreased significantly ($p < 0.05$) in the GBH-treated *L. terrestris* units. A similar response was not observed for *A. caliginosa* ($p > 0.05$). Production of cocoons of both species decreased and hatching decreased as well and was significant for *A. caliginosa* (from 71% to 32%; $p < 0.05$) but could not be tested for *L. terrestris* (from 43% to 17% but too few replicates for statistical analysis because

[7]https://www.duenger-shop.de/Pflanzenschutz/Unkrautbekaempfung/Roundup-Speed-1-Liter.html.

of total absence of cocoons in some test units). Concentrations of nitrate and phosphate increased in the herbicide treated units regardless of the presence of earthworms but whether this was from breakdown of the herbicides was not investigated. Because only one excessive rate of application of a non-agricultural formulation of glyphosate intended for residential settings and containing a second herbicide was studied, causality for the statistically significant responses of earthworms could not be specifically assigned to glyphosate alone, thus rendering the studies unusable for risk assessment. In addition, too many variables were included in the comparisons and number of replicates was too small to adequately estimate precision. At best, these studies are preliminary and firm conclusions cannot be drawn.

Overall, there is no compelling evidence that use of glyphosate under good agricultural practices is harmful to soil-dwelling organisms.

Microbiota in Soil In its evaluation of glyphosate, EFSA (2015) noted only small effects on mineralization of nitrogen (N) and carbon (C) after 28-d exposures to glyphosate acid and a formulated product applied at higher than FAR. Values of 6% reduction in mineralization of N at 33 mg glyphosate a.e./kg dry soil and 8% for the formulated product at 94 mg glyphosate a.e./kg dry soil. Corresponding values for mineralization of C were 9.3% reduction at 6.4 mg/kg dry soil for the acid and 15% reduction at 94 mg a.e./kg dry soil. In a recent review, Thiour-Mauprivez et al. (2019) summarized the results of several studies that used various biochemical tools to characterize microbiota and concluded that treatment of soil with GBH at FAR resulted in little or no effects on soil microorganisms. They also noted that rates > FAR stimulate the microbial community. They also pointed out that some studies using newer genomic techniques showed negative effects on some microbiota while others did not. They attributed this to the communities containing different proportions of glyphosate-tolerant, glyphosate-sensitive, and glyphosate-degrading guilds of microbiota. Different responses such as these might be a result of the experimental design and choice of plants. For example, the results of a greenhouse study on the effect treatment of GR corn and soybean with GBH showed that some RNA transcripts in bacteria associated with the rhizosphere were upregulated and others were downregulated when compared to the control (Newman et al. 2016). There were two replicates per treatment. The soil was Blount silt loam (fine, illiticmesic Aeric Epiaqualf) and the GBH used was PowerMAX™ (Monsanto, MO, USA), applied pre-plant and at 31 and 51 days after planting at a nominal rate of 0.163 kg a.e./ha (less than the recommended rate of 0.9 kg/ha Monsanto Canada Inc. 2016). Above-ground plant material was harvested 7 days after the last treatment. Treatment was repeated for four growth periods and the rhizosphere samples collected on day-58 of the last period. Based on the results of analysis of phospholipid fatty acids and RNA sequencing, the authors suggested that long-term use of glyphosate can affect rhizosphere bacterial activity. They suggested that this could potentially shift composition of the bacterial community to favoring more glyphosate-tolerant bacteria; however, tolerance to glyphosate by the microbial community was not measured in the study. Different results were obtained by Lu et al. (2018) in a field study on the effects of glyphosate on nitrogen-fixation, suppression of pathogens and disease, and diversity of the rhizobacterial community in GR soybeans. The soybean line ZUTS31 was planted in 6 × 2 m plots and the triplicate treatment plots were sprayed with GBH (an unspecified formulation from Monsanto Malaysia containing 41% glyphosate IPA) at a nominal rate of 0.6 kg a.e./ha 18 days after seeding. Controls were sprayed with water and hand weeded. At seed-filling stage, the plants were removed from the soil and the rhizosphere soil analyzed for 16S rRNA genes as well as metagenomic DNA. Based on analysis of the results, the authors concluded that treatment with glyphosate did not significantly affect the alpha and beta diversity of the rhizobacterial community of the tested soybean line. However, it significantly influenced some functional genes involved in plant growth-promoting traits in the rhizosphere. The fraction of glyphosate sprayed on the plants that reached the rhizosphere was not measured so it is possible that the lack of response was due to lack of exposure.

A field study was conducted that investigated the effects of use of GBH on soil (Dundee silt loam) in which transgenic glyphosate-resistant soybean (*Glycine max* (L.) Merr.) were grown (Weaver et al. 2007). This study was conducted in Southern Mississippi and the GBH, Roundup Ultra® (Monsanto, St Louis, USA), was applied at 28 d after planting (two- to three-trifoliate leaf stage) and again at 42 days after planting (six- to seven-trifoliate leaf stage) at a nominal rate of 2.5 kg a.e./ha (1.5-times greater than the FAR). Control plots were untreated. Structure of the microbial community was characterized with the ester-linked fatty acid methyl ester (EL-FAME) procedure and changes were analyzed by analysis of dendrograms. Only small differences were observed between treated and control plots and the authors concluded that, even at the greater than normal rates tested, effects on the microbial community were small and transient. A recent characterization of microbiological communities in soils from two regions of the US, Maryland and Mississippi, each with two crops (corn and soybeans) reported that there were no differences in the communities between plots were glyphosate was applied as compared to where it was not (Kepler et al. 2020). Also, the authors reported no effects on prokaryotic and fungal communities in soil microbial communities associated with GR varieties of corn and soybeans across a range of farming systems. In addition, no effect of the use of glyphosate on populations of pathogenic species, such as *Fusarium* spp. was detected.

In a meta-analysis, Nguyen et al. (2016) concluded that, at normal FARs (< 10 mg a.e./kg soil), GBHs had no effect on soil microbial biomass and soil microbial respiration but biomass was decreased at greater rates of application (equivalent to 10–100 mg a.e./kg soil). Using fatty acid methyl ester (FAME) analysis and characterization of 16S rRNA genes Lancaster et al. (2010) showed that gram-negative bacteria were increasingly represented in the bacterial community with increasing applications (1–5) of GBH (Roundup WeatherMAX, 480 g/L glyphosate a.e./L, from Monsanto St Louis, MO, USA) to soil in laboratory conditions. Spacing between applications was 2 weeks which is shorter than normal field applications so it is not clear if this response would occur under field conditions. A field study on the effect of two GBHs from Monsanto Europe (Roundup Max™ (a granulated formulation containing 68% glyphosate a.e. by weight and POEA) and Roundup Quick™ (a premixed spray product containing 0.72% glyphosate a.e. by weight) and no POEA) showed that applications of these products at rates of 1,000- and 300-times the FAR did not cause decreases in numbers of heterotrophic bacteria in soils (Sihtmäe et al. 2013). They also noted an increase in the population of heterotrophic bacteria, which was most pronounced in the first 45 d after treatment of the soil. Overall, these studies indicate that applications of GBH to crops do not cause declines in populations of microbiota. Changes in the diversity and biomass of microbiota were observed but this might be expected because the addition of glyphosate provided a potential source of organic matter, nitrogen, and phosphorus to the microbiota.

Other studies have characterized the dissipation of glyphosate in treated soils. In a review of the effects of pesticides on the microbiota of soil, Jacobsen and Hjelmsø (2014) pointed out that pesticides may have effects on soil microbiota, one of which being the selection or induction of enzymes and metabolic processes that increase the rate of metabolism of pesticides. An example of this is the enhancement of the mineralization of fluometuron by GBH in soil and in a culture-medium in laboratory studies (Lancaster et al. 2008). The formulation of GBH was Roundup WeatherMAX and it was added to the soil (Weswood silty clay loam) at a nominal rate equivalent to 1.25 and 2.5 kg a.e./ha or 146 and 292 μg a.e./mL medium inoculated with *Rhizoctonia solani*. Mineralization of fluometuron was increased by a factor of 1.9 at a rate of 2.5 kg a.e./ha.

Mineralization of glyphosate in soils is most easily measured using ^{14}C-labeled glyphosate which is added to soil incubated in sealed flasks fitted with device to absorb $^{14}CO_2$. Usually a solution of NaOH or KOH is used to capture any $^{14}CO_2$ released by microbial activity. In a laboratory study, the mineralization of $^{14-}$C-labeled glyphosate was measured using this approach in an Ultisol soil from Brazil (Andréa et al. 2003). Samples of soil were treated with a mixture of GHB (Nortox®, Brazil) and glyphosate labeled on the phosphonomethyl carbon) 1-, 2-, 3-, and 4-times at intervals of 14 days and formation of $^{14}CO_2$ measured every week. The interval between treatments was not environmentally realistic and results were again counterintuitive, with greater proportion of $^{14}CO_2$ released in the single as compared to the multiple treatments. This was probably the result dilution from the repeated addition of more radiolabeled parent material. When the total $^{14-}$C-budget was measured, between 25 and 45% was mineralized as CO_2, 15–50% as ^{14}C extractable with 0.35 mol/L H_3PO_4, and 15–25% was bound as non-extractable residue. Total recovery of ^{14}C ranged from 80 to 100% after two months indicating total dissipation of glyphosate via mineralization and/or the formation of strongly sorbed or inactive substances.

In another study in soil (Lancaster et al. 2010), ^{14}C-labeled glyphosate (labeled on the phosphonomethyl carbon) was applied to soil (Weswood silt loam) to monitor mineralization via the release of $^{14}CO_2$. At the start of the experiment, all soils were treated with non-labeled Roundup WeatherMAX at a nominal rate of 49 μg a.e./g. This treatment was repeated every 14 days for 1, 2, 3, 4, and 5 treatments. The

^{14}C-glyphosate was added on the day of the last treatment with unlabeled GBH and the evolution of $^{14}CO_2$ monitored for 14 days. The results were counterintuitive; rate of mineralization of the labeled glyphosate decreased with increasing treatments with GBH; however, this could have been because the labeled glyphosate applied during the last treatment was diluted with the unlabeled material applied previously and this slowed the mineralization of the labeled material. Also, the intervals between treatment were short and not environmentally realistic. The amount of ^{14}C incorporated into microbial biomass did increase with number of treatments with GBH with up to 60% incorporated in soil with 5 treatments of GBH after 3 days, but this decreased to ca. 25% for all treatments with GBH after 14 days of incubation.

An assessment of the structure of the microbial community in soils with a history of the use of GBH and those where GBHs had not been used indicated that functional diversity was not different (Allegrini et al. 2015). This study was conducted in the La Pampa region of Argentina in four locations to with 19–20 years of use of GBH and two without. Community function in the soils was characterized with Community-Level Physiological Profiling (CLPP). Tolerance to glyphosate (potassium salt) was characterized with Pollution Induced Community Tolerance (PICT) assay using a control and concentrations of glyphosate (potassium salt) of 2,440–1,224,000 μg a.e./L of growth medium (unrealistic concentrations). Structure of the microbial community was further characterized with Denaturant Gradient Gel Electrophoresis (DGGE). While there were differences in utilization three amino acids in the CLPP assay, analysis of catabolic evenness indicated no significant differences between history and non-history soils. Likewise, the PICT assay indicated that tolerance to glyphosate was not consistent with previous history of herbicide exposure. Characterization of community structure using DGGE indicated > 90% similarity of 16S rDNA fingerprints between history and non-history soils.

Overall, the results of most of the studies on the effects of glyphosate and GBHs on microbiota in soils do not indicate adverse effects on the structure of the soil microbial community structure and function under field conditions. In some cases, increases in the numbers or activity of several soil bacteria were observed after application. Most researchers regard these increases as non-adverse but that might not always be the case; for example, increased rates of degradation of other soil-active pesticides might result in decreases in biological activity of these other pesticides. However, this is common to many pesticides and is not an issue with glyphosate alone.

Microbiota in Water-Sediment Systems Microbes in aquatic ecosystems have been reported to be affected by glyphosate (assumed to be acid). In a study on brackish-water, 12-L microcosm, Janßen et al. (2019) reported biofilm communities of microbes to be less affected by a pulse of 14×10^3 μg glyphosate/L than those living freely in water, but the effects were of longer duration. Exposures to glyphosate and AMPA were confirmed by analysis using isotopically labeled internal standards, derivatization with FMOC-Cl, and LC-MS-MS analysis. However, the

LOD and LOQ were not reported in the paper or in the separately published method of analysis (Skeff et al. 2016). The concentration of glyphosate was unrealistic (see discussion of residues in surface waters above) but the effect on the biofilm microbes was minor and similar to that in earlier studies that found little or no effects of glyphosate on biofilm microbiota (Khadra et al. 2018; Lozano et al. 2018b). In the Janßen et al. study, total cell counts were initially increased in the treated microcosms with a return to control numbers at 60 days after treatment. Evidence of rapid dissipation of glyphosate and AMPA from the water in the microcosms was interpreted to suggest that the increases in certain microbiota were due to their use of glyphosate and AMPA as a source of nutrient; however, adsorption to organic matter and sediments could also have been involved; Janßen et al. (2019) did not measure concentrations in the sediments in the cosms. Also, in many brackish ecosystems, there is tide-driven exchange of water, so the effects observed in this study without water exchange and dilution may be greater than those that might occur in nature. Studies on the effects of glyphosate on the microbiological community in freshwater microcosms did not alter the physical and chemical properties of the water or the composition of the major species of microbiota (Lu et al. 2020). This was a laboratory study in pesticide-free water inoculated with plankton from a lake. Cosms were 2-L flasks containing 1.2 L of modified BG11 medium. There were three control cosms and three treated with 2.5×10^3 μg glyphosate a.e./L Exposures were not confirmed by analysis, so concentrations are nominal. Additional cosms were inoculated with the cyanobacteria, *Synechococcus* 7,942 and *Pseudanabaena* sp. After 7 days, RNA was extracted from algae collected on 0.2 μm filters for analysis. The experiment was repeated three times to allow other data to be measured at times up to 15 days. Physicochemical parameters changed over the 15-day period but there were few significant differences between control and treated cosms. The metatranscriptomic analyses indicated that the transcription in some cyanobacteria was influenced by glyphosate, particularly in genes associated with translation, biosynthesis of secondary metabolites, transport, and catabolism. Phosphorus from the glyphosate was utilized for growth by the cyanobacterium, *Synechococcus,* which increased in numbers, while numbers of *Pseudanabaena* declined. In a study of microbiota in biofilms from the Artière River in France, phosphorus and glyphosate decreased the richness and diversity of eukaryotes species in biofilms (Carles and Artigas 2020). This was observed in microcosms as described in Carles et al. (2019, discussed in Sect. 2.2.2). Species richness and diversity in bacterial communities were not affected by glyphosate, although the structure of these communities shifted in relation to degradation of glyphosate. Increases in the relative abundance of certain Bacteroidetes, Chloroflexi, Cyanobacteria, Planctomycetes, and alpha-Proteobacteria were observed in bacterial communities, presumed to be capable of using glyphosate as a source of phosphorus.

A field study in Argentina (Berman et al. 2020) compared communities of picocyanobacterial in lakes from a region of intensive agriculture (Pampaen region with predominant crops of soybean, wheat, maize, and sunflower) where GBHs are used with those on a remote non-agricultural area (Patagonia, mainly livestock farming) where little, if any GBHs are used. They reported that abundance of picocyanobacteria (small autotrophic cyanobacteria < 2 µm) was greater in Pampean lakes and suggested that this was related to the use of GBHs. However, in an earlier paper (Castro Berman et al. 2018), the frequency of detection of glyphosate was low in the Pampean region (7 of 52 lakes > LOD see Sect. 3.1.2) and concentrations were < 4.5 µg/L. The authors did consider some potential confounders such as double crops and total dissolved organic nitrogen, which showed significant correlations with populations of picocyanobacteria, but omitted nutrients such as phosphorus. Given the intensive agricultural activity in the Pampaen region, runoff of nutrients, such as phosphorus, into the lakes is a more logical cause of the observations.

Overall, changes in the structure of communities of aquatic microbiota in response to exposure to glyphosate were small and were mostly observed at concentrations equivalent to or greater than worst-case exposures in the environment.

Microbiome of the Gut Microbiota of the gut are essential to the health of most animals. This has led to the theory that glyphosate might interfere with some members of the bacterial community in the gut of animals by selectively inhibiting this pathway in bacteria with sensitive Class I EPSPS and not in those with insensitive Class II EPSPS (Nielsen et al. 2018). Several studies have investigated effects of glyphosate on the gut microbiome; however, an important question is "are these microbes sensitive to glyphosate at realistic concentrations?" and "are changes in the microbiota of the gut relevant to apical endpoints or are they adaptive?" Studies have reported effects of glyphosate on gut microbiota of various animals, but the concentrations used have seldom been shown to be those that these animals might be exposed to in the environment. For example, Motta et al. (2018) reported that feeding solutions of sugar[8] containing 5,000 and 10×10^3 µg glyphosate a.e./L to honeybees influenced the bacterial species and strains of microbiota in the gut of the bee. Exposures were not confirmed by analysis and it is not clear if the glyphosate used was a.e., but this is assumed. They cite Herbert et al. (2014) to support the view that this is a realistic dose of glyphosate to which honeybees might be exposed, but Herbert et al. do not report concentrations in honey and cites other papers regarding concentrations of glyphosate found in surface and groundwater but not in honey, pollen, or nectar. In fact, concentrations of glyphosate in honey are small; in the low µg/kg range (see Sect. 3.2.1). Furthermore, dose responses were inconsistent (Motta et al. Fig. 1B) for responses of total bacteria and *Lactobacillus* Firm-5, *Lactobacillus* Firm-4, and *Bifidobacterium* on day-3. None of these papers made any effort to determine the maximum concentrations of glyphosate that might be consumed by bees in nectar or found in honey, pollen, or beebread and royal jelly (that is fed to the larvae) where glyphosate is used. A later paper on the effects of

[8]The authors report that oral exposures to glyphosate were in 0.5 M sugar syrup which has a density of 1.072 g/kg. Exposure solutions of 5×10^3 and 10×10^3 µg glyphosate a.e./L were equivalent to 4.7×10^3 and 9.4×10^3 µg/kg of syrup.

glyphosate on gut microbiota in honeybees was published in 2020 (Motta et al. 2020). Newly emerged worker bees were exposed in the laboratory to 1 mM glyphosate and GBH (Roundup PROMAX®, containing 660 g potassium salt of glyphosate/L, equivalent to 540 g/L glyphosate a.e.) sugar syrup for five days. The concentration of glyphosate (a.e. assumed) in the syrup was equivalent to 169×10^3 µg/L or 158×10^3 µg/kg (unrealistic concentrations). Using analysis of 16S rRNA, significant reductions ($p < 0.05$) in all bacteria in the gut were reported. Mean number of gene copies decreased from about 3.2×10^7 in the controls to 2×10^7 in exposed worker bees. Similar reductions in gene copies for the bacteria *Snodgrassella alvi*, *Gilliamella spp.*, and *Bifidobacterium spp.* were also reported (Fig. 1B to F in Motta et al. 2020). However, apical endpoints at the level of the hive were not reported and the relevance of effects at this unrealistic exposure is not useful for risk assessment. Measurements of responses of the same bacteria and *Lactobacillus* Firm-5 in the gut and subsequent recovery after a 3-day exposure to 0.1% GBH in sucrose syrup (assumed to be 934×10^3 µg glyphosate a.e./kg) showed decreases in the number of gene copies for all bacteria after a recovery period of three days with partial recovery to levels similar to the controls after an additional 2 days. Again, because of the unrealistic exposures and use of a single exposure concentration, these data are not useful for risk assessment. These authors also assessed hive-level exposures to the same formulation on the response of the hive to a challenge with *Serratia marcescens* kz19 under field conditions. Only the study at site 1 in 2019 is discussed here. Two concentrations of GBH were used: 0.001 and 0.1% in sucrose syrup. These concentrations were unrealistic and equivalent to 934×10^3 and 934 µg glyphosate per kg of syrup (a.e. assumed). Exposures were for 4 weeks, Bees were sampled and gene copies of 16S rRNA for *Snodgrassella* measured at the end of week 0, 3, 4,and 7. Honey was sampled at the end of weeks 0, 1, 3, 4, and 7. Concentrations of glyphosate in honey were analyzed using HPLC-MS. The preparation of the samples and analytical conditions were well described. Use of internal standards was not reported and, although the calculation of the LOQ was described, it and the recovery were not reported. Worker bees were collected from the hives at the end of week-4 and challenged with *Serratia* under laboratory conditions in groups of about 25 bees each with six replicates from each of at least three hives. Bees were exposed to the bacteria suspended in sugar syrup and observed for 10 days for effects on mortality. Survival was adversely affected at both concentrations of GHB with mortality between 30 and 40% at ten days. Mortality in the controls ranged from 5 to 10%. Over the seven-week study, mean concentrations of glyphosate in samples of honey increased from 0 to 1.5×10^6 µg/L (0–1.4×10^6 µg/kg) in the high-exposed group and up to 40×10^3 µg/L (37.3×10^3 µg/kg). Concentrations of glyphosate were orders of magnitude greater than those reported from honey collected from agricultural areas where glyphosate is used in weed management.

Two studies have reported that exposures to glyphosate and AMPA, singly or in a mixture causes changes in physiological parameters and in the microbiome of the hepatopancreas of the Mediterranean mussel *Mytilus galloprovincialis* L. (Iori et al. 2020; Matozzo et al. 2018). Mussels were collected from the Lagoon of Venice and maintained in the laboratory where they were exposed in two replicates of 35 each to glyphosate (acid) at nominal concentrations of 10, 100, and 1,000 μg/L for 7, 14 and 21 days. Exposure concentrations were confirmed at time zero and just before renewal of the exposure solution at 48 h by analysis by LC-MS after derivatization with FMOC-Cl. Isotopically labeled internal standards were used to normalize for matrix effects. The LODs were 1.5 μg/L and 50 μg/L for glyphosate and 50 μg/L for AMPA, respectively. Exposures were verified only once. Several physiological parameters were measured, and some showed significant changes. The authors suggested that these responses were indicative of adverse effects; however, apical responses related to survival growth, development, and reproduction were not reported, so these data are not useful for risk assessment. In addition, the greater exposures and times of exposure were not realistic. Based on mussels exposed to glyphosate and AMPA and a mixture of the two, all at 100 μg a.e./L for 7 and 21 days in the Matozzo et al. (2018) study, Iori et al. (2020) characterized the response of the microbiome in *M. galloprovincialis* using RNA gene expression analyses. Compared to control, several genes were upregulated, and others downregulated; however, since no apical endpoints were measured in the study, the relevance of these changes in relation to environmental exposures at realistic concentrations and for realistic durations cannot be determined. The observed changes could be compensatory or only biomarkers of exposure.

Many of the other studies on the effects of glyphosate on mammalian gut microbiota are also problematic owing to unrealistic exposure levels. An in vitro study on interactions between the gut bacteria, *Enterococcus* spp. and *Clostridium botulinum*, showed effects at the unrealistic concentrations of 1×10^3 and 10×10^3 μg/L with no observed effects at 100 μg/L in the growth medium (Krüger et al. 2013). In another example, rats fed 500 mg/kg of GBH were reported to have altered abundance and composition of gut microbiota (Aitbali et al. 2018). The same microbiological responses as well as inflammation of the gut were also reported in another study in rats gavaged with 5, 50, or 500 mg glyphosate a.e./kg (Tang et al. 2020). Other similar studies in the literature include the findings of Lozano et al. (2018a) that GBH alters the gut microbiota composition of rats. In this study, rats were dosed via drinking water containing Roundup, presumably containing formulants. Final measured concentrations in water were 50 ng/L, 0.1 g/L and 2.25 g/L of glyphosate. Clearly the latter two concentrations are not environmentally realistic, and neither is the use of formulated product if one assumes dietary exposure via food and/or water. Although not clear from the paper, it appears that the only statistically significant responses were from the unrealistic doses. Another study (Qiu et al. 2020) used formulated GBH added to food (doses of 10, 20 and 40 mg a. e./kg over 35 days) to characterize effects on intestinal physiology and morphology in weaned piglets.

Studies such as these are almost meaningless; some cannot differentiate between the effect of glyphosate and its formulants. Furthermore, the concentrations of glyphosate are far above what is found in any food, and the formulants would not be present in food.

A study on bacteria in the gut of Sprague Dawley rats showed that oral administration of glyphosate had very limited effects on composition of the community (Nielsen et al. 2018). Rats were exposed via oral gavage to glyphosate acid at 2.5 and 25 mg a.e./kg/day for 2 weeks. The higher dose was 50-fold greater than the Acceptable Daily Intake under European Union regulations. In addition, a group of rats was exposed to a GBH (Glyfonova®, FMC, Denmark) at 25 mg glyphosate a. e./kg/day. To minimize the effects of acidity of the formulation (pH $\approx$ 2) on the rats and the bacteria in the gut, the authors adjusted the pH of all solutions to 5 using NaOH. DNA was isolated from the fecal pellets of the rats and from the ileum, cecum, and colon, at terminal sacrifice. The composition of the bacterial community was determined by sequencing of the hypervariable V3-region of the 16S rRNA gene in the extracted bacterial DNA. There was no difference in weight-gain over the 2-week exposure between any of the treatments. Regardless of treatment, there were no physiological abnormalities observed in any of the organs examined at the end of the study; however, a blood protein, acute phase protein haptoglobin (involved in regulating immune response), was slightly but significantly increased (1 vs 0.85 mg/mL) in the rats dosed with Glyfonova compared to those dosed with glyphosate acid at the same concentration of glyphosate a.e. as in the formulation. The authors concluded that glyphosate and Glyfonova had very limited effects on composition of the gut microbial and suggested that this was because there were enough dietary aromatic amino acids in the gut environment and that this mitigated the effect of inhibition of the shikimate pathway in some bacteria.

The one study that adjusted the pH of the glyphosate before dosing (Nielsen et al. 2018) showed no or very small effects even at reasonable worst-case exposures (25 mg a.e./kg/day). That others who observed adverse effects at greater exposures did not report controlling for pH suggests that pH might be a confounding stressor in experiments on composition of gut bacteria in mammals dosed with unrealistically large amounts of glyphosate or GBHs. We have not noted this pH adjustment in other toxicity studies, but the pH has been shown to influence toxicity in aquatic organisms (Mann and Bidwell 1999; Tsui and Chu 2003). The chelation of cations by glyphosate has been discussed in relation interference with micronutrients required in plants (Mertens et al. 2018) and has been suggested to promote adverse effects in mammals by interfering with availability of manganese, an essential mineral, from the gut (Samsel and Seneff 2015). The evidence for this at small realistic dietary exposures to glyphosate is weak (Mesnage and Antoniou 2017) but at the heroic doses of glyphosate that are given to mammals in some toxicity tests, this may be responsible for the observed effects in the animals as well as on the microbiota of the gut. Because of inappropriate and unrealistic exposure doses, and/or the use of formulated material, the results of most of the studies on effects of glyphosate on bacteria of the gut cannot be used to assess risks.

4 Fate, Ecotoxicology, and Risks of Formulants Used with Glyphosate

Adjuvants are frequently added to pesticides to enhance efficacy, to make application easier, or to change the size distribution of droplets in order to mitigate drift (Foy 1992). The most common formulants in GBHs are surfactants added to facilitate the penetration of the polar glyphosate molecule through the waxy plant cuticles. Over the years, a great variety of surfactants have been used in different formulations of glyphosate. The use of a different surfactant in a particular formulation will yield different abilities to enhance efficacy of weed control (Leaper and Holloway 2000); however, different surfactants can also display different ecotoxicological profiles (Currie et al. 2015; Mesnage et al. 2019).

Historically, the most widely used surfactants in commercial glyphosate-products have been those based in alkylamine ethoxylates (ANEOs), also known as polyoxyethylene amines (POEAs), particularly those derived from animal tallow. The acronym POEA has been commonly used to refer to a particular mixture of POEAs: polyoxyethylene(15) tallow amine (POE-tallow amine, POE-T or POE-15, CAS no. 61791-26-2, see Fig. 6). Roundup branded herbicide products, at least the earlier versions, are typically formulated with POE-T, which was used in the original formulations of Roundup Original®. POE-T is a tallow-based mixture of alkylamine ethoxylates (ANEOs). The distribution of alkyl/alkene chain lengths, R, resembles that of tallow's fatty acids with varying length between C_{14} and C_{18}, but predominantly saturated or unsaturated C_{16} and C_{18} (Corbera et al. 2010; Visek 2000). The two ethoxylate chains in POEAs can also vary in length but, POE-T has an average length of 15 ethoxy groups (EO). As a reference, the molecular weight for a molecule with a saturated C_{18} alkyl chain and an average of 14 ethoxylates in the EO chain ($C_{18}H_{37}N(EO)_{14}$, one of the most common homologs in this mixture) would be 886.3.

Through this chapter, we use the acronym POE-T to refer to the specific polyoxyethylene (15) tallow amine and POEAs when talking about other mixtures of polyoxyethylene amines or individual POEA homologs. The composition of the mixtures of adjuvants in commercial formulations of pesticides is commonly not disclosed to the public; it is considered confidential business information (USEPA

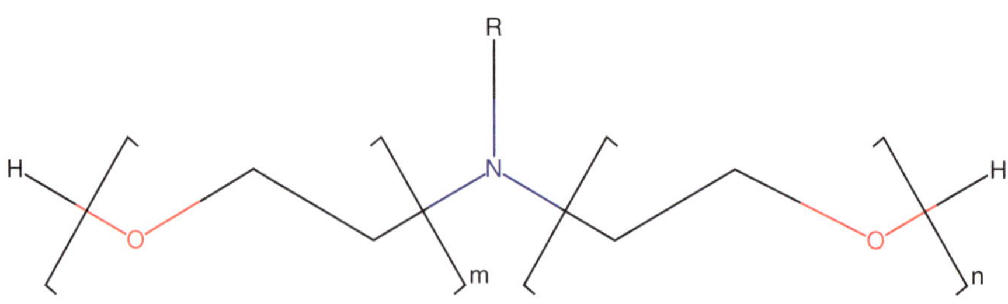

Fig. 6 General structure of on a polyoxyethylene amine (POEA). R = number of carbons in the alkyl chain (R) (tallow moiety) and n + m (total number of EO groups)

2015). Despite this, some of the original formulations are, by now, well understood. For example, early Monsanto (now Bayer Crop Sciences) formulations, such as Original Roundup® and its forestry-use equivalent, Vision® are known to contain ~15% (by weight) of a surfactant mixture with internal Monsanto code MON 0818 (Edginton et al. 2004). Later chemical analysis of this MON 0818 mixture has shown it to be made up of around 70% POE-T (Rodriguez-Gil et al. 2016).

POE-T-containing glyphosate formulations have not been registered for use on over-water applications since the mid-1990s in many jurisdictions including Canada, the US, and Australia (NRA 1996; PMRA 2015a; USEPA 2015) and no-spray buffer zones may be required for use near water bodies (PMRA 2015a), which limits the potential for exposure to aquatic systems. In addition, in the USA and many other jurisdictions, no aquatic use of any pesticide is allowed without a National Pollution Discharge Elimination System Permit (NPDES) or similar permit. Additionally, in 2017 the European Union completely banned the use of POE-T surfactants in GBH for use within the Union (EC 2017). POE-T-containing GBHs are still used; however, in large amounts around the world and, as such, its ecotoxicology and possible risks, as well as those of other surfactant alternatives also used in GBHs, are reviewed herein.

4.1 *Fate of POE-T in the Environment*

POEAs, are known to rapidly sorb to soil, sediments, and particulates in the water column (Krogh et al. 2003b; Tush and Meyer 2016). The mechanisms responsible for this sorption are presented in (Krogh et al. 2003b) an, in summary, involve hydrophobic interactions between the alkyl chain and organic material, hydrogen or polar interactions between the negatively charged clay particles and the ethoxy chains, as well as ionic binding with the central nitrogen when it is protonated (Fig. 7). The relative importance of each of these processes will depend on the relative lengths of the alkyl and ethoxy chains, as well as the pH of the media. With a pKa around 7, the central nitrogen will appear protonated at most common pH values in soil and water.

The assessment of the fate and behavior of POE-T in the environment has traditionally been limited by the lack of analytical methods able to measure the surfactant mixture at environmentally relevant concentrations. To our knowledge, only four different analytical methods able to quantify POE-T at environmentally relevant concentrations have been used in the peer review literature. Even with the improvements in analytical techniques over the past decade and with an increased availability of the surfactant mixture to researchers, quantification of POE-T has remained a challenge.

The two main challenges associated with the analysis and quantification of POE-T are the facts that it is a complex mixture and its strong adsorption to soils and sediment. Strong adsorption to soil and sediment requires the use of intense extraction methods for analysis. Accelerated solvent extraction (ASE), where the

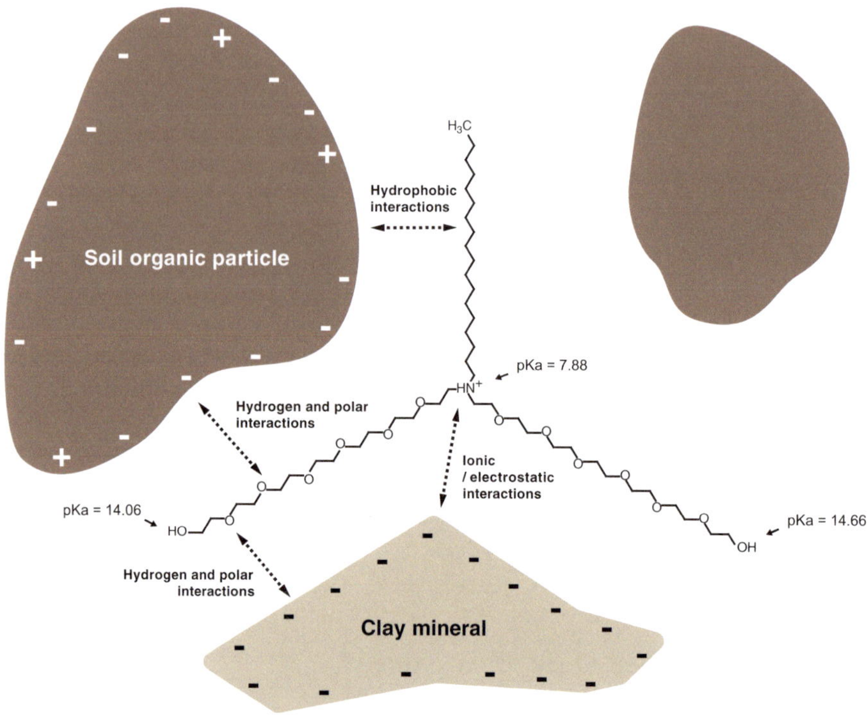

Fig. 7 Diagram summarizing the main chemical interactions between a typical POEA molecule and soil components. The represented chemical structure is that of $C_{18}H_{37}N(EO)_{14}$, a representative polyethoxylated tallow amine (POEA) homologue with an 18-carbon alkyl chain and two 7-ethoxyl chains (14 EO groups in total)

sample is exposed to solvent at high temperature and pressure, is currently the preferred method (Krogh et al. 2002; Rodriguez-Gil et al. 2016; Ross and Liao 2015; Tush et al. 2013). Even with these intense methods recoveries are still low (~50% (Rodriguez-Gil et al. 2016).), with high percentages of non-extractable material remaining in the sample.

Once extracted, reliable separation techniques are needed to detect and quantify the hundreds of individual homologs. These have traditionally followed two approaches: reverse phase HPLC based on differential hydrophobicity associated to the length of the alkyl chain, and normal phase chromatography based on the average length of the ethoxy chain. Reverse phase separation is currently the most common approach. In order to quantify homologs with the same alkyl chain length but different degrees of ethoxylation which elute at the same time, tandem mass spectrometry or time-of-flight mass spectrometry are employed (Krogh et al. 2002; Rodriguez-Gil et al. 2016; Ross and Liao 2015; Tush et al. 2013). Quantification of each individual homolog; however, is not possible and, typically, a subset of representative homologs is often monitored in order to quantify the mixture. For example, nine homologs are monitored in the method employed by Ross and Liao

(2015) and Rodriguez-Gil et al. (2016). Homologs with an average of 10, 12 and 14 ethoxy groups were considered for each of three different alkyl chains, unsaturated C_{16} and C_{18} and a saturated C_{18} (yielding the nine combinations). More recent methods have calculated the concentration of POE-T based on the sum of the areas of all detected homologs. However, two assumptions are made in this approach; that each homolog gives the same molar response as every other homolog and that the molar spike concentration is a known quantity based on the mass added and the average molecular mass of the distribution (Tush et al. 2018).

The choice of homologs to quantify is not trivial. As pointed out by Tush and Meyer (2016), the fate and behavior of each of the individual POEA homologs in the environment will be different, due to their slightly different chemical properties and susceptibility to degradation. This often results in different mixture profiles present in different environmental samples, limiting the applicability of quantification methods based in a subset of homologs. For example, unsaturated C_{18} homologs have been shown to dissipate faster than saturated C_{18} and C_{16} homologs, and as such, using them for the quantification of an aged mixture in environmental matrices could lead to an underestimation of the total mixture concentration. When used in short duration experiments, such as those by Rodriguez-Gil et al. (2016), where degradation is unlikely, the approach can still provide reliable measures.

Based on these analytical methodologies, several studies have reported measured concentrations of POE-T in environmental matrices. In a study of two agricultural soils Krogh et al. (2003a) measured concentrations of POE-T before and at several times after the application of a GBH (Eranca®) up to 14 days post-application. Variability between the two fields was large, but they observed increases in concentration of POE-T 4 days after application of the product and slow dissipation at 14 days post-application in one of the sampled fields. Initial concentrations (as mean concentration of C_{16}–C_{18}N(EO)13–18) were 97 and 170 μg/kg dry weight for each of the fields increasing up to 524 μg/kg dry weight 14 d after application.

Using a qualitative analytical method Tush and Meyer (2016) detected POEAs in soil samples collected between February and March, before the planting season, from a number of corn and/or soybean fields in five US states (Iowa, Illinois, Indiana, Mississippi, and Missouri). POEAs were detected in all samples from agricultural fields ($n = 20$), but they were not detected in samples from three reference soils. As mentioned earlier, a preferential loss of the unsaturated C_{18} homologs was observed, which could be an indication, the authors noted, of degradation of the alkyl chain at the double bond due to photo- or biodegradation.

With an improved, quantitative, analytical method Tush et al. (2018) monitored the concentration of POE-T in soil collected from an active silt loam, tile-drained agricultural field in Indiana to which glyphosate had been applied over the period of a year. The study site was planted with corn in 2003 and rotated into Roundup Ready soybeans in 2004. Samples were collected in April 2004 before glyphosate application, in May 2004 after the first glyphosate application, in July 2004 before a second glyphosate application, in October 2004 after harvest, and in April 2005 after winter. Soil samples were collected from three depths (0–15, 15–30, and 30–45 cm). The GBH applied to this field was not named in the study. POE-T was detected in all samples from the uppermost depth at concentrations ranging between 77 and 420 μg/kg dry weight. POE-T was present at 98 μg/kg in the preapplication sample despite the authors indicating no previous application of glyphosate for at least 2 years prior to the collection. Measured concentrations increased by about four times after glyphosate application to the field, decreasing afterwards. POE-T was still present at 230 μg/kg dry weight a year after application. In relation to the vertical distribution of the compound, it was observed that POE-T was preferentially found in the 0–15 cm segment, which is consistent with its strong sorption to soil and sediments and expected low mobility.

In this study, the authors also analyzed 16 sediment samples from 13 US streams and rivers draining areas where glyphosate is applied. These samples had been collected as part of previous studies extending from 2006 to 2014 in Georgia, North Carolina, South Carolina, Hawaii, Iowa, and Mississippi. POE-T was found in all the analyzed samples in concentrations ranging from 1.3 to 160 μg/kg dry weight. The authors note that, while concentrations of POE-T in soil samples were generally higher than those of glyphosate and its metabolite AMPA, this pattern reversed in sediment samples. The differences in transport of POEAs from the application site to the streambed sediment is still unexplained but is likely the result of differences in the intensive properties of the two chemicals, such as K_{OC}, K_{OM}, and Kd.

As noted earlier, one aspect common to all current analytical methods employed for soil/sediment analysis is the use of Accelerated Solvent Extraction (ASE) for the extraction of POEAs from the sample. This technique employs high temperature and pressure to increase extraction yields. While this approach results in the best recoveries and most precise estimates of the actual amount of compound in the analyzed matrix, the use of high temperatures and pressures likely results in extraction of larger amounts of the compound than otherwise would not be bioavailable to the organisms in the soil or sediment. Even under these intense conditions, these methods currently only achieve extraction recoveries around 50% (Krogh et al. 2003a; Rodriguez-Gil et al. 2016; Tush and Meyer 2016), highlighting the strength of the sorption of the POEAs to soil/sediment particles, were they can remain as non-extractable residue. The low bioavailability of sorbed POEAs is consistent with results of aquatic toxicity testing in the presence of sediment (covered in the next section).

In addition to the field surveys described above; several experimental studies have tried to determine parameters related to the fate of POE-T in the environment. Wang et al. (2005) measured the changes on concentrations of MON 0818 in water in laboratory-based microcosms (72-L) containing sediments with different amounts of total organic carbon (TOC). The authors observed shorter water column half-lives in aquaria containing sediment with higher TOC (18 and 13 h for 1.5 and 3% TOC, respectively), while no significant changes in concentration in water were observed in the aquaria with no sediment over the duration of the study (96 h).

Rodriguez-Gil et al. (2016) characterized the effect of sediment total organic carbon (0.05–2.05% TOC), and water depth (15, 30, and 90 cm) on the fate of MON 0818, in outdoor microcosms (4 m diameter) monitored over 28 days. Consistent with the field observations, the surfactant showed strong affinity for sediment materials. Under microcosm conditions, water depth or sediment characteristics did not significantly affect the water column $t\frac{1}{2}$ of POE-T, which was fairly short and ranged from 3.2–5.3 h. Sorption of POEAs to suspended solids was observed, which dissipated via one- or two-phase exponential decay; when two-phase decay occurred, fast phase $t\frac{1}{2}$ values ranged from 0.71–1.3 h and slow-phase values ranged from 18 to 44 h. Concentrations of POE-T increased in sediment shortly after application and decreased over the study period with a $t\frac{1}{2}$ of 5.8–71 d. The concentrations of POE-T in the sediment of the shallow (15 cm) mesocosms dissipated following a two-phase exponential decay model with an initial fast phase $t\frac{1}{2}$ of 1.1–8.9 d and a slower second-phase $t\frac{1}{2}$ of 21 d.

These observations from field surveys and micro/mesocosm experiments indicate a strong affinity of POE-T to sediment and soil particles that is supported by results from more traditional laboratory studies (reviewed in Krogh et al. 2003b). Van Ginkel (1993) carried out a series of test of biodegradability of ethoxylated fatty amines, including POE-T using a prolonged version of the Closed Bottle test (OECD 301D). The authors incubated POE-T in closed bottles together with secondary activated sludge and observed that the biodegradation curved for POE-T were characterized by a two-phase growth. From their results, the authors noted that the most likely biodegradation pathway is the central fission of the molecule cleaving it into the alkyl chain and a secondary ethoxylated amine. This way, the early rapid first phase of the biodegradation would correspond to the oxidation of the alkyl chain followed by a slower oxidation of the secondary ethoxylated amine. The authors also noted lower percentages of degradation at study termination (day 28) for POEAs with increasing a number of EO groups, with degradation percentages of 60, 28, and 22% for POEAs with 2, 15, and 50 EO groups, respectively (Van Ginkel et al. 1993).

Results from a number of fate studies with radiolabeled ^{14}C-MON 0818 (labeled on the ethylene oxide moiety) were provided by Monsanto company as part of a review of the registration of glyphosate in Australia and are summarized in NRA (1996). These studies included aerobic shake flasks tests with sterile and non-sterile soil of varying organic and clay contents. These studies showed POEA to be relatively stable under sterile conditions compared to non-sterile conditions where 24–31% of the applied radiocarbon was recovered as $^{14}CO_2$ after 7 weeks. Sorption to soil particles was high in both systems, with only 3% of the applied radiocarbon remaining in the water column after 7 weeks in the non-sterile treatment. In addition to these shake-flask tests, additional flasks were set up containing natural water and sediment from three different US locations and the radiolabeled surfactants and monitored over 14 weeks. Similarly, 40–50% of the applied radiocarbon evolved to $^{14}CO_2$ over the duration of the study. Sorption to the sediment under these conditions; however, was lower with 21–53% of the applied radiocarbon remaining in the aqueous phase and 7–29% becoming sorbed to suspended sediment over the study.

These laboratory fate studies have been recently updated and extended (Mitchell Kurtzweil, Bayer CropScience, personal communication, 2020) to include aerobic soil degradation, hydrolysis, adsorption/desorption, and aerobic aquatic degradation studies conducted according to U.S. EPA and OECD pesticide guidelines (USEPA 2016). In the same way as those presented in NRA (1996), these studies made use of ^{14}C-POE-T, which was shown to be hydrolytically stable at pHs ranging from 4–9. POE-T sorbed strongly to soil (K_{Foc}^{ads} = 17,600–114,000) and was not readily desorbed. Aerobic soil DT_{50}s were determined to range from 20 to 166 days and increased with the organic content of the soil. Water column $t½$ values in a natural water-sediment system under aerobic conditions were consistent with previous data (NRA 1996) and were estimated to be between 2–3 h with POE-T dissipating from water through a combination of metabolism and adsorption to sediment. The DT_{50}s for POE-T in the whole water-sediment systems, however, were much longer and ranged between 14 and 29 days. The studies with radiolabeled POE-T indicated that the ethylene oxide (EO) moiety of the molecule is degraded, thus shortening the length of the EO chain. Shorter EO chain length is associated with reduced toxicity to non-target aquatic organisms (Krogh et al. 2003b).These results are consistent with the above-mentioned mesocosm studies (Rodriguez-Gil et al. 2016; Wang et al. 2005) and field survey data (Tush and Meyer 2016; Tush et al. 2018).

In addition to their survey of soils collected around the US, Tush and Meyer (2016) also developed adsorption isotherms for three POEA homologs in the POE-T mixture. Their measured Freundlich constant values are consistent with those from the updated ^{14}C-labeled POE-T (Mitchell Kurtzweil, Bayer CropScience, personal communication, 2020). In addition, because Tush and Meyer (2016) noted sorption isotherms were not linear, they hypothesized cooperativity as a sorption mechanism wherein greater fractions of POEA were sorbed as the concentration of the surfactant increased owing to formation of bilayers and micelles.

In summary, the available data indicate that POEAs have a strong affinity for soil and sediment materials, where they become strongly sorbed, resulting in reduced exposure to organisms in the water column, as well as those in soil and/or sediment.

4.2 Toxicity of POEAs and Risks to Non-Target Aquatic Organisms

As for the exposure assessment, the direct assessment of the toxicity of the surfactants used in GBHs has traditionally been limited by the confidential business information nature of their composition and the difficulties of access to these surfactant mixtures for testing. In general, two ways of characterizing the toxicity POE-Ts have been used. One approach that has been used is comparing the toxicity of glyphosate alone to that of the GBHs (e.g., Demetrio et al. 2014; Janssens and Stoks 2017; Mann and Bidwell 1999; Mayer and Ellersieck 1986; Perkins et al. 2000; Tatum et al. 2012), another is by evaluating POE-T alone at nominal concentrations, without confirmation by chemical analysis (e.g., Brausch et al. 2007; Brausch and Smith 2007; Bringolf et al. 2007; Folmar et al. 1979; Frontera et al. 2011; Guilherme et al. 2012; Moore et al. 1986; Moore et al. 2012; Perkins et al.

2000; Servizi et al. 1987; Tsui and Chu 2003; Wan et al. 1989). These earlier studies noted that, due to the low toxicity of glyphosate itself, the surfactants, and not the active ingredient, were the main drivers of the toxicity observed in aquatic organisms (Bidwell and Gorrie 1995; Folmar et al. 1979; Mayer and Ellersieck 1986; Servizi et al. 1987; Wan et al. 1989).

Worst-case scenario calculations indicating a small margin of safety between expected environmental concentrations and the toxicity data from the early laboratory studies, initiated an early move away from the use of formulations containing POE-T for over-water applications as early as the 1980s. This can be observed in the introduction of Rodeo®, a GBH without POE-T, in the early 1990s and early research in the topic (Paveglio et al. 1996; Simenstad et al. 1996).

In the late 1990s, the Australian Environmental Protection Agency officially exclude over-water applications of herbicides containing POE-T (NRA 1996). In time, other jurisdictions (e.g., the USA, Canada, and the EU), implemented similar policies which are still in effect today (EC 2017; PMRA 2015a; USEPA 2015). Discussion of the toxicity of POE-T alone had been included in regulatory environmental risk assessments of glyphosate and glyphosate-containing formulations (NRA 1996; SERA 1996, 1997; USEPA 1993) and stimulated trials with Rodeo and alternative surfactants to control invasive species in overwater uses in Washington State (Paveglio et al. 1996; Simenstad et al. 1996).

In the most recent instances of the environmental and human-health risk as-assessments of GBHs related to their re-registration in the different jurisdictions, regulatory authorities have begun to include POEA-specific sections with larger toxicity data sets and POEA-specific assessments (PMRA 2015a; SERA 2011; USEPA 2015). The PMRA (2015a) included a methodological assessment of the hazard posed by MON 0818 and other mixtures of POEAs to aquatic organisms, noting that no GBHs registered in Canada contain more than 20% POEAs by weight. For their assessment, PMRA created individual SSDs for aquatic invertebrates (14 data points), amphibians (7 data points) and saltwater fish (21 data points) and calculated EC50-based HC5s (the concentration at which the lowest fifth centile of species in the SSD would be exposed to a hazardous concentration) of 4.1 μg POEA/L for invertebrates, 350 μg POEA/L for amphibians and 2060 μg POEA/L for saltwater fish. Parallel HC5 endpoints for formulated herbicide products containing POEA were also calculated as 190 μg glyphosate a.e./L for invertebrates, 930 μg glyphosate a.e./L for amphibians, 100 μg glyphosate a.e./L for marine invertebrates, and 3.04×10^3 μg glyphosate a.e./L for saltwater fish, reflecting the substantially greater toxicity of formulated products compared to POEA alone.

In this context, Rodriguez-Gil et al. (2017a), published a refinement to the aquatic risk assessment of POE-T. This assessment made use of data from POE-T-only studies available in the literature, as well as newly generated data from standard toxicity test with 15 additional species exposed to MON 0818. This new dataset was used to generate a species-sensitivity distribution based on a total of 37 aquatic species. The calculated EC50-based HC_5 for this data set was 170 (95%CI 150–200) μg POE-T/L (Fig. 8a), like that proposed by the Canadian PMRA.

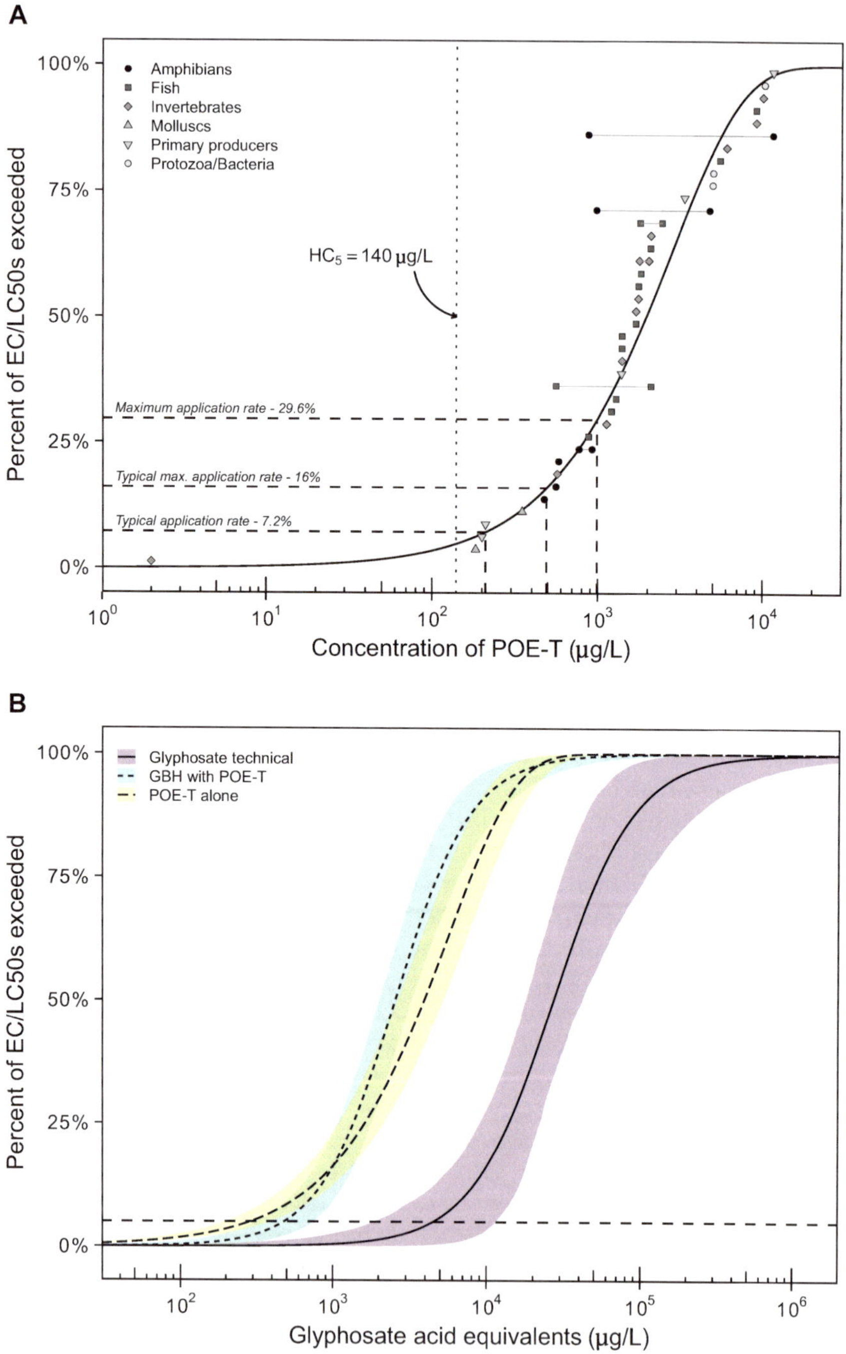

Fig. 8 (**a**) Species sensitivity distribution (SSD) of acute toxicity values for POE-T in aquatic organisms. Percent of EC50s exceeded under three theoretical worst-case exposure scenarios (See text for full description) are also indicated. Shaded area represents the 95% confidence interval of the distribution. (**b**) Species sensitivity distributions (SSDs) of toxicity values for glyphosate technical (purple curve), POE-T (yellow curve) and GBHs containing POEA as a formulant (blue curve). The POE-T data from panel B was normalized to glyphosate acid equivalents (a.e.) to allow for comparison in panel B. All data from (Rodriguez-Gil et al. 2017a). Raw data and sources of information are provided in SI Table 1. Distribution parameters and centile intercepts are provided in SI Table 2 and SI Table 3

Rodriguez-Gil et al. (2017a), also compared the generated SSD to 3 calculated worst-case scenario POE-T water concentrations based on the concentration expected from an unintentional direct over-spray of a water body of 15 cm depth at three different application rates (maximum rate, typical use maximum, and typical use) of the forestry-use formulation Vision® (999, 490, 150 μg POEA/L respectively). In addition to these, a fourth scenario was calculated from maximum glyphosate concentrations measured in inland waters and the amount of POE-T that would be expected if both compounds originated from a formulation containing 15% MON 0818 (with 70% POE-T) and POE-T and glyphosate remained in the same ratio in the environment (140 μg POEA/L). From this analysis, the percentage of species in the SSD that would be exposed to POE-T concentrations in water in exceedance of their EC50 would be 43% under the selected worst-case scenario of unintentional over-spray of a 15-cm-deep water body (maximum forestry application rate scenario), whereas 21% would be in this situation for the typical maximum rate scenario, 7% for the typical scenario, and decreasing to 4% when compared to theoretical maximum POE-T concentrations in water derived from the maximum measured glyphosate concentrations. These observations are consistent with those that led to the removal of POE-T from GBHs for over-water uses in the late nineties. For this review, we updated the dataset used in Rodriguez-Gil et al. (2017a) to include new data published since then, and to incorporate new analysis techniques, such as the ability to fit the SSD to interval and censored data as described in previous sections and the extended methods section of the SI. With this new updated data set, containing data for 40 species, we generated an SSD (Fig. 8, Panel A) which was compared to the theoretical worst-case scenarios presented in Rodriguez-Gil et al. (2017a). This new updated SSD provided an HC5 of 140 μg/L (95% CI 104–223 μg/L), very close to the 170 μg/L originally presented. Exceedances of the theoretical worst-case scenarios for this updated SSD are slightly lower at 29.6% for the maximum application rate scenario, 16.0% for the more typical maximum application rate and 7.2% for the typical application rate. A comparison of this POE-T-only SSD and one based on toxicity data for POEA-containing GBHs (after adjusting the POE-T to glyphosate a.e.) is presented in Fig. 8b.

In addition to the above-mentioned standard toxicity tests, Rodriguez-Gil et al. (2017b) assessed the effect of other environmental conditions, such as the presence or absence of sediment, on the toxicity of POE-T to aquatic organisms. Consistent with the observations on sorption of POE-T to sediment presented in Sect. 4.1 above, the toxicity of POE-T to 4 sediment-duelling organisms (*Hexagenia sp.,* Hyalella azteca Saussure, Lumbriculus variegatus Müller) and fathead minnow, (Pimephales promelas Rafinesque) was substantially reduced (between 2- and 10-fold increases in the measured LC50s) in the presence of sediment. These observations support the theory that POE-T's high affinity for the sediment results in a reduction of the exposure to aquatic organisms. Of particular interest was the case of fathead minnow, where the measured LC50 in the absence of sediment (880 μg/L, as MON 0818) was lower than the calculated worst-case water column concentration (1.41×10^3 μg/L as MON 0818), but crossed this threshold into the no-risk levels when tested in the presence of sediment ($LC50_{\text{with sediment}} = 3.5 \times 10^3$ μg/L as MON

0818). These results, the authors point out, would indicate that the high exceedances observed when comparing worst-case scenario concentrations to toxicity data generated from standard laboratory data are not representative of real-world scenarios where the exposure of aquatic organisms to POE-T would be substantially reduced by rapid sorption to soil and sediment.

A similar toxicity-mitigation phenomenon has been observed to take place when the toxicity of a number of surfactants (including ANEOs) has been assessed in the presence of other organic materials, such as humic acids, in the water column (Andersson 2012; Chen et al. 2014; Deese et al. 2016). These results further point to the need to account for these effects when assessing the toxicity of compounds with a tendency for rapid and strong sorption to organic materials.

A recent Draft Screening Risk Assessment of the Poly(alkoxylates/ethers) chemical group (which includes ANEOs, including POEAs) generated by Environment and Climate Change Canada and Health Canada (GC 2019) also agrees with this conclusion and notes that "Considering all available lines of evidence presented in this draft screening assessment, there is low risk of harm to the environment from the 21 substances considered in this assessment." This conclusion is based on the reductions in exposure expected due to sorption to sediment and other organic materials.

Very little information is available on the toxicity of POEAs to terrestrial organisms, despite observations such as those by Tush and Meyer (2016) reporting on the common presence of POEAs in agricultural soils. It has recently been pointed out that, due to changes in leaf porosity, reductions in drought-tolerance could occur in plants exposed via spray to commonly used surfactants, including POEAs and alkylpolyglucosides (Raesch et al. 2018). Because that spray would contain surfactant and glyphosate, the herbicidal effects of glyphosate would likely exceed any impairment due to the surfactant. Thus, this observation is likely of academic interest only.

Mendonca et al. (2019) exposed three strains of soil *Pseudomonas* species to POE-T alone or a combination of POE-T and glyphosate in an in vitro study. The form of glyphosate (acid or salt), source, and purity were not reported and the concentrations of glyphosate and POE-T (purity not reported) were not confirmed. Only a single concentration of glyphosate was used (0.5 mM equivalent to 84×10^3 μg/L) and was unrealistic. POE-T was used at 30% of this concentration (25.4×10^3 μg/L), also unrealistic. They observed a 60% reduction in growth when the bacteria where exposed to POE-T alone. Addition of glyphosate to the exposure medium resulted in no or small (5–15%) additional decrease in growth. Additionally, exposure was in liquid growth media, representing a worst-case scenario without any possible exposure reduction due to sorption of the compounds to soil. Smaller concentrations were not tested, and the results of the study are unusable for risk assessment.

Only one study was found related to terrestrial insects. Bednářová et al. (2020) exposed fruit fly cultures (*Drosophila melanogaster*) to glyphosate, POE-T, or a GBH via application of the diluted product in a sucrose solution to filter paper in the holding vial to characterize the effects of dermal and oral exposures. The authors did not relate these concentrations to what would be expected on plants sprayed with GBH and, because the dimensions of the filter paper were not reported, it was not possible to estimate a relationship to field exposures. Also, the addition of sucrose to the test solutions introduced another uncontrolled confounder that might increase rates of oral consumption and that is not representative of field use.

Acute exposure (24 h) to POE-T (45×10^3 μg/L in the solution of 5% sucrose) reduced lifespan of female flies by approximately 30% compared to untreated controls or glyphosate (100×10^3 μg/L) alone. Fecundity was also reduced by about 50% after a 24-h exposure to 45×10^3 μg/L of POE-T. Additionally, the authors also chronically (72 h) exposed the flies to varying concentrations of the POE-T and reported LC50 values of 1.04×10^6 μg/L for male and 1.3×10^6 μg/L for female flies (Bednářová et al. 2020). To our knowledge, no data exist on measured concentrations of POEAs on plant surfaces after application, making it hard to compare these data to concentrations that are to be expected in the field. This route of exposure would be limited to the sprayed crop and the immediate edge-of-field environment. These observations are inconsistent with other studies using application of GBH to terrestrial insects, which did not result in direct adverse effects (see Sect. 3.2.2 above) probably because the concentrations used by Bednářová et al. (2020) were very large and not environmentally relevant.

4.3 *Toxicity and Risks from Other Adjuvants*

The removal of POE-T from GBHs labeled for overwater use in several countries (discussed above) resulted in the use other surfactants that were less toxic to aquatic organisms (e.g., Paveglio et al. 1996; Simenstad et al. 1996). Several other formulants have been tested for addition to formulations of glyphosate that contain no surfactant and are used for over-water applications (reviewed in Solomon and Thompson 2003). In this use, appropriate adjuvants that will provide efficacious control of the target plants (usually emergent vegetation such as the invasive species *Phragmites australis* and smooth cordgrass (*Spartina alterniflora*)) and are of low risk to non-target organisms in the water are selected for use. The removal of POE-T from all GBHs in the European Union (EC 2017) resulted in the adoption of other surfactants in the GBH used within the European market.

Based on a study on the acute (48-h) toxicity of several commercial pesticide adjuvants to *Daphnia pulex* under laboratory conditions Stark and Walthall (2003) reported the 48-h LC50s summarized in Table 3. Based on LC50 data, these products are all less toxic than POE-T to the related species *Ceriodaphnia dubia* (1.18×10^3 μg/L) and *D. magna* (1.84×10^3 μg/L) as reported by Rodriguez-Gil et al. 2017a. However, when extinction concentrations (where the intrinsic rate of increase [r_i] in the population is < 1 and the population will go extinct) were modeled from population growth in 10-day chronic bioassays with *D. pulex*, there was no obvious relationship to 48-h toxicity. Extinction concentration is a novel concept in estimating risk to organisms in general and has not yet been used in a regulatory

Table 3 LC50 values and extinction concentrations for several agricultural adjuvants tested in *Daphnia pulex* under laboratory conditions

Product[a]	48-h LC50 (μg/L)[b]	Extinction concentration (μg/L)[b]
R-11®	13.2×10^3	0.9×10^3
X-77®	16.4×10^3	1.6×10^6
Sylgard 309®	22.9×10^3	18×10^3
Silwet L-77®	23.4×10^3	28×10^3
Kinetic®	111×10^3	25×10^3
Bond®	614×10^3	450×10^3
Plyac®	2.6×10^6	610×10^3
Water Maxx®	16.3×10^6	13×10^3

[a]Where available, data on constituents, CAS numbers and distributors are provided in SI Table 4
[b]Toxicity values and extinction concentrations from data in Stark and Walthall (2003)

context. Chronic tests conducted under laboratory conditions in plastic test vessels do not necessarily represent the natural aquatic environment where environmental fate would be expected to be different, thus, extrapolation to responses in the environment would be difficult.

In their review of overwater use of glyphosate, Solomon and Thompson (2003) characterized the toxicity of several surfactants that could be used with glyphosate. Because the toxicity data were sparse and were mostly limited to three standard test organisms (bluegill sunfish, *Lepomis macrochirus* Rafinesque; rainbow trout, *Oncorhynchus mykiss* Walbaum; and daphnia, *Daphnia magna* Straus) data were characterized using an SSD and the tenth centile of the toxicity data was used as a point of departure rather than the more usual fifth centile. These data (Fig. 9) show that of all the surfactants tested, POEA is the most toxic and LI700 the least toxic.

More commonly, GBHs already contain the required formulants for the desired herbicidal performance. The original formulations of Roundup contained POE-T. However, more recent formulations can include any number of other surfactant options (or mixtures of) such as etheramine alkoxylates which can allow for more concentrated formulations (resulting in less expensive packaging, shipping and storage costs), especially those of the potassium salt of glyphosate which is known to be incompatible with traditional POEA surfactants (Wright et al. 1998). In recent years, newer formulations developed for European markets include other surfactant alternatives such as propoxylated quaternary ammonium surfactant (Prill and Richardson 1997). Recently Mesnage et al. (2019), reviewed the toxicity of a number of these new surfactant options, mainly from publicly available toxicity data from registration materials and product Safety Data Sheets (SDS). This review indicates that, while the toxicity of etheramine alkoxylates to aquatic organisms is generally similar to that of traditional POEA-based formulations, the newer surfactant options available in the European markets show lower toxicity, with LC50 values 1 to 2 orders of magnitude higher. Mesnage et al. (2019), have discussed the surfactants in newer GHBs and point to the recent introduction of several classes of surfactants such as the propoxylated quaternary ammonium (in Roundup Biactive®), alkypolyglycocide (in Roundup Pro Bio®), polyethylene ether amine (Roundup

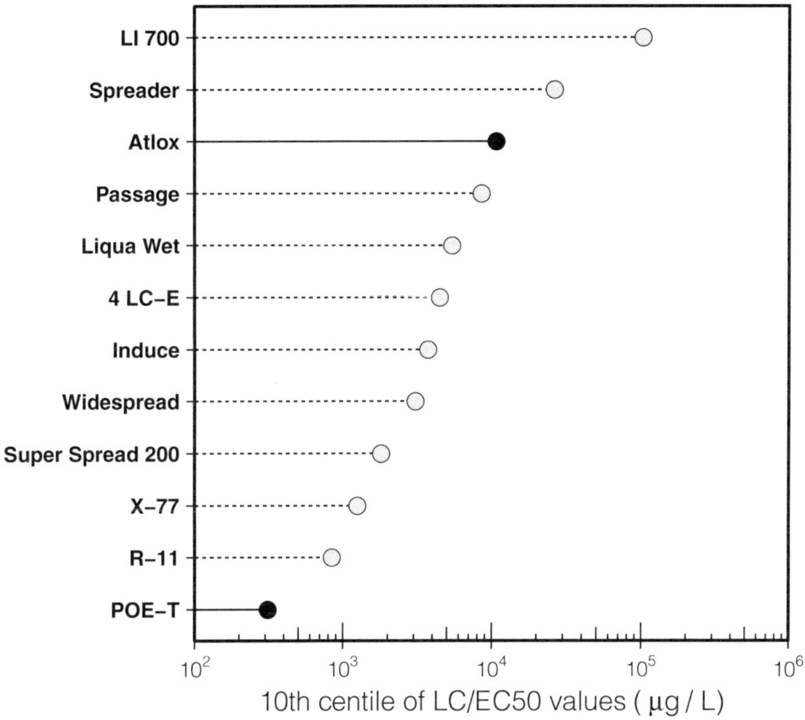

Fig. 9 Toxicity of several adjuvants considered for use with POEA-free glyphosate. The tenth centile value for POE-T (highlighted) was obtained from an SSD plotted from data in (Rodriguez-Gil et al. 2017a), the tenth centile for Atlox (an alkylpolysacharide surfactant, highlighted) was based on an SSD generated from toxicity tests on five aquatic organisms, and the tenth centile values for the other adjuvants are from (Solomon and Thompson 2003). Where available, data on constituents, CAS numbers, and distributors are provided in SI Table 5

GT®), alkoxylated etheramine oxide (in Roundup Full®), and 'N-N-dimethyl-C12-C14-(even numbered)-alkyl-1-amine (in Roundup Pro 360®) surfactants in the EU and Argentina, a trend that they state is not limited to agricultural products but can also be observed in industrial, household, and personal care products. The authors also point out that there is little ecotoxicological information on these surfactants but that many are less toxic than POE-T when tested in cell lines in vitro (Mesnage et al. 2019).

The use of alkypolyglycocide surfactants in agricultural applications, and in glyphosate formulations in particular, has been previously assessed in a number of studies (Fig. 9). Of these surfactants, Atlox AL 2575W (an alkyl alkypolyglycocide marketed by Croda Crop Care), has previously been used in Australia as a component of Crop Care's Touchdown® (NRA 1996). The toxicity of this surfactant is compared to those evaluated by (Solomon and Thompson 2003) and that of POE-T (MON 0818) (Rodriguez-Gil et al. 2017a) in Fig. 9, highlighting the low toxicity of this group of surfactants. GBHs containing glyphosate and alkypolysaccharide

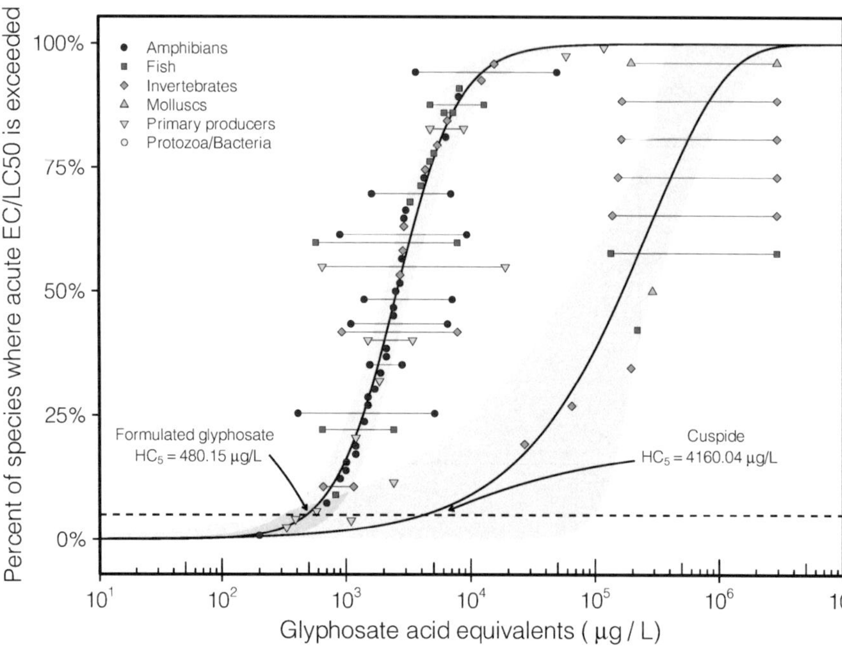

Fig. 10 Acute toxicity of glyphosate formulated with POEAs as compared to Cúspide 480SL® formulated with an alkypolysaccharide surfactant and contained the adjuvant Cosmo-Flux F411. Data from (Currie et al. 2015). Distribution parameters and centile intercepts are provided in SI Table 2 and SI Table 3

surfactants have also been tested. This is the case of Cúspide 480SL®, a Colombian GBH formulation that was used for a short period for the control of *Erythroxylum coca* L., the source of cocaine, in Colombia, South America. This control program has recently resumed with the use of Cúspide 480SL® (Talanú Chemical Ltda, Ibagué, Colombia). Toxicity testing with this product in the laboratory (Currie et al. 2015) showed that most aquatic organisms were considerably less sensitive to this formulation than to those containing POEAs. An updated SSD (Fig. 10) was generated employing the same methodologies used for all other datasets in this review. This SSD suffers from large variability due to a large number of the species showing right-censored toxicity data (i.e., the EC50 was above the highest concentration tested). Despite our re-analysis, which is able to handle this censored data, the high proportion of these cases results in increased variability. For visualization purposes the range showed for these data points has a maximum value chosen as ten times the maximum measured of the distribution. In addition, the two most sensitive points in the SSD were primary producers (algae), which were likely directly affected by the herbicidal action of glyphosate in the Cúspide formulation.

The use of alkypolysaccharide surfactants is expected to increase, as its implementation in different formulation products becomes more common among the main manufacturers of GBH. In fact, in addition to practical aspects (e.g., they work well in formulations containing triethanolamine salts of glyphosate), mammalian toxicology of these products was reviewed in Knaak et al. (1997) and environmental aspects are also mentioned as part of recent formulation development (e.g., Yeritsyan 2016).

5 Ecotoxicology of the Degradation Product, AMPA

Only one degradation product of glyphosate in plants and in the environment is of potential concern is aminomethylphosphonic acid AMPA (Fig. 3). This degradate is mainly of concern because of its mobility in the environment and its relatively longer persistence (Bento et al. 2016; Grandcoin et al. 2017; Levine et al. 2015) as compared to the parent and its relatively greater bioavailability in soils (smaller K_{OC}).

The toxicity of AMPA to aquatic organisms has been investigated in a well conducted study (Levine et al. 2015). Chronic toxicity of AMPA to fathead minnow (*P. promelas*) was evaluated in a fish early life stage study. The NOEC based on larval survival, growth, and development $\geq 12 \times 10^3$ µg/L. Chronic toxicity to *D. magna* was evaluated in a 21-d reproduction test and, based on the endpoints of survival, growth, and reproduction, the NOEC was 15×10^3 µg/L. The authors noted that the maximum reported concentrations of AMPA in surface waters from ten different countries were less than 50 µg/L with only two above that value (397 µg/L in the USA and > 200 µg/L in Ireland). All these values were at least two orders of magnitude less than the NOECs from the toxicity tests. In another study, the toxicity of AMPA was evaluated in the saltwater oyster (*Crassostrea gigas* Thunberg), the freshwater (FW) alga (*Raphidocelis subcapitata* (Korshikov) Nygaard, 1987 [formerly known as *Selenastrum capricornutum* and *Pseudokirchneriella subcapitata*] and the FW crustacean *D. magna*) (Di Poi et al. 2018). The EC10s (used as an approximation to the NOEC) for embryotoxicity and metamorphosis in *C. gigas* were 39×10^3 µg/L and 27×10^3 µg/L, respectively. The EC10 for growth in *R. subcapitata* was 85×10^3 µg/L and that for immobility in *D. magna* was $> 100 \times 10^3$ µg/L. The results of these two papers are consistent and indicate that risks from exposure to AMPA in surface waters are *de minimis*.

Because AMPA is found in soils of fields treated with glyphosate, its potential risks to soil organisms were evaluated. In its evaluation of glyphosate, EFSA (2015) noted low toxicity in the soil mite (*H. aculeifer*) and the springtail (*F. candida*) after 14 and 28 d exposures to AMPA (NOECs = 320×10^3 and 315×10^3 µg/kg dry soil, respectively). Similar low toxicity was noted for the earthworms *Eisenia andrei* Bouché (14-d LC50 > 1×10^6 µg AMPA/kg dry weight of soil) and for *E. fetida* a 56-day NOEC of 132×10^3 µg/kg dry weight (EFSA 2015). For microbial activity in soil, EFSA (2015) noted values of 21% reduction in mineralization of N at 160×10^3 µg AMPA/kg dry weight of soil and, for mineralization of C and 18% reduction at 94×10^3 µg AMPA/kg dry weight. In another study on soil biota, (von Merey et al. 2016) no effects were observed on transformation of nitrogen at a concentration 160×10^3 μg AMPA/kg dry soil. At greater concentrations (320×10^3 and 640×10^3 μg/kg dry weight) rates of transformation were 27 and 43% above control. The authors suggested that these increases were a result of stimulation of the microbiota by the increased availability of nitrogen and phosphorus from the degradation of AMPA. In tests on the predatory mite, (*H. aculeifer*), the reproductive $NOEC_{reproduction}$ was $\geq 320 \times 10^3$ μg/kg dry weight (von Merey et al. 2016). These authors concluded that the there was a very low likelihood of adverse effects of AMPA on soil biota or soil processes.

Overall, the toxicity data indicate that risks from AMPA to aquatic and soil organisms and soil processes are *de minimis*.

6 Conclusions

Globally, glyphosate is the most widely used pesticide. There have been many studies conducted on glyphosate for regulatory purposes as well as a large number in the open scientific literature. In addition, it has recently been reviewed by several national and global regulatory agencies. Its chemistry and fate in the environment are well characterized. It is essentially non-volatile and is not transported in air other than in the form of spray drift or sorbed to dust. It is soluble in water but sorbs strongly to organic matter and clay particles in soil, thus limiting its transport away from soil in treated fields. It is metabolized in plants and soils to several products that are of lower toxicity and not of toxicological concern. One of these products (AMPA) is commonly found in surface waters but can also be formed from aminophosphonates used for water treatments such as descaling and from use of some organophosphate detergents. Thus, not all AMPA in surface waters is necessarily derived from glyphosate and it is inappropriate to stoichiometrically sum AMPA and glyphosate to derive loadings to surface waters. In animals, glyphosate is rapidly cleared from the body and does not bioaccumulate or biomagnify in trophic food webs. Because of its short biological *t*½ in water, sediment, and tissues, exposures of organisms in the environment are acute, and therefore acute toxicity data are the most relevant for assessment of direct effects on non-target organisms.

Because of its high solubility in water and low octanol–water partition coefficient, glyphosate does not penetrate well into target plants and has low efficacy unless a formulant, such as a surfactant, is added to the formulation and/or spray tank. In addition to lowering the surface tension of water to promote rapid spreading on leaf surfaces, the surfactant also disrupts cuticular waxes and speeds up penetration into the tissues of the leaf from which it moves via the phloem to rest of the plant.. The most used formulant in GBHs is a mixture of polyoxyethylene amines (POEAs); however, some newer GBHs contain different formulants. By design, formulants have very different intensive properties from the pesticides they are combined with and the mixture has different physical properties from the active ingredient. Because of this, the active ingredient and the formulants in the commercial products have different fates in the environment and do not necessarily move at the same rate through the environment as the active pesticidal substance. Thus, for GBHs (and other pesticides) formulants are not likely to be found in surface waters in the same ratio as in the commercial product unless there is a spill or a deliberate overspray.

Technical glyphosate is of low toxicity to aquatic and terrestrial animals because, at permissible environmental use rates it inhibits EPSPS in the biochemical pathway for the synthesis of aromatic amino acids that is only known to be expressed in plants and some bacteria and fungi and not animals. Toxicity values for glyphosate for many species have been reported in the literature but some of these report concentrations without specifying the form (acid, or salt) of the test substance. All concentrations of glyphosate should be normalized to acid equivalents when characterizing environmental concentrations and toxicity so that data from different studies can be more easily compared.

The fifth centile (i.e., HC5) is a commonly used metric to characterize the distribution of toxicity endpoints among a diversity of test organisms. The HC5 for technical glyphosate in aquatic organisms is 4.5×10^3 µg glyphosate a.e./L, whereas the HC5 for formulations containing POEAs is 480 µg glyphosate a.e./L. Clearly, the POEAs significantly contribute to the toxicity of the commercial product across a diversity of taxa (This is clearly presented in Fig. 8b). As such, toxicological data from tests conducted with formulated product cannot be directly used to characterize the risk of glyphosate alone. The use of off-the-shelf home and garden products for toxicity testing is also inappropriate as these products sometimes contain a second herbicide, such as a fatty acid, that provides rapid kill of the plants (Bugg et al. 1999). These mixtures are not representative of commercial products used in agriculture and forestry, which represents the bulk of the herbicide use. This is the source of considerable confusion and wasted effort and resources in the plethora of toxicity tests conducted with commercial formulations of glyphosate with unknown formulants. GBHs containing POEAs are not registered for aquatic uses to control emergent vegetation and are illegal when sprayed directly on water or allowed to drift to water. Therefore, non-target organisms are unlikely to be exposed to POEAs unless applied illegally.

The distribution of toxicity values for aquatic organisms were compared to the concentrations in surface waters of the US reported in the National Water Quality Monitoring Council database as well as data from the French Naïdes database. The resulting environmental exposure distributions for both datasets did not exceed (the estimated exceedance, and associated confidence interval was zero) the fifth centile of the distribution of toxicity values (SSD) of the glyphosate technical product (4.49×10^3 µg/L). In fact, the highest measured concentration (940 µg/L) was over 4 times lower than this value and still lower than the first centile of the SSD (1.64×10^3 µg/L). It is important to note that the 99th centiles of the exposure distributions were even lower at 1.54 and 8.45 µg/L for the French and US datasets respectively, increasing the margin of safety between exposure and acute effects.

Insufficient numbers and diversity of terrestrial animal studies yielding toxicological endpoints (e.g., LD50) have been published to conduct a risk assessment using species-sensitivity distributions and terrestrial residues. However, the available data do not suggest that the use of GBHs is likely cause adverse effects in terrestrial insects, such as honeybees, terrestrial vertebrates, arthropods and invertebrates in soil and microbiota that process nutrients and carbon-containing compounds in soil.

The toxicity of AMPA, one of the major degradates of glyphosate in plants and the environment has been investigated. Overall, the toxicity data indicate that risks from AMPA to aquatic and soil organisms and soil processes are *de minimis*. However, exposures in aquatic systems might result from use of other substances that also degrade into AMPA. Currently, concentrations of AMPA in surface waters present a *de minimis* risk. These conclusions are similar to those reached by regulatory agencies such as the USEPA in the United States, EFSA in Europe, PMRA in Canada, the AVPMA in Australia, the Food and Agricultural Organization of the United Nations, and the FSCJ in Japan.

Acknowledgments The authors thank Bayer Crop Science for providing complete copies of several toxicity studies that were submitted to regulatory agencies as well as data from recently conducted studies on the environmental fate of POE-T. We also thank the anonymous reviewers of the manuscript for their useful suggestions, all of which have improved the chapter.

Declaration of Interest J-LR-G and KRS contributed equally to two-thirds of this chapter and RSP and SOD contributed equally the other third. All authors declare no conflict of interest. The authors of this chapter were all self-supported and no funding was provided by any of the registrants of glyphosate. The opinions expressed in this chapter are those of the authors alone and do not necessarily reflect those of their respective institutions. Mention of a trade name of a product does not constitute a recommendation for its use.

References

Aitbali Y, Ba-M'hamed S, Elhidar N, Nafis A, Soraa N, Bennis M (2018) Glyphosate based-herbicide exposure affects gut microbiota, anxiety and depression-like behaviors in mice. Neurotox Teratolog 67:44–49. https://doi.org/10.1016/j.ntt.2018.04.002

Allegrini M, Zabaloy MC, Gómez EV (2015) Ecotoxicological assessment of soil microbial community tolerance to glyphosate. Sci Tot Environ 533:60–68. https://doi.org/10.1016/j.scitotenv.2015.06.096

Anadón A et al (2009) Toxicokinetics of glyphosate and its metabolite aminomethyl phosphonic acid in rats. Toxicol Lett 190:91–95. https://doi.org/10.1016/j.toxlet.2009.07.008

Andersson M (2012) Acute toxicity to *Daphnia magna* in River Water; investigating mitigation and bioavailability of pure cationic surfactants and mixtures with SPME. M.Sc. thesis University of Gothenburg

Andréa MMD, Peres TB, Luchini LC, Bazarin S, Papini S, Matallo MB, Savoy VLT (2003) Influence of repeated applications of glyphosate on its persistence and soil bioactivity Pesqu Agropecu. Bras 38:1329–1335

Annett R, Habibi HR, Hontela A (2014) Impact of glyphosate and glyphosate-based herbicides on the freshwater environment. J Appl Toxicol 34:458–479. https://doi.org/10.1002/jat.2997

APVMA (2017) Final regulatory position: Consideration of the evidence for a formal reconsideration of glyphosate. Australian Pesticides and Veterinary Medicines Authority, Canberra, ACT, Australia

Artigas J, Batisson I, Carles L (2020) Dissolved organic matter does not promote glyphosate degradation in auto-heterotrophic aquatic microbial communities. Environ Pollut 259:113951. https://doi.org/10.1016/j.envpol.2020.113951

ATSDR (2019) Toxicological Profile for Glyphosate (Draft for public comment). U.S. Department of Health and Human Services, Public Health Service. Agency for Toxic Substances and Disease Registry, Atlanta, GA, USA

Balbuena MS, Tison L, Hahn M-L, Greggers U, Menzel R, Farina WM (2015) Effects of sublethal doses of glyphosate on honeybee navigation. J Exp Biol 218:2799–2805. https://doi.org/10.1242/jeb.117291

BCPC (2012) The pesticide manual. 16. British Crop Protection Council, Alton, Hampshire, UK

Bednářová A, Kropf M, Krishnan N (2020) The surfactant polyethoxylated tallowamine (POEA) reduces lifespan and inhibits fecundity in *Drosophila melanogaster*-in vivo and in vitro study. Ecotoxicol Environ Safety 188:109883. https://doi.org/10.1016/j.ecoenv.2019.109883

Bento CPM et al (2016) Persistence of glyphosate and aminomethylphosphonic acid in loess soil under different combinations of temperature, soil moisture and light/darkness. Sci Tot Environ 572:301–311. https://doi.org/10.1016/j.scitotenv.2016.07.215

Berg CJ, King HP, Delenstarr G, Kumar R, Rubio F, Glaze T (2018) Glyphosate residue concentrations in honey attributed through geospatial analysis to proximity of large-scale agriculture and transfer off-site by bees. PLoS One 13:e0198876. https://doi.org/10.1371/journal.pone.0198876

Berman MC et al (2020) Field evidence supports former experimental claims on the stimulatory effect of glyphosate on picocyanobacteria communities. Sci Tot Environ 701:134601. https://doi.org/10.1016/j.scitotenv.2019.134601

Bernal MH, Solomon KR, Carrasquilla G (2009) Toxicity of formulated glyphosate (Glyphos®) and Cosmo-Flux® to larval and juvenile Colombian frogs 2. Field and laboratory microcosm acute toxicity. J Toxicol Environ Hlth A 72:966–973. https://doi.org/10.1080/15287390902929717

Berry C (2020) Glyphosate and cancer: the importance of the whole picture. Pest Manag Sci. https://doi.org/10.1002/ps.5834

Bidwell JR, Gorrie JR (1995) Acute toxicity of a herbicide to selected frog species: Final Report. Western Australian Dept. of Environmental Protection, Perth, W.A

Blake R, Pallett K (2018) The environmental fate and ecotoxicity of glyphosate. Outlooks Pest Manag 29:266–269. https://doi.org/10.1564/v29_dec_08

Bo Nielsen J, Ahm Sørensen J, Nielsen F (2009) The usual suspects—influence of physicochemical properties on lag time, skin deposition, and percutaneous penetration of nine model compounds. J Toxicol Environ Health A 72:315–323. https://doi.org/10.1080/15287390802529872

Boily M, Sarrasin B, DeBlois C, Aras P, Chagnon M (2013) Acetylcholinesterase in honey bees (*Apis mellifera*) exposed to neonicotinoids, atrazine and glyphosate: laboratory and field experiments. Environ Sci Poll Res 20:5603–5614. https://doi.org/10.1007/s11356-013-1568-2

Bonfleur EJ, Lavorenti A, Tornisielo VL (2010) Mineralization and degradation of glyphosate and atrazine applied in combination in a Brazilian Oxisol. J Environ Sci Hlth B 46:69–75. https://doi.org/10.1080/03601234.2011.534384

Borggaard OK, Gimsing AL (2008) Fate of glyphosate in soil and the possibility of leaching to ground and surface waters: a review. Pest Manag Sci 64:441–456. https://doi.org/10.1002/ps.1512

Bott S, Tesfamariam T, Kania A, Eman B, Aslan N, Römheld V (2011) Phytotoxicity of glyphosate soil residues re-mobilised by phosphate fertilization. Plant and Soil 342:249–263. https://doi.org/10.1007/s11104-010-0689-3

Botta F et al (2009) Transfer of glyphosate and its degradate AMPA to surface waters through urban sewerage systems. Chemosphere 77:133–139. https://doi.org/10.1016/j.chemosphere.2009.05.008

Brausch J, Smith P (2007) Toxicity of three polyethoxylated tallowamine surfactant formulations to laboratory and field collected fairy shrimp, *Thamnocephalus platyurus*. Arch Environ Contam Toxicol 52:217–221. https://doi.org/10.1007/s00244-006-0151-y

Brausch J, Beall B, Smith P (2007) Acute and sub-lethal toxicity of three POEA surfactant formulations to *Daphnia magna*. Bull Environ Contam Toxicol 78:510–514. https://doi.org/10.1007/s00128-007-9091-0

Bringolf RB, Cope WG, Mosher S, Barnhart MC, Shea D (2007) Acute and chronic toxicity of glyphosate compounds to glochidia and juveniles of *Lampsilis siliquoidea* (Unionidae). Environ Toxicol Chem 26:2094–2100. https://doi.org/10.1897/06-519R1.1

Bugg MW, Arnold KA, White RJ (1999) Method of preparing glyphosate herbicide formulations. United States of America Patent 5,994,269 Patent

Byer JD, Struger J, Klawunn P, Todd A, Sverko E (2008) Low cost monitoring of glyphosate in surface waters using the ELISA method: an evaluation. Environ Sci Technol 42:6052–6057. https://doi.org/10.1021/es8005207

Carles L, Artigas J (2020) Interaction between glyphosate and dissolved phosphorus on bacterial and eukaryotic communities from river biofilms. Sci Tot Environ 719:137463. https://doi.org/10.1016/j.scitotenv.2020.137463

Carles L, Gardon H, Joseph L, Sanchis J, Farre M, Artigas J (2019) Meta-analysis of glyphosate contamination in surface waters and dissipation by biofilms. Environ Int 124:284–293. https://doi.org/10.1016/j.envint.2018.12.064

Carretta L, Cardinali A, Marotta E, Zanin G, Masin R (2019) A new rapid procedure for simultaneous determination of glyphosate and AMPA in water at sub μg/L level. J Chromatogr A 1600:65–72. https://doi.org/10.1016/j.chroma.2019.04.047

Castro Berman M, Marino DJG, Quiroga MV, Zagarese H (2018) Occurrence and levels of glyphosate and AMPA in shallow lakes from the Pampean and Patagonian regions of Argentina. Chemosphere 200:513–522. https://doi.org/10.1016/j.chemosphere.2018.02.103

Cedergreen N, Streibig JC (2005) The toxicity of herbicides to non-target aquatic plants and algae: Assessment of predictive factors and hazard. Pest Manag Sci 61:1152–1160. https://doi.org/10.1002/ps.1117

Chen Y, Geurts M, Sjollema SB, Kramer NI, Hermens JLM, Droge STJ (2014) Acute toxicity of the cationic surfactant C12-benzalkonium in different bioassays: how test design affects bioavailability and effect concentrations. Environ Toxicol Chem 33:606–615. https://doi.org/10.1002/etc.2465

Connors KA et al (2019) Creation of a curated aquatic toxicology database: EnviroTox. Environ Toxicol Chem 38:1062–1073. https://doi.org/10.1002/etc.4382

Corbera M, Simonet B, Salvado V, Hidalgo M (2010) Characterisation of alkylamine ethoxylates (ANEOs) in commercial herbicide formulations using liquid chromatography/electrospray ionisation mass spectrometry. Rapid Comm Mass Spectrom 24:2931–2937

Coupe RH, Kalkhoff SJ, Capel PD, Gregoire C (2012) Fate and transport of glyphosate and aminomethylphosphonic acid in surface waters of agricultural basins. Pest Manag Sci 68:16–30. https://doi.org/10.1002/ps.2212

Crump K (2020) The potential effects of recall bias and selection bias on the epidemiological evidence for the carcinogenicity of glyphosate. Risk Anal 40:696–704. https://doi.org/10.1111/risa.13440

Crump K, Crouch E, Zelterman D, Crump C, Haseman J (2020) Accounting for multiple comparisons in statistical analysis of the extensive bioassay data on glyphosate. Toxicol Sci 175:156–167. https://doi.org/10.1093/toxsci/kfaa039

Currie Z, Prosser RS, Rodriguez-Gil JL, Mahon K, Poirier D, Solomon KR (2015) Toxicity of Cúspide 480SL® spray mixture formulation of glyphosate to aquatic organisms. Environ Toxicol Chem 34:1178–1184. https://doi.org/10.1002/etc.2913

Dai P et al (2018) The herbicide glyphosate negatively affects midgut bacterial communities and survival of honey bee during larvae reared in vitro. J Agric Food Chem 66:7786–7793. https://doi.org/10.1021/acs.jafc.8b02212

Deese RD, LeBlanc MR, Cook RL (2016) Surfactant toxicity to *Artemia franciscana* and the influence of humic acid and chemical composition. Environ Chem 13:507–516. https://doi.org/10.1071/EN15108

Demetrio PM, Bonetto C, Ronco AE (2014) The effect of cypermethrin, chlorpyrifos, and glyphosate active ingredients and formulations on *Daphnia magna* (Straus). Bull Environ Contam Toxicol 93:268–273. https://doi.org/10.1007/s00128-014-1336-0

Di Fiori E, Pizarro H, dos Santos AM, Cataldo D (2012) Impact of the invasive mussel *Limnoperna fortunei* on glyphosate concentration in water. Ecotoxicol Environ Safety 81:106–113. https://doi.org/10.1016/j.ecoenv.2012.04.024

Di Poi C, Costil K, Bouchart V, Halm-Lemeille M-P (2018) Toxicity assessment of five emerging pollutants, alone and in binary or ternary mixtures, towards three aquatic organisms. Environ Sci Poll Res 25:6122–6134. https://doi.org/10.1007/s11356-017-9306-9

Dill GM et al (2010) Glyphosate: discovery, development, applications, and properties. In: Nandula VK (ed) Glyphosate resistance in crops and weeds: history, development, and management. Wiley, New York, pp 1–33

Druart C, Millet M, Scheifler R, Delhomme O, De Vaufleury A (2011) Glyphosate and glufosinate-based herbicides: Fate in soil, transfer to, and effects on land snails. J Soils Sediments 11:1373–1384. https://doi.org/10.1007/s11368-011-0409-5

Duke SO (2018) The history and current status of glyphosate. Pest Manag Sci 74:1027–1034. https://doi.org/10.1002/ps.4652

Duke SO (2020) Glyphosate: uses other than in glyphosate-resistant crops, mode of action, degradation in plants, and effects on non-target plants and agricultural microbes. Rev Environ Contam Toxicol 255. https://doi.org/10.1007/398_2020_53

EC (2017) Commission implementing regulation (EU) 2017/2324 of 12 December 2017 renewing the approval of the active substance glyphosate in accordance with Regulation (EC) No 1107/2009 of the European Parliament and of the Council concerning the placing of plant protection products on the market, and amending the Annex to Commission Implementing Regulation (EU) No 540/2011. Text with EEA relevance Official Journal of the European Union 10:L333

Edge C, Thompson D, Hao C, Houlahan J (2014) The response of amphibian larvae to exposure to a glyphosate-based herbicide (Roundup WeatherMax) and nutrient enrichment in an ecosystem experiment. Ecotoxicol Environ Safety 109:124–132. https://doi.org/10.1016/j.ecoenv.2014.07.040

Edge CB, Thompson DG, Hao C, Houlahan JE (2012) A silviculture application of the glyphosate-based herbicide VisionMax to wetlands has limited direct effects on amphibian larvae. Environ Toxicol Chem 31:2375–2383. https://doi.org/10.1002/etc.1956

Edge CB et al (2020) Compensatory indirect effects of an herbicide on wetland communities. Sci Tot Environ 718:137254. https://doi.org/10.1016/j.scitotenv.2020.137254

Edginton A, Sheridan P, Boermans H, Thompson D, Holt J, Stephenson G (2004) A comparison of two factorial designs, a complete 3X3 factorial and a central composite rotatable design, for use in binomial response experiments in aquatic toxicology. Arch Environ Contam Toxicol 46:216–223. https://doi.org/10.1007/s00244-003-2176-9

EFSA (2015) Conclusion on the peer review of the pesticide risk assessment of the active substance glyphosate. EFSA J 13:4302. https://doi.org/10.2903/j.efsa.2015.4302

EFSA (2018) Scientific Report on evaluation of the impact of glyphosate and its residues in feed on animal health. EFSA J 16:5283

El Agrebi N, Tosi S, Wilmart O, Scippo M-L, de Graaf DC, Saegerman C (2020) Honeybee and consumer's exposure and risk characterisation to glyphosate-based herbicide (GBH) and its degradation product (AMPA): residues in beebread, wax, and honey. Sci Tot Environ 704:135312. https://doi.org/10.1016/j.scitotenv.2019.135312

Faghani M (2018) Effect of glyphosate on honey bee (*Apis mellifera*) performance. Arthropods 7:77–81. https://doi.org/

Faita MR, de Medeiros OE, Júnior VVA, Orth AI, Nodari RO (2018) Changes in hypopharyngeal glands of nurse bees (*Apis mellifera*) induced by pollen-containing sublethal doses of the herbicide Roundup®. Chemosphere 211:566–572. https://doi.org/10.1016/j.chemosphere.2018.07.189

Farina WM, Balbuena MS, Herbert LT, Mengoni Gonalons C, Vazquez DE (2019) Effects of the herbicide glyphosate on honey bee sensory and cognitive abilities: Individual impairments with implications for the hive. Insects 10. https://doi.org/10.3390/insects10100354

Folmar LC, Sanders HO, Julin AM (1979) Toxicity of the herbicide glyphosate and several of its formulations to fish and aquatic invertebrates. Arch Environ Contam Toxicol 8:269–278. https://doi.org/10.1007/BF01056243

Foy CL (ed) (1992) Adjuvants for agrichemicals. CRC Press, Boca Raton, FL

Franz JE, Mao MK, Sikorski JA (eds) (1997) Glyphosate: a unique global herbicide vol 189. ACS Monograph. American Chemical Society, Washington, DC

Fritz-Wallace K et al (2020) Quantification of glyphosate and aminomethylphosphonic acid from microbiome reactor fluids Rapid. Comm Mass Spectrom 34:e8668. https://doi.org/10.1002/rcm.8668

Frontera J, Vatnick I, Chaulet A, Rodríguez E (2011) Effects of glyphosate and polyoxyethylenamine on growth and energetic reserves in the freshwater crayfish *Cherax quadricarinatus* (Decapoda, Parastacidae). Arch Environ Contam Toxicol 61:590–598. https://doi.org/10.1007/s00244-011-9661-3

Gaupp-Berghausen M, Hofer M, Rewald B, Zaller JG (2015) Glyphosate-based herbicides reduce the activity and reproduction of earthworms and lead to increased soil nutrient concentrations. Sci Rep 5:12886. https://doi.org/10.1038/srep12886

GC (2019) Draft screening assessment – poly(alkoxylates/ethers) group. Environment and climate change Canada. Health Canada, Ottawa

Giesy JP, Dobson S, Solomon KR (2000) Ecotoxicological risk assessment for Roundup® herbicide. Rev Environ Contam Toxicol 167:35–120. https://doi.org/10.1007/978-1-4612-1156-3_2

Goñalons CM, Farina WM (2018) Impaired associative learning after chronic exposure to pesticides in young adult honey bees. J Exp Biol 221:jeb176644. https://doi.org/10.1242/jeb.176644

Grandcoin A, Piel S, Baurès E (2017) AminoMethylPhosphonic acid (AMPA) in natural waters: Its sources, behavior and environmental fate. Water Res 117:187–197. https://doi.org/10.1016/j.watres.2017.03.055

Gros P, Meissner R, Wirth MA, Kanwischer M, Rupp H, Schulz-Bull DE, Leinweber P (2020) Leaching and degradation of $^{13}C_2$-^{15}N-glyphosate in field lysimeters. Environ Monit Assess 192:127. https://doi.org/10.1007/s10661-019-8045-4

Guilherme S, Santos MA, Barroso C, Gaivão I, Pacheco M (2012) Differential genotoxicity of Roundup® formulation and its constituents in blood cells of fish (*Anguilla anguilla*): considerations on chemical interactions and DNA damaging mechanisms. Ecotoxicol 21:1381–1390. https://doi.org/10.1007/s10646-012-0892-5

Hao C, Morse D, Morra F, Zhao X, Yang P, Nunn B (2011) Direct aqueous determination of glyphosate and related compounds by liquid chromatography/tandem mass spectrometry using reversed-phase and weak anion-exchange mixed-mode column. J Chromatogr A 1218:5638–5643. https://doi.org/10.1016/j.chroma.2011.06.070

HC (2019) Statement from Health Canada on Glyphosate. Health Canada, Government of Canada. https://www.canada.ca/en/health-canada/news/2019/01/statement-from-health-canada-on-glyphosate.html. Accessed January 2019

Herbert LT, Vázquez DE, Arenas A, Farina WM (2014) Effects of field-realistic doses of glyphosate on honeybee appetitive behaviour. J Exp Biol 217:3457–3464. https://doi.org/10.1242/jeb.109520

Leaper C, Holloway PJ (2000) Adjuvants and glyphosate activity. Pest Manag Sci 56:313–319. https://doi.org/10.1002/(SICI)1526-4998(200004)56:4<313::AID-PS147>3.0.CO;2-3

IARC (2015) Some Organophosphate Insecticides and Herbicides IARC Monographs on the Evaluation of Carcinogenic Risks to Humans Volume 112, Glyphosate vol 112–112. International Agency for Research on Cancer, Leon, France

Iori S et al (2020) The effects of glyphosate and AMPA on the mediterranean mussel *Mytilus galloprovincialis* and its microbiota. Environ Res 182:108984. https://doi.org/10.1016/j.envres.2019.108984

Jacobsen CS, Hjelmsø MH (2014) Agricultural soils, pesticides and microbial diversity. Curr Opin Biotechnol 27:15–20. https://doi.org/10.1016/j.copbio.2013.09.003

Janßen R, Skeff W, Werner J, Wirth MA, Kreikemeyer B, Schulz-Bull D, Labrenz M (2019) A glyphosate pulse to brackish long-term microcosms has a greater impact on the microbial diversity and abundance of planktonic than of biofilm assemblages. Front Mar Sci 6:758. https://doi.org/10.3389/fmars.2019.00758

Janssens L, Stoks R (2017) Stronger effects of Roundup than its active ingredient glyphosate in damselfly larvae. Aquat Toxicol 193:210–216. https://doi.org/10.1016/j.aquatox.2017.10.028

JMPR (2016) Summary Report for Diazinon, Glyphosate, Malathion. Food and Agriculture Organization of the United Nations / Geneva, World Health Organization (WHO). Joint FAO/WHO Meeting on Pesticide Residues (JMPR). 6 p Available from: http://www.who.int/foodsafety/jmprsummary2016.pdf?ua=1

Kepler RM et al (2020) Soil microbial communities in diverse agroecosystems exposed to the herbicide glyphosate. Appl Environ Microbiol 86:e01744–e01719. https://doi.org/10.1128/AEM.01744-19

Khadra M, Planas D, Girard C, Amyot M (2018) Age matters: submersion period shapes community composition of lake biofilms under glyphosate stress. FACETS 3:934–951. https://doi.org/10.1139/facets-2018-0019

Kishore G, Jacob GS (1987) Degradation of glyphosate by Pseudomonas sp. PG2982 via a sarcosine intermediate. J Biol Chem 262:12164–12168

Klátyik S et al (2017) Dissipation of the herbicide active ingredient glyphosate in natural water samples in the presence of biofilms. Int J Environ Anall Chem 97:901–921. https://doi.org/10.1080/03067319.2017.1373770

Knaak J, Leung H-W, Stott W, Busch J, Bilsky J (1997) Toxicology of mono-, di-, and triethanolamine. Rev Environ Contam Toxicol 149:1–86. https://doi.org/10.1007/978-1-4612-2272-9_1

Kolpin DW, Thurman EM, Lee EA, Meyer MT, Furlong ET, Glassmeyer ST (2006) Urban contributions of glyphosate and its degradate AMPA to streams in the United States. Sci Tot Environ 354:191–197. https://doi.org/10.1016/j.scitotenv.2005.01.028

Krogh K, Vejrup K, Mogensen B, Halling-Sorensen B (2002) Liquid chromatography-mass spectrometry method to determine alcohol ethoxylates and alkylamine ethoxylates in soil interstitial water, ground water and surface water samples. J Chromatogr A 957:45–57. https://doi.org/10.1016/S0021-9673(02)00077-8

Krogh K, Mogensen B, Halling-Sorensen B, Cortes A, Vejrup K, Barceló D (2003a) Analysis of alcohol ethoxylates and alkylamine ethoxylates in agricultural soils using pressurised liquid extraction and liquid chromatography-mass spectrometry. Anal Bioanal Chem 376:1089–1097. https://doi.org/10.1007/s00216-003-2062-3

Krogh KA, Halling-Sørensen B, Mogensen BB, Vejrup KV (2003b) Environmental properties and effects of nonionic surfactant adjuvants in pesticides: a review. Chemosphere 50:871–901. https://doi.org/10.1016/S0045-6535(02)00648-3

Krüger M, Shehata AA, Schrödl W, Rodloff A (2013) Glyphosate suppresses the antagonistic effect of *Enterococcus spp.* on *Clostridium botulinum*. Anaerobe 20:74–78. https://doi.org/10.1016/j.anaerobe.2013.01.005

Laitenen P, Rämö S, Siimes K (2007) Glyphosate translocation from plants to soil – does it constitute a significant proportion of residues in soil? Plant Soil 300:51–60. https://doi.org/10.1007/s11104-007-9387-1

Lajmanovich RC et al (2015) Harmful effects of the dermal intake of commercial formulations containing chlorpyrifos, 2,4-D, and glyphosate on the common toad *Rhinella arenarum* (Anura: Bufonidae). Water Air Soil Pollut 226:427. https://doi.org/10.1007/s11270-015-2695-9

Lancaster SH, Haney RL, Senseman SA, Kenerley CM, Hons FM (2008) Microbial degradation of fluometuron is influenced by Roundup WeatherMAX. J Agric Food Chem 56:8588–8593. https://doi.org/10.1021/jf801648w

Lancaster SH, Hollister EB, Senseman SA, Gentry TJ (2010) Effects of repeated glyphosate applications on soil microbial community composition and mineralization of glyphosate. Pest Manag Sci 66:59–64. https://doi.org/10.1002/ps.1831

Lanctot C, Robertson C, Navarro-Martin L, Edge C, Melvin SD, Houlahan J, Trudeau VL (2013) Effects of the glyphosate-based herbicide Roundup WeatherMax(R) on metamorphosis of wood frogs (Lithobates sylvaticus) in natural wetlands. Aquat Toxicol 140–141:48–57. https://doi.org/10.1016/j.aquatox.2013.05.012

Lanctôt C, Navarro-Martín L, Robertson C, Park B, Jackman P, Pauli BD, Trudeau VL (2014) Effects of glyphosate-based herbicides on survival, development, growth and sex ratios of wood frog (*Lithobates sylvaticus*) tadpoles. II: Agriculturally relevant exposures to Roundup WeatherMax® and Vision® under laboratory conditions. Aquat Toxicol 154:291–303. https://doi.org/10.1016/j.aquatox.2014.05.025

Levine SL, von Merey G, Minderhout T, Manson P, Sutton P (2015) Aminomethylphosphonic acid has low chronic toxicity to *Daphnia magna* and *Pimephales promelas*. Environ Toxicol Chem 34:1382–1389. https://doi.org/10.1002/etc.2940

Lozano VL et al (2018a) Sex-dependent impact of Roundup on the rat gut microbiome. Tox Rep 5:96–107. https://doi.org/10.1016/j.toxrep.2017.12.005

Lozano VL et al (2018b) Effects of glyphosate and 2,4-D mixture on freshwater phytoplankton and periphyton communities: a microcosms approach. Ecotoxicol Environ Safety 148:1010–1019. https://doi.org/10.1016/j.ecoenv.2017.12.006

Lu G-H et al (2018) Impact of glyphosate on the rhizosphere microbial communities of an EPSPS-transgenic soybean line ZUTS31 by metagenome sequencing. Curr Genomics 19:36–49. https://doi.org/10.2174/1389202918666170705162405

Lu T et al (2020) Understanding the influence of glyphosate on the structure and function of freshwater microbial community in a microcosm. Environ Pollut 260:114012. https://doi.org/10.1016/j.envpol.2020.114012

Lupi L, Bedmar F, Puricelli M, Marino D, Aparicio VC, Wunderlin D, Miglioranza KSB (2019) Glyphosate runoff and its occurrence in rainwater and subsurface soil in the nearby area of agricultural fields in Argentina. Chemosphere 225:906–914. https://doi.org/10.1016/j.chemosphere.2019.03.090

Lutri V et al (2020) Hydrogeological features affecting spatial distribution of glyphosate and AMPA in groundwater and surface water in an agroecosystem. Córdoba, Argentina Sci Tot Environ 711:134557. https://doi.org/10.1016/j.scitotenv.2019.134557

Mac Loughlin TM, Peluso ML, Aparicio VC, Marino DJ (2020) Contribution of soluble and particulate-matter fractions to the total glyphosate and AMPA load in water bodies associated with horticulture. Sci Tot Environ 703:134717. https://doi.org/10.1016/j.scitotenv.2019.134717

Mamy L, Barriuso E, Gabrielle B (2016) Glyphosate fate in soils when arriving in plant residues. Chemosphere 154:425–433. https://doi.org/10.1016/j.chemosphere.2016.03.104

Mann RM, Bidwell JR (1999) The toxicity of glyphosate and several glyphosate formulations to four species of southwestern australian frogs. Arch Environ Contam Toxicol 36:193–199. https://doi.org/10.1007/s002449900460

Marek LJ, Koskinen WC (2014) Simplified analysis of glyphosate and aminomethylphosphonic acid in water, vegetation and soil by liquid chromatography–tandem mass spectrometry. Pest Manag Sci 70:1158–1164. https://doi.org/10.1002/ps.3684

Matozzo V et al (2018) Ecotoxicological risk assessment for the herbicide glyphosate to non-target aquatic species: a case study with the mussel *Mytilus galloprovincialis*. Environ Pollut 233:623–632. https://doi.org/10.1016/j.envpol.2017.10.100

Matozzo V, Fabrello J, Marin MG (2020) The effects of glyphosate and its commercial formulations to marine invertebrates: a review. J Mar Sci Engin 8:399. https://doi.org/10.3390/jmse8060399

Mayer FL, Ellersieck MR (1986) Manual of acute toxicity: interpretation and data base for 410 chemicals and 66 species of freshwater animals. U.S. Dept. of the Interior, Fish and Wildlife Service., Washington, DC

Medalie L, Baker NT, Shoda ME, Stone WW, Meyer MT, Stets EG, Wilson M (2020) Influence of land use and region on glyphosate and aminomethylphosphonic acid in streams in the USA. Sci Tot Environ 707:136008. https://doi.org/10.1016/j.scitotenv.2019.136008

Melo K, De Nucci G, Trape A, Jacobucci S, Rosa P (2018) Brief review analytical methods for the determination of glyphosate. MOJ Toxicol 4:39–42

Mendonca CM, Reed ML, Kukurugya MA, Aristilde L (2019) Adverse metabolic outcomes in soil *Pseudomonas* species exposed to polyethoxylated tallow amine and glyphosate. Environ Sci Technol Lett 6:448–455. https://doi.org/10.1021/acs.estlett.9b00363

Mertens M, Höss S, Neumann G, Afzal J, Reichenbecher W (2018) Glyphosate, a chelating agent—relevant for ecological risk assessment? Environ Sci Pollut Res 25:5298–5317. https://doi.org/10.1007/s11356-017-1080-1

Mesnage R, Antoniou MN (2017) Facts and fallacies in the debate on glyphosate toxicity. Front Publ Hlth 5:316. https://doi.org/10.3389/fpubh.2017.00316

Mesnage R, Benbrook C, Antoniou MN (2019) Insight into the confusion over surfactant co-formulants in glyphosate-based herbicides. Food Chem Toxicol 128:137–145. https://doi.org/10.1016/j.fct.2019.03.053

Meza-Joya FL, Ramírez-Pinilla MP, Fuentes-Lorenzo JL (2013) Toxic, cytotoxic, and genotoxic effects of a glyphosate formulation (Roundup® SL–Cosmoflux® 411F) in the direct developing frog *Eleutherodactylus johnstonei*. Environ Mol Mutagen 54:362–373. https://doi.org/10.1002/em.21775

Moermond C, Beasley A, Breton R, Junghans M, Laskowski R, Solomon K, Zahner H (2017) Assessing the reliability of ecotoxicological studies: an overview of current needs and approaches. Integr Environ Assess Manag 13:640–651. https://doi.org/10.1002/ieam.1870

Monsanto Canada Inc. (2016) Monsanto Canada Inc., Winnipeg

Monsanto Europe (2017a) MON 0139: Acute Contact Toxicity to the Solitary Bee, *Osmia bicornis* under Laboratory Conditions. Monsanto Europe S.A. on behalf of Glyphosate Task Force, Antwerp, Belgium

Monsanto Europe (2017b) MON 0139: Acute Oral and Contact Toxicity to the Bumble Bee, *Bombus terrestris L.* under Laboratory Conditions. Monsanto Europe S.A. on behalf of Glyphosate Task Force, Antwerp, Belgium

Monsanto USA (2017) MON 0139: Chronic Oral Toxicity Test on the Honey Bee (*Apis mellifera L.*) in the Laboratory. Monsanto Company on behalf of the Glyphosate Task Force, St. Louis, MO

Montiel-León JM et al (2019) Widespread occurrence and spatial distribution of glyphosate, atrazine, and neonicotinoids pesticides in the St. Lawrence and tributary rivers. Environ Pollut 250:29–39. https://doi.org/10.1016/j.envpol.2019.03.125

Moore LJ, Fuentes L, Rodgers JH Jr, Bowerman WW, Yarrow GK, Chao WY, Bridges WC Jr (2012) Relative toxicity of the components of the original formulation of Roundup® to five North American anurans. Ecotoxicol Environ Safety 78:128–133. https://doi.org/10.1016/j.ecoenv.2011.11.025

Moore SB, Avery GB, Barnhardt JM, Diehl RA (1986) Acute and chronic aquatic toxicities of textile surfactants. In: National Technical Conference, USA. pp 290–293

Motta EV, Raymann K, Moran NA (2018) Glyphosate perturbs the gut microbiota of honey bees. Proc Natl Acad Sci 115:10305–10310. https://doi.org/10.1073/pnas.1803880115

Motta EV et al (2020) Oral and topical exposure to glyphosate in herbicide formulation impact the gut microbiota and survival rates of honey bees. Appl Environ Microbiol 86:e01150–e01120. https://doi.org/10.1128/AEM.01150-20

Nagy K, Duca RC, Lovas S, Creta M, Scheepers PTJ, Godderis L, Ádám B (2020) Systematic review of comparative studies assessing the toxicity of pesticide active ingredients and their product formulations. Environ Res 181:108926. https://doi.org/10.1016/j.envres.2019.108926

NAIADES (2020) Données sur la qualité des eaux de surface. NAIADES. http://www.naiades.eaufrance.fr/acces-donnees#/physicochimie. Accessed June 2020

Nakamura A, Catterall CP, Kitching RL, House AP, Burwell CJ (2008) Effects of glyphosate herbicide on soil and litter macro-arthropods in rainforest: implications for forest restoration. Ecol Manag Restor 9:126–133. https://doi.org/10.1016/j.cois.2015.05.006

Nandula VK, Reddy KN, Rimando AM, Duke SO, Poston DH (2007) Glyphosate-resistant and-susceptible soybean (*Glycine max*) and canola (*Brassica napus*) dose response and metabolism relationships with glyphosate. J Agric Food Chem 55:3540–3545. https://doi.org/10.1021/jf0635681

Navarro-Martin L, Lanctot C, Jackman P, Park BJ, Doe K, Pauli BD, Trudeau VL (2014) Effects of glyphosate-based herbicides on survival, development, growth and sex ratios of wood frogs (*Lithobates sylvaticus*) tadpoles. I: chronic laboratory exposures to VisionMax(R). Aquat Toxicol 154:278–290. https://doi.org/10.1016/j.aquatox.2014.05.017

Newman MM et al (2016) Changes in rhizosphere bacterial gene expression following glyphosate treatment. Sci Tot Environ 553:32–41. https://doi.org/10.1016/j.scitotenv.2016.02.078

Nguyen DB, Rose MT, Rose TJ, Morris SG, Van Zwieten L (2016) Impact of glyphosate on soil microbial biomass and respiration: a meta-analysis. Soil Biol Biochem 92:50–57. https://doi.org/10.1016/j.soilbio.2015.09.014

Nguyen NK, Dörfler U, Welzl G, Munch JC, Schroll R, Suhadolc M (2018) Large variation in glyphosate mineralization in 21 different agricultural soils explained by soil properties. Sci Tot Environ 627:544–552. https://doi.org/10.1016/j.scitotenv.2018.01.204

Niedobová J, Skalský M, Ouředníčková J, Michalko R, Bartošková A (2019) Synergistic effects of glyphosate formulation herbicide and tank-mixing adjuvants on Pardosa spiders. Environ Pollut 249:338–344. https://doi.org/10.1016/j.envpol.2019.03.031

Nielsen LN et al (2018) Glyphosate has limited short-term effects on commensal bacterial community composition in the gut environment due to sufficient aromatic amino acid levels. Environ Pollut 233:364–376. https://doi.org/10.1016/j.envpol.2017.10.016

NRA (1996) NRA Special Review of Glyphosate. National Registration Authority for Agricultural and Veterinary Chemicals, Canberra, Australia

Nuutinen V, Hagner M, Jalli H, Jauhiainen L, Rämö S, Sarikka I, Uusi-Kämppä J (2020) Glyphosate spraying and earthworm *Lumbricus terrestris* L. activity: Evaluating short-term impact in a glasshouse experiment simulating cereal post-harvest. Eur J Soil Biol 96:103148. https://doi.org/10.1016/j.ejsobi.2019.103148

NWQMC (2019) Water Quality Data. National Water Quality Monitoring Council. http://waterqualitydata.us/portal/. Accessed May 2019

NZ EPA (2016) Review of the evidence relating to glyphosate and carcinogenicity. New Zealand Environmental Protection Authority, Wellington NZ

OECD (2020) OECD Guidelines for the Testing of Chemicals. Organization for Economic Cooperation and Development. http://www.oecd-ilibrary.org/environment/oecd-guidelines-for-the-testing-of-chemicals-section-2-effects-on-biotic-systems_20745761. Accessed May 2020

Okada E, Costa JI, Bedmar F (2016) Adsorption and mobility of glyphosate in different soils under no-till and conventional tillage. Geoderma 263:78–85. https://doi.org/10.1016/j.geoderma.2015.09.009

Okada E, Coggan T, Anumol T, Clarke B, Allinson G (2019) A simple and rapid direct injection method for the determination of glyphosate and AMPA in environmental water samples. Anal Bioanal Chem 411:715–724. https://doi.org/10.1007/s00216-018-1490-z

Okada E, Allinson M, Barral MP, Clarke B, Allinson G (2020) Glyphosate and aminomethylphosphonic acid (AMPA) are commonly found in urban streams and wetlands of Melbourne. Aust Water Res 168:115139. https://doi.org/10.1016/j.watres.2019.115139

Paveglio FL, Kilbride KM, Grue CE, Simenstad CA, Fresh KL (1996) Use of Rodeo® and X-77® spreader to control smooth cordgrass (*Spartina alterniflora*) in a southwestern Washington esturay I. Environmental fate. Environ Toxicol Chem 15:961–968. https://doi.org/10.1002/etc.5620150619

Pekár S (2012) Spiders (Araneae) in the pesticide world: an ecotoxicological review. Pest Manag Sci 68:1438–1446. https://doi.org/10.1002/ps.3397

Pérez GL, Vera MS, Miranda L (2011) Effects of herbicide glyphosate and glyphosate-based formulations on aquatic ecosystems. In: Kortekamp A (ed) Herbicides and environment. IntechOpen, Rijeka, Croatia, pp 343–368

Perkins PJ, Boermans HJ, Stephenson GR (2000) Toxicity of glyphosate and triclopyr using the frog embryo teratogenesis assay—Xenopus. Environ Toxicol Chem 19:940–945. https://doi.org/10.1002/etc.5620190422

Pessagno RC, Torres Sánchez RM, dos Santos AM (2008) Glyphosate behavior at soil and mineral–water interfaces. Environ Pollut 153:53–59. https://doi.org/10.1016/j.envpol.2007.12.025

Pinto E, Soares AG, Ferreira IM (2018) Quantitative analysis of glyphosate, glufosinate and AMPA in irrigation water by in situ derivatization–dispersive liquid–liquid microextraction combined with UPLC-MS/MS. Anal Methods 10:554–561. https://doi.org/10.1007/s00216-018-1490-z

PMRA (2015a) Proposed Re-evaluation Decision, Glyphosate. Canadian Pest Management Regulatory Agency, Ottawa, ON

PMRA (2015b) Proposed Re-evaluation Decision: Glyphosate. Pest Management Regulatory Agency, Health Canada, Ottawa, ON

Pochron S, Simon L, Mirza A, Littleton A, Sahebzada F, Yudell M (2020) Glyphosate but not Roundup® harms earthworms (*Eisenia fetida*). Chemosphere 241:125017. https://doi.org/10.1016/j.chemosphere.2019.125017

Poiger T, Buerge IJ, Bächli A, Müller MD, Balmer ME (2017) Occurrence of the herbicide glyphosate and its metabolite AMPA in surface waters in Switzerland determined with on-line solid phase extraction LC-MS/MS. Environ Sci Poll Res 24:1588–1596. https://doi.org/10.1007/s11356-016-7835-2

Portier CJ (2020) A comprehensive analysis of the animal carcinogenicity data for glyphosate from chronic exposure rodent carcinogenicity studies. Environ Health 19:18. https://doi.org/10.1186/s12940-020-00574-1

Prill E, Richardson R (1997) European Patent EP, 378985 – Improved glyphosate formulations

Prosser RS, Anderson J, Hanson ML, Solomon KR, Sibley PK (2016) Indirect effects of herbicides on biota in edge-of-field habitats: a critical review of the literature. Agric Ecosys Environ 232:59–72. https://doi.org/10.1016/j.agee.2016.07.009

Qiu S, Fu H, Zhou R, Yang Z, Bai G, Shi B (2020) Toxic effects of glyphosate on intestinal morphology, antioxidant capacity and barrier function in weaned piglets. Ecotoxicol Environ Safety 187:109846. https://doi.org/10.1016/j.ecoenv.2019.109846

Raesch A, Hunsche M, Mail M, Burkhardt J, Noga G, Pariyar S (2018) Agricultural adjuvants may impair leaf transpiration and photosynthetic activity. Plant Physiol Biochem 132:229–237. https://doi.org/10.1016/j.plaphy.2018.08.042

Richmond ME (2018) Glyphosate: a review of its global use, environmental impact, and potential health effects on humans and other species. J Environ Stud Sci 8:416–434. https://doi.org/10.1007/s13412-018-0517-2

Rittman S, Wrinn KM, Evans SC, Webb AW, Rypstra AL (2013) Glyphosate-based herbicide has contrasting effects on prey capture by two co-occurring wolf spider species. J Chem Ecol 39:1247–1253. https://doi.org/10.1007/s10886-013-0353-5

Rodriguez-Gil JL, Lissemore L, Solomon K, Hanson M (2016) Dissipation of a commercial mixture of polyoxyethylene amine surfactants in aquatic outdoor microcosms: effect of water depth and sediment organic carbon. Sci Tot Environ 550:449–458. https://doi.org/10.1016/j.scitotenv.2016.01.140

Rodriguez-Gil JL et al (2017a) Aquatic hazard assessment of MON 0818, a commercial mixture of akylamine ethoxylates commonly used in glyphosate-containing herbicide formulations. Part 1: species sensitivity distribution from laboratory acute exposures. Environ Toxicol Chem 36:501–511. https://doi.org/10.1002/etc.3559

Rodriguez-Gil JL et al (2017b) Aquatic hazard assessment of MON 0818, a commercial mixture of akylamine ethoxylates commonly used in glyphosate-containing herbicide formulations. Part 2: roles of sediment, temperature, and capacity for recovery following a pulsed exposure. Environ Toxicol Chem 36:512–521. https://doi.org/10.1002/etc.3558

Rodríguez-Gil JL, Cáceres N, Dafouz R, Valcárcel Y (2018) Caffeine and paraxanthine in aquatic systems: global exposure distributions and probabilistic risk assessment. Sci Tot Environ 612:1058–1071. https://doi.org/10.1016/j.scitotenv.2017.08.066

Ross AR, Liao X (2015) A novel method for the rapid determination of polyethoxylated tallow amine surfactants in water and sediment using large volume injection with high performance liquid chromatography and tandem mass spectrometry. Anal Chim Acta 889:147–155. https://doi.org/10.1016/j.aca.2015.06.046

Rott E, Steinmetz H, Metzger JW (2018) Organophosphonates: a review on environmental relevance, biodegradability and removal in wastewater treatment plants. Sci Tot Environ 615:1176–1191. https://doi.org/10.1016/j.scitotenv.2017.09.223

Rubio F, Guo E, Kamp L (2014) Survey of glyphosate residues in honey, corn and soy products. J Environ Anal Toxicol 249. https://doi.org/10.4172/2161-0525.1000249

Samsel A, Seneff S (2015) Glyphosate, pathways to modern diseases III: Manganese, neurological diseases, and associated pathologies. Surg Neurol Int, 6. https://doi.org/10.4103/2152-7806.153876

Santadino M, Coviella C, Momo F (2014) Glyphosate sublethal effects on the population dynamics of the earthworm *Eisenia fetida* (Savigny, 1826). Water Air Soil Pollut 225:2207. https://doi.org/10.1007/s11270-014-2207-3

Schönbrunn E, Eschenburg S, Shuttleworth WA, Schloss JV, Amrhein N, Evans JNS, Kabsch W (2001) Interaction of the herbicide glyphosate with its target enzyme 5-enolpyruvylshikimate 3-phosphate synthase in atomic detail. Proc Natl Acad Sci 98:1376–1380. https://doi.org/10.1073/pnas.98.4.1376

Seide VE, Bernardes RC, Pereira EJG, Lima MAP (2018) Glyphosate is lethal and Cry toxins alter the development of the stingless bee *Melipona quadrifasciata*. Environ Pollut 243:1854–1860. https://doi.org/10.1016/j.envpol.2018.10.020

SERA (1996) Selected commercial formulations of glyphosate – Accord, Rodeo, Roundup and Roundup Pro – Risk Assessment. Final Report. Technical report prepared by Syracuse Environmental Research Associates, Inc. for the United States. Dept. of Agriculture, Forest Services, Syracuse, NY

SERA (1997) Effects of surfactants on the toxicitiy of glyphosate, with specific reference to Rodeo®. Technical report prepared by Syracuse Environmental Research Associates, Inc. for the United States. Dept. of Agriculture., Syracuse, NY

SERA (2011) Glyphosate human health and ecological risk assessment (FINAL REPORT). Technical report prepared by Syracuse Environmental Research Associates, Inc. for the United States. Dept. of Agriculture., Syracuse, NY

Servizi JA, Gordon RW, Martens DW (1987) Acute toxicity of Garlon 4 and Roundup herbicides to salmon, daphnia, and trout. Bull Environ Contam Toxicol 39:15–22. https://doi.org/10.1007/BF01691783

Sihtmäe M, Blinova I, Künnis-Beres K, Kanarbik L, Heinlaan M, Kahru A (2013) Ecotoxicological effects of different glyphosate formulations. Appl Soil Ecol 72:215–224. https://doi.org/10.1016/j.apsoil.2013.07.005

Silva V, Montanarella L, Jones A, Fernández-Ugalde O, Mol HGJ, Ritsema CJ, Giessen V (2018) Distribution of glyphosate and aminomethylphosphonic acid (AMPA) in agricultural topsoils of the European Union. Sci Tot Environ 621:1352–1359. https://doi.org/10.1016/j.scitotenv.2017.10.093

Simenstad CA et al (1996) Use of Rodeo® and X-77® spreader to control smooth cordgrass (Spartina alterniflora) in a southwestern Washington esturay II. Effects on benthic microflora and invertebrates. Environ Toxicol Chem 15:969–978. https://doi.org/10.1002/etc.5620150620

Skeff W, Recknagel C, Schulz-Bull DE (2016) The influence of salt matrices on the reversed-phase liquid chromatography behavior and electrospray ionization tandem mass spectrometry detection of glyphosate, glufosinate, aminomethylphosphonic acid and 2-aminoethylphosphonic acid in water. J Chromatogr A 1475:64–73. https://doi.org/10.1016/j.chroma.2016.11.007

Solomon KR (2016) Glyphosate in the general population and in applicators: A critical review of studies on exposures. Crit Rev Toxicol 46(S1):21–27. https://doi.org/10.1080/10408444.2016.1214678

Solomon KR (2020) Estimated exposure to glyphosate in humans via environmental, occupational, and dietary pathways: an updated review of the scientific literature. Pest Manag Sci 76:2878–2885. https://doi.org/10.1002/ps.5717

Solomon KR, Stephenson GL (2017) Quantitative weight of evidence assessment of higher tier studies on the toxicity and risks of neonicotinoid insecticides in honeybees 1: methods. J Toxicol Environ Health B 20:316–329. https://doi.org/10.1080/10937404.2017.1388563

Solomon KR, Thompson DG (2003) Ecological risk assessment for aquatic organisms from over-water uses of glyphosate. J Toxicol Environ Health B 6:211–246. https://doi.org/10.1080/15287390390155571

Solomon KR, Anadón A, Carrasquilla G, Cerdeira A, Marshall J, Sanin L-H (2007) Coca and poppy eradication in Colombia: environmental and human health assessment of aerially applied glyphosate. Rev Environ Contam Toxicol 190:43–125. https://doi.org/10.1007/978-0-387-36903-7_2

Solomon KR, Marshall EJP, Carrasquilla G (2009) Human health and environmental risks from the use of glyphosate formulations to control the production of coca in Colombia: overview and conclusions. J Toxicol Environ Health A 72:914–920. https://doi.org/10.1080/15287390902929659

Stalikas CD, Konidari CN (2001) Analytical methods to determine phosphonic and amino acid group-containing pesticides. J Chromatogr A 907:1–19. https://doi.org/10.1016/S0021-9673(00)01009-8

Stark JD, Walthall WK (2003) Agricultural adjuvants: Acute mortality and effects on population growth rate of Daphnia pulex after chronic exposure. Environ Toxicol Chem 22:3056–3061. https://doi.org/10.1897/02-504

Stephenson GL, Solomon KR (2017) Quantitative weight of evidence assessment of higher tier studies on the toxicity and risks of neonicotinoid insecticides in honeybees 2: Imidacloprid. J Toxicol Environ Health B 20:330–345. https://doi.org/10.1080/10937404.2017.1388564

Székács A, Darvas B (2018) Re-registration challenges of glyphosate in the European Union. Front Environ Sci 6:78. https://doi.org/10.3389/fenvs.2018.00078

Tang Q, Tang J, Ren X, Li C (2020) Glyphosate exposure induces inflammatory responses in the small intestine and alters gut microbial composition in rats. Environ Pollut 261:114129. https://doi.org/10.1016/j.envpol.2020.114129

Tatum VL, Borton DL, Streblow WR, Louch J, Shepard JP (2012) Acute toxicity of commonly used forestry herbicide mixtures to *Ceriodaphnia dubia* and Pimephales promelas. Environ Toxicol 27:671–684. https://doi.org/10.1002/tox.20686

Thiour-Mauprivez C, Martin-Laurent F, Calvayrac C, Barthelmebs L (2019) Effects of herbicide on non-target microorganisms: towards a new class of biomarkers? Sci Tot Environ 684:314–325. https://doi.org/10.1016/j.scitotenv.2019.05.230

Thompson HM, Levine SL, Doering J, Norman S, Manson P, Sutton P, von Merey G (2014) Evaluating exposure and potential effects on honeybee brood (*Apis mellifera*) development using glyphosate as an example. Integr Environ Assess Manag 10:463–470. https://doi.org/10.1002/ieam.1529

Tsui MTK, Chu LM (2003) Aquatic toxicity of glyphosate-based formulations: comparison between different organisms and the effects of environmental factors. Chemosphere 52:1189–1197. https://doi.org/10.1016/S0045-6535(03)00306-0

Tush D, Meyer MT (2016) Polyoxyethylene tallow amine, a glyphosate formulation adjuvant: soil adsorption characteristics, degradation profile, and occurrence on selected soils from agricultural fields in Iowa, Illinois, Indiana, Kansas, Mississippi, and Missouri. Environ Sci Technol 50:5781–5789. https://doi.org/10.1021/acs.est.6b00965

Tush D, Loftin KA, Meyer MT (2013) Characterization of polyoxyethylene tallow amine surfactants in technical mixtures and glyphosate formulations using ultra-high performance liquid chromatography and triple quadrupole mass spectrometry. J Chromatogr A 1319:80–87. https://doi.org/10.1016/j.chroma.2013.10.032

Tush D, Maksimowicz MM, Meyer MT (2018) Dissipation of polyoxyethylene tallow amine (POEA) and glyphosate in an agricultural field and their co-occurrence on streambed sediments. Sci Tot Environ 636:212–219. https://doi.org/10.1016/j.scitotenv.2018.04.246
USEPA (1993) Reregistration Eligibility Decision (RED) Glyphosate, EPA 738-R-93-014. U.S. Environmental Protection Agency, Washington, DC, USA
USEPA (2014) Guidance for Assessing Pesticide Risks to Bees. Office of Chemical Safety and Pollution Prevention, Office of Pesticide Programs. United States Environmental Protection Agency, Washington, DC
USEPA (2015) Preliminary Ecological Risk Assessment in Support of the Registration Review of Glyphosate and Its Salts. United States Environmental Protection Agency, Office of Chemical Safety and Pollution Prevention Washington DC
USEPA (2016) Series 835-Fate, Transport and Transformation Test Guidelines. United States Environmental Protection Agency. https://www.epa.gov/test-guidelines-pesticides-and-toxic-substances/series-835-fate-transport-and-transformation-test. Accessed May 2020
USEPA (2017) Revised Glyphosate Issue Paper: Evaluation of Carcinogenic Potential. United States Environmental Protection Agency, Office of Chemical Safety and Pollution Prevention Washington DC
USEPA (2019) EPA takes next step in review process for herbicide glyphosate, reaffirms no risk to public health. United States Environmental Protection Agency. https://www.epa.gov/newsreleases/epa-takes-next-step-review-process-herbicide-glyphosate-reaffirms-no-risk-public-health. Accessed May 2019
USGS (2010) National Field Manual for the Collection of Water-Quality Data. US Geological Survey Techniques of Water-Resources Investigations. http://pubs.water.usgs.gov/twri9A. Accessed September 2010
Van Bruggen AHC, He MM, Shin K, Mai V, Jeong KC, Finckh MR, Morris JG Jr (2018) Environmental and health effects of the herbicide glyphosate. Sci Tot Environ 616–617:255–268. https://doi.org/10.1016/j.scitotenv.2017.10.309
Van Der Kraak GJ, Hosmer AJ, Hanson ML, Kloas W, Solomon KR (2014) Effects of atrazine in fish, amphibians, and reptiles: an analysis based on quantitative weight of evidence. Crit Rev Toxicol 44(S5):1–66. https://doi.org/10.3109/10408444.2014.967836
Van Ginkel C, Stroo C, Kroon A (1993) Biodegradability of ethoxylated fatty amines: detoxification through a central fission of these surfactants. Sci Tot Environ 134:689–697. https://doi.org/10.1016/S0048-9697(05)80072-5
Vázquez DE, Ilina N, Pagano EA, Zavala JA, Farina WM (2018) Glyphosate affects the larval development of honey bees depending on the susceptibility of colonies. PLoS ONE 13: e0205074. https://doi.org/10.1371/journal.pone.0205074
Vázquez DE, Latorre-Estivalis JM, Ons S, Farina WM (2020) Chronic exposure to glyphosate induces transcriptional changes in honey bee larva: a toxicogenomic study. Environ Pollut 261:114148. https://doi.org/10.1016/j.envpol.2020.114148
Villamar-Ayala CA, Carrera-Cevallos JV, Vasquez-Medrano R, Espinoza-Montero PJ (2019) Fate, eco-toxicological characteristics, and treatment processes applied to water polluted with glyphosate: a critical review. Crit Rev Environ Sci Technol 49:1476–1514. https://doi.org/10.1080/10643389.2019.1579627
Visek K (2000) Amines, fatty. In: Kirk-Othmer Encyclopedia of chemical technology, vol 2. Wiley, New York, pp 518–537
von Merey G, Manson PS, Mehrsheikh A, Sutton P, Levine SL (2016) Glyphosate and aminomethylphosphonic acid chronic risk assessment for soil biota. Environ Toxicol Chem 35:2742–2752. https://doi.org/10.1002/etc.3438
Wagner N, Reichenbecher W, Teichmann H, Tappeser B, Lötters S (2013) Questions concerning the potential impact of glyphosate-based herbicides on amphibians. Environ Toxicol Chem 32:1688–1700. https://doi.org/10.1002/etc.2268
Wan MT, Watts RG, Moul DJ (1989) Effects of different dilution water types on the acute toxicity to juvenile pacific salmonids and rainbow trout of glyphosate and its formulated products. Bull Environ Contam Toxicol 43:378–385. https://doi.org/10.1007/BF01701872

Wang N et al (2005) Influence of sediment on the fate and toxicity of a polyethoxylated tallowamine surfactant system (MON 0818) in aquatic microcosms. Chemosphere 59:545–551. https://doi.org/10.1016/j.chemosphere.2004.12.009

Wang S et al (2016) (Bio)degradation of glyphosate in water-sediment microcosms – a stable isotope co-labeling approach. Water Res 99:91–100. https://doi.org/10.1016/j.watres.2016.04.041

Weaver MA, Krutz LJ, Zablotowicz RM, Reddy KN (2007) Effects of glyphosate on soil microbial communities and its mineralization in a Mississippi soil. Pest Manag Sci 63:388–393. https://doi.org/10.1002/ps.1351

Williams GM et al (2016) A review of the carcinogenic potential of glyphosate by four independent expert panels and comparison to the IARC assessment. Crit Rev Toxicol 46(S1):3–20

Williams GM et al (2018) Corrigendum to Williams et al. (2016) Crti Rev Toxicol 46. Crit Rev Toxicol 48:893–894. https://doi.org/10.1080/10408444.2018.1522175

Wright DR, Brinker RJ, Sandbrink JJ, Wideman AS (1998) U.S. Patent No. 5,750,468 – Glyphosate formulations containing etheramine surfactants. USA Patent

Xu J, Smith S, Smith G, Wang W, Li Y (2019) Glyphosate contamination in grains and foods: an overview. Food Control 106:106710. https://doi.org/10.1016/j.foodcont.2019.106710

Yale RL, Sapp M, Sinclair CJ, Moir JWB (2017) Microbial changes linked to the accelerated degradation of the herbicide atrazine in a range of temperate soils. Environ Sci Res Int 24:7359–7374. https://doi.org/10.1007/s11356-017-8377-y

Yeritsyan K (2016) European Patent EP 2 230 916 B1 – Herbicidal formulations for triethanolamine salts of glyphosate

Zaller JG, Heigl F, Ruess L, Grabmaier A (2014) Glyphosate herbicide affects belowground interactions between earthworms and symbiotic mycorrhizal fungi in a model ecosystem. Sci Rep 4:5634. https://doi.org/10.1038/srep05634

Zhu YC, Adamczyk J, Rinderer T, Yao J, Danka R, Luttrell R, Gore J (2015) Spray toxicity and risk potential of 42 commonly used formulations of row crop pesticides to adult honey bees (Hymenoptera: Apidae). J Econ Entomol 108:2640–2647. https://doi.org/10.1093/jee/tov269

Zoller O, Rhyn P, Zarn JA, Dudler V (2020) Urine glyphosate level as a quantitative biomarker of oral exposure. Int J Hyg Environ Health 228:113526. https://doi.org/10.1016/j.ijheh.2020.113526

GPSR Compliance

The European Union's (EU) General Product Safety Regulation (GPSR) is a set of rules that requires consumer products to be safe and our obligations to ensure this.

If you have any concerns about our products, you can contact us on ProductSafety@springernature.com

In case Publisher is established outside the EU, the EU authorized representative is:

Springer Nature Customer Service Center GmbH
Europaplatz 3
69115 Heidelberg, Germany

Batch number: 10777422

Printed by Printforce, the Netherlands